Pergamon Series in Analytical Chemistry
Volume 6
General Editors: R. Belcher† (Chairman), D. Betteridge & L. Meites

Computers
in Analytical Chemistry

Related Pergamon Titles of Interest

BOOKS

Earlier Volumes in the Pergamon Series in Analytical Chemistry

OTHER BOOKS

JOURNALS

*Full details of all Pergamon publications/free specimen copy of any Pergamon journal
available on request from your nearest Pergamon Office.*

Computers
in Analytical Chemistry

by

PHILIP G. BARKER, FBCS, FRSC

Principal Lecturer in Computer Science,
Teesside Polytechnic

Sometime Lecturer in Computing,
University of Durham

Sometime SRC Research Fellow,
Department of Chemistry (Swansea),
University of Wales

PERGAMON PRESS

OXFORD · NEW YORK · TORONTO · SYDNEY · PARIS · FRANKFURT

U.K.	Pergamon Press Ltd., Headington Hill Hall, Oxford OX3 0BW, England
U.S.A.	Pergamon Press Inc., Maxwell House, Fairview Park, Elmsford, New York 10523, U.S.A.
CANADA	Pergamon Press Canada Ltd., Suite 104, 150 Consumers Rd., Willowdale, Ontario M2J 1P9, Canada
AUSTRALIA	Pergamon Press (Aust.) Pty. Ltd., P.O. Box 544, Potts Point, N.S.W. 2011, Australia
FRANCE	Pergamon Press SARL, 24 rue des Ecoles, 75240 Paris, Cedex 05, France
FEDERAL REPUBLIC OF GERMANY	Pergamon Press GmbH, Hammerweg 6, D-6242 Kronberg-Taunus, Federal Republic of Germany

First edition 1983

Library of Congress Cataloging in Publication Data

Barker, Philip G.
Computers in analytical chemistry.
(Pergamon series in analytical chemistry ; v. 6)
Includes bibliographical references and indexes.
1. Chemistry, Analytic—Data processing. I. Title.
II. Series.
QD75.4.E4B37   1983   543'.0028'54   82-22276

British Library Cataloguing in Publication Data

Barker, Philip G.
Computers in analytical chemistry—(Pergamon
series in analytical chemistry; 6)
1. Chemistry, and analytical—Data processing
I. Title
543'.0028'5   QD75.4.E4
ISBN 0-08-024008-9

Printed in Great Britain by A. Wheaton & Co. Ltd., Exeter

**To God
Science and Humanity**

Organisation of the Book

Chemical analysis and analytical chemistry are domains of scientific endeavour that are fundamental to many areas of applied science and technology. Taken together these subjects form a mass of knowledge which has its own structure and subdivisions, theories and folklore, experts, practitioners and manufacturing technology. Analytical chemistry thus covers an extremely complex web of human activity. During the last century Society has experienced many changes in technology that have affected everybody's way of life - particularly the scientist and the way in which scientific chemical analysis is undertaken. Moreover, the last ten to fifteen years has seen the rapid growth of computer technology. This too is slowly having its impact on analytical chemistry in a variety of ways ranging from large scale automation through to laboratory machines that show some degree of 'intelligence'. It is the intention of this book to cover some of the ways in which computers are used in analytical chemistry - the latter term being interpreted in its widest sense. An understanding or background of chemistry/analytical chemistry is assumed.

Because of the all-pervasive nature of the subject an attempt is made in Chapter One to answer the question, 'What is Analytical Chemistry?. It does this by looking at some application areas in which analysis forms a fundamental part of the day to day activity of those involved in its practice. For the purpose of illustration, some techniques of analysis are described in considerable detail. Obviously, the space devoted to this chapter does not enable it to be comprehensive. Hopefully, though, it will at least be representative of what happens in a wide variety of analytical chemistry laboratories - if not in academia, then, at least, in the commercial and industrial worlds.

In Chapter Two the author tries to formulate the basic steps involved in the analytical process. Various illustrations are used to describe how a computer system might assist the analyst in the sometimes formidable task of analysing an unknown material. Here, fact occasionally turns to fancy when a description is given of some of the ways in which computers may be used in the not too distant future. On this front, some of the important applications include the use of industrial robots, the use of sophisticated mathematical modelling and simulation techniques, the creation of large data bases and the application of artificial intelligence concepts to the design of computer programs known as 'expert systems'. This chapter also describes the influences that recent developments in computer input and output devices are having on the way in which the analyst is able to record and display the results of his/her work.

Chapter Three is devoted to a cursory study of some of the instruments that are currently being used by analysts throughout a wide variety of laboratories. It pays particular attention to those instruments that for one reason or another require the application of computer technology. As early as 1964 Joshua Lederberger at Stanford University was probably one of the first to pioneer the use of real-time computer systems within the laboratory environment. Since this early work, which was devoted to mass spectral analysis, many significant advances have taken place. Nowadays there are five important application areas of computers within analytical instrumentation: (1) data acquisition, (2) data processing and enhancement, (3) storage and retrieval of data, (4) automatic interpretation of results, and (5) control uses. After presenting a simple model for a typical analytical instrument and outlining some of its important properties this chapter discusses some of the applications of computers in three of the most important analytical methods: gas chromatography, infrared and mass spectrometry. Obviously, many of the general principles outlined in the specific examples chosen apply to a much wider range of instrumentation. The chapter concludes with a description of how smaller analytical units such as GC, IR, MS can be linked together to form more 'sophisticated' instruments.

Any form of technical description of computers is delayed until Chapter Four. Here a description of small computer systems (microprocessors and microcomputers), medium sized computers (minicomputers) and large computers (mainframes and super-computers) is presented. The chapter introduces the concepts of computer hardware and software. The architecture of a simple integrated circuit (or 'chip') is discussed in the context of its being the building block for larger systems. Examples of minicomputer, mainframe and super-computer systems are also briefly outlined.

Control applications are an important facet of the computer. Equally important - perhaps, more so - is its ablility to store and retrieve data. In Chapter Five, the concepts of data collection (or data acquisition) are discussed. Theoretical considerations relevant to sampling theory are outlined and techniques for storing analogue and digital data are briefly described. A summary of the various types of storage device, such as magnetic tape and disks, is also given in this section.

The inter-connection of instruments - both to each other and to computer hardware - is a major area of interest within all branches of analytical chemistry. Consequently, Chapter Six is devoted to the topic of interfacing. Although the main emphasis is on electrical interfacing, other forms are briefly outlined. Various types of standard electrical and electronic interfaces such as the IEEE-488, CAMAC and S-100 are described. In addition, some of the problems associated with the design, implementation and use of non-standard interfaces are mentioned.

In Chapter Seven, as a natural follow-on from interfacing, the author turns his attention to the basic types of communication channel that are necessary to enable instrument/computer inter-connection and the flow of data between one device and another. Several different types of communication link are discussed - couriers, wired links, optical fibres, infrared data links, microwaves and satellite systems. Fundamental to present day data communication are the world's telecommunication networks. Consequently, in view of its importance, this chapter contains a brief section devoted to basic telecommunications; the basic principles of communication networks are introduced in order to provide the necessary background for a later chapter on computer networks. Because of the need to ensure the privacy of data and information that is transmitted over the available communication links, the chapter concludes with a short discussion of data encryption.

Progress in electronics, computing and precision engineering have made possible significant advances in automation. Chemistry, in particular applied analytical chemistry, has been unable to avoid the onward march of this new area of technology. Automatic analysis is now a rapidly growing sector of chemical science. Because of its importance, Chapter Eight is devoted to a study of the application of automation to various aspects of laboratory analysis. Several approaches are considered including the use of in-house development teams and the purchase or lease of turnkey packages. A brief overview of the future potential role of the analyst within this area is then given.

Chapter Nine deals with various aspects of data processing. Most of the activities that take place in the laboratory produce data in one form or another. Before information can be derived from experimental or analytical measurements, the raw data obtained in the laboratory (or on site) has to be processed by methods that are appropriate to the investigation concerned. Several different approaches to computerised data processing are outlined. The chapter discusses both numeric and non-numeric techniques as well as some of the more advanced applications such as pattern recognition, modelling and optimisation.

The storage of experimental results and other important information in computer based systems requires the design and implementation of appropriate storage mechanisms. Computer data bases are often used to provide the required facilities. In view of their growing importance, Chapter Ten is used to present an introduction to this topic.

Because of the existing large volume and continual high growth rate of information that is of relevance to analytical science - both its literature and scientific results - the need for computerised information services is of vital importance. Such services provide a variety of computer based (or generated) tools to enable the analytical scientist to gain access to and, hence, make use of the valuable work of other scientists. The wide range of tools are grouped together under the title of 'information services' and are described in Chapter Eleven.

Many large companies have geographically dispersed laboratories, processing plants and administrative centres. The sharing of data, information and control is of vital importance, as is the general coordination of an organisation's activity. Because of the growing interest in the use of distributed communication systems for achieving these goals the final chapter of the book is devoted to a study of computer networks. Various types of system are described: local area, national and international networks. The fundamental modes of operation of these systems are outlined and likely future directions of development summarised.

Acknowledgements

I would like to express my sincere thanks to Professor D. Betteridge for suggesting that this work should be undertaken. I am also indebted to the University of Durham and Teesside Polytechnic for providing many of the resources that enabled much of the background research to be undertaken. Mr. Peter Henn and the staff of Pergamon Press contributed invaluable help and advice. To all those who have helped me with the preparation of this book I am deeply grateful.

Contents

1
What is Analysis?

INTRODUCTION

Chemistry is one of the oldest sciences known to Mankind. Introductory text books on the subject often describe its domain as: the science of the properties of elementary and compound substances and the nature of the laws that describe how these combine together and react one with another. Chemistry is thus concerned with a wide range of topics. It deals with the everyday objects and events that take place around us and with the more esoteric entities and happenings on distant planets, in flames, in ionization chambers and within the human body. It is concerned with the discovery and manufacture of new materials (food, fuel, plastics, clothes - to mention just a few) and the safety of the environment in which we live - land, sea, river and the atmosphere. Above all it is a well documented, progressive science that offers an exciting intellectual framework for those wishing to embark upon its more detailed study.

For a variety of reasons, most people at some stage in their lives become acquainted with the subject or with one or other of its multifarious applications. Children often encounter the subject for the first time at school - as one of the many options contained in an over-laden science curriculum. Others gain their knowledge as a result of some less formal interest - assaying the alchohol content of a wine they have been fermenting; trying to ascertain why the soil in their garden will not sustain the growth of roses; or, by means of a suitable dye, attempting to restore the colour of a garment whose attractiveness has faded with old age. Once kindled, an interest in chemistry is often difficult to extinguish. Consequently, for a large number of people chemistry becomes a professional discipline that forms the basis of a life long career.

Those who undertake a deeper, more involved, study of the subject do so as a consequence of several possible factors. Some are motivated by the financial rewards of their employment as professional chemists. Many others are endowed with an insatiable desire to discover new knowledge and cherish the accolades that this can bring. However, there are also those who have a less professional interest and who treat the subject as a hobby - just like stamp collecting, bird watching or painting. These are the amateur chemists, of whom there are a great number. No matter how their fundamental motivations may differ, those who pursue the discipline of chemistry have much in common. This common ground undoubtedly embraces: (a) an understanding of the basic principles of science and its application to the bringing together of cause and effect; (b) some knowledge of engineering principles and a desire to build things - the chemist often constructs molecules, theories, experimental rigs and measuring apparatus; and, (c) a working knowledge of the basic ideas inherent in design - of synthetic pathways, of equipment and of analytical techniques. Like other scientists, perhaps, the most fundamental attribute that chemists have in common is that of being an observer. In this context the feature that distinguishes the chemist from other scientists lies in the nature of the systems that are dealt with and the types of observations that are made.

Fundamental to chemistry - indeed, to all branches of science - are the basic principles involved in the 'analytical method' (Ber68). Using this approach a scientist sets about enumerating those factors which influence a particular situation or state of affairs. An attempt is then made to measure in a quantitative way the influences of the forces or factors that are deemed to be important within the confines of the system that has been chosen for investigation. All scientists are thus primarily concerned with observations made on some particular, carefully selected, system.

Because of its subsequent importance, the concept of a system needs to be more firmly established. In the broadest possible sense such an entity may be defined as:

> a set of objects together with relationships between
> the objects and between their attributes (Wei75)

For the chemist, the objects of interest will be atoms and molecules. The relationships will be the numerous laws that define how these entities interact and react together. The attributes will be the various properties that the atoms and molecules themselves possess. Such was the definition of chemistry that might have been promoted years ago. Today, however, it is much more than this. Nowadays the definition of chemistry and the scope of chemical systems must include the very process of life and the mechanisms of living. The relationships mentioned above must be extended to include a study of the ways in which atoms and molecules (and their properties) affect people and other living species - if only roses in an undernourished garden.

Provided one accepts the importance of gaining a knowledge of the relationships and attributes that dominate a chemical system one must also accept the need for appropriate methods to enable these to be elucidated and recorded. Most often these objectives are realised through 'controlled' experiments that incorporate proven measuring techniques that form the basis of the scientist's interest in analytical chemistry. However, despite the longevity of chemistry itself, it is only in relatively recent times that chemical analysis has gained the hallmarks of a precise and sophisticated science. Indeed, fewer than 100 years ago its deficiencies were a cause for public scandal (Cam79).

As time progresses new measuring tools become available which are capable of providing the scientist with previously unknown measuring capabilities.

Undoubtedly, during the last twenty years one of the most significant factors that has influenced modern instrumentation and analytical techniques has been the rapid developments that have taken place in electronics. Particularly noticable has been the explosive evolution of the electronic digital computer in its various forms - ranging from microchip through to super-computer. Combined with other developments in instrumentation and technique, the analytical scientist of today has significantly more observational power than his/her counterpart of 100 years ago.

Despite the changes that have taken place over the years, the goals and objectives of chemical analysis have not changed. What has changed are the ways in which these objectives are realised, the problems that remain to be solved and the seemingly never ending plethora of new areas of development that advances within any one sphere seem to bring. As the boundaries of knowledge are pushed further and further over new terrain, so the limitations of our previous ideas and expectations become apparent. Often many of the previously accepted theories and beliefs become suspect as more refined and sophisticated analysis presents us with many new phenomena and observations that we are often unable to explain. Such has been the influence of the computer in analytical chemistry. Its existence has opened up many new avenues to the practice and theory of analytical science.

THE NATURE OF ANALYTICAL SCIENCE

Chemical analysis is a science of measurement that forms a basic component of a large number of other areas of scientific endeavour. Particularly important are those areas that overlap with processes that influence our daily lives. Indeed, our modern way of life would not proceed as smoothly as it does were it not for much of the basic chemical analysis that goes on "behind the scenes". One needs only to mention such areas as the food and confectionary industries, the brewing industry and the many different pharmaceutical producers. These, taken together, are responsible for providing most of the materials that we ingest into our bodies. Without proper quality control via suitable chemical analysis our existence could well be a chance affair. When we become ill, very many medical and bio-chemical tests may be performed in an attempt to locate the cause of illness. Sudden death as a result of an unknown cause can often result in detailed post-mortem examinations in a pathology laboratory. Many of the routine tests used in medical and pathology labs often originate from methods that have been developed in the chemistry laboratory - either directly or as a by-product of some other investigation. Just as the standard of our lives is influenced by analytical chemistry, so is our continued existence. Many industries exist only because of their ability to produce products to a pre-defined customer standard - steel having a specific composition; plastics having particular types of properties; drugs that are safe to use and which are free of long term side effects. In each of these, and indeed most other process industries, there will usually be a significant number of analyses taking place in order to ensure the quality and safety of the goods that are manufactured. Some of these will be discussed in more detail in the case studies that are presented later in this chapter.

The Five Basic Questions

Analytical science is a discipline whose origins are deeply embedded in the history of chemical analysis. It is thus concerned with observations of and measurements made on chemical moities that are present within some chemical system that is of interest to us. In general, when considering a chemical system there will be three major factors to take into account. Firstly, the system itself. Secondly, its environment, and, thirdly, the boundary that separates the two. Figure 1.1 shows the relationship between these three items.

The boundary that separates the system under study from its surrounding environment may not always be sharply defined. Furthermore, depending upon the exact nature of this boundary, there will be various degrees of interaction between the system and the environment. Based upon a detailed

Fig. 1.1 Basic components of a system.

consideration of the type of energy/material flow that takes place, systems may be defined as open or closed. Closed systems are usually much simpler to understand than those of the open variety. Consequently, many of the early theories of chemistry related mainly to chemical systems which were of a closed nature. More detailed discussions of these and of open systems are presented in the reference texts (Ber68, Eme69, Bei72, Wei75) listed in the bibliography at the end of this chapter. For the present discussion the model presented above is adequate.

Fundamentally, within analysis one attempts to ask the following questions of a system S:

(1) What measuring tools are available to measure and observe the system?
(2) What are the constituent components of the system at a particular instant in time, T, say?
(3) What are the relative/absolute proportions of these components present in the system at that instant in time?
(4) How accurately can the measurements be made? How reliable are they and what confidence can one place in the results?
(5) Are the results that are reported time variant - as in the case of a kinetics experiment whose reaction half-life is relatively short?

The answers to the above questions can be represented by means of a simple mathematical expression of the following form:

$$\langle \text{result set} \rangle \quad ::= \quad \left\{ S, \; \left[C_i \right]_1^n, \; \left[Q_i \right]_1^n, \; \left[A_i \right]_1^n, \; T \right\}$$

where S defines the system under investigation, C_i is the list of chemical components within S, Q_i are the quantities of these components and A_i denote the confidence limits of the results that are reported. All results refer to a particular time, T, at which the analyses were made.

Obviously, from the above five basic questions a multitude of others are likely to arise. Some of the more important of these will be outlined briefly in order to present a framework for some of the material that is to follow in subsequent chapters of this book.

One of the reasons why chemical analysis is so important is that ultimately the activity involved will lead to some form of decision making. From this point of view, the answers to the questions listed above are all that are necessary. Thus, for someone having to make a decision on whether to accept or reject a delivery of a particular chemical used in a manufacturing process, the quality analysis and date are frequently all that are required. However, for those interested in being able to reproduce the measurements the detailed nature of the analytical methods that were employed will be of significant importance. Similarly, the cost of conducting an analysis should also be of concern - particularly if it is expensive as a consequence of its dependence upon costly chemicals or equipment. An expensive analysis performed at frequent intervals soon eats into a laboratory budget. Analysis time is yet another factor that must be taken into consideration and about which questions need to be posed. If analysis time is lengthy this might mean that other processes whose initiation are dependent upon the result might be held up pending its availability. In such a situation it would obviously be desirable to search for an alternative analytical procedure capable of eliminating the delays inherent in the method being used.

There are many other important questions that need to be asked. Amongst these will be those which relate to the analytical methods used. Of vital concern should be the question of whether or not the method of observing the system influences or changes it in any way. Thus, in the case of analysis by gas chromatography, reaction of one or more of the sample's components with the carrier gas used for elution could conceivably mask their existence unless their presence is detected by some alternative method. Similarly, addition of reagents necessary to conduct a particular 'wet' analysis might significantly affect many of the equilibria that exist in a system thereby giving false results. These two simple examples serve only to illustrate the care that needs to be taken by the analyst when designing methods for detecting the presence of particular materials and formulating procedures for quantitatively estimating their abundance.

Analytical science can thus be an extremely complex business. Despite this, it has a valuable role to play. By its very nature it is an applied branch of science. In many ways it is totally dependent upon many of the results of the pure sciences (physics, chemistry, biology, etc) and of technology. It uses the endeavours of these fundamental disciplines in order to construct a framework of analytical methods capable of supporting

both the local and the global affairs upon which society is based.   This
relationship is shown in Fig. 1.2.

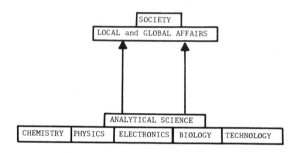

Fig. 1.2   The role of analytical science in Society.

In the following section the nature of the framework that analytical
science provides will be briefly described by means of some examples taken
from a few of the more important process industries, medicine and
research.

SOME CASE STUDIES

In order to illustrate the wide range of 'application areas' that are so
dependent upon analytical chemistry some introductory case studies will be
briefly described.   The examples that have been selected will serve the
purpose of introducing a small section of the many different analytical
methods that are used in present day situations.   Each case study will be
considered under two headings:

(A)   Application Description

(B)   Analytical Methods

In the application description an attempt is made to present a brief
overview of the functional role of the appliction area, the processes
involved and the role of analysis in controlling these processes.   The
descriptions of the analytical methods employed will not be
comprehensive.   They will be presented only in sufficient depth to give
the reader a flavour of the analytical procedures involved.   Their purpose
is to set the scene for subsequent chapters rather than to provide
detailed technical commentaries on the methods employed.   The illustrative
examples that have been singled out for further discussion are listed
below:

(1)   The Petrochemical Industry
         - as an example of chemical manufacture,
(2)   A Clinical Laboratory
         - as an illustration of health screening,
(3)   The Food Industry
         - as a research/quality assurance example.

Example 1 - The Petrochemical Industry

Application Description

The major objective of the petrochemical industry is to transform crude petroleum oils into materials of a more useful nature. These include petrol and other light fuels, aromatics and a variety of other hydrocarbons. The oils that are used originate from many different sources. They are converted by a process called thermal cracking (or pyrolysis) into other chemicals that may be used as such or which may form the essential intermediates in some other manufacturing process. Figure 1.3 shows a simplified version of the nature of the material flow involved in a typical petrochemical plant.

The diagram reflects the idea of inputs and outputs (shown by arrows) and processes (shown as boxes). During the routine operation of a production plant that implements processes of this type there will be a call for three categories of analysis:

   (a) that of input raw materials,
   (b) that of products, by-products, waste materials, and
   (c) that of intermediate products, processes and equipment.

The efficiency of a manufacturing process may be critically dependent on the purity of its raw materials - catalysts may be deactivated by traces of particular impurities; some reactions become explosive when certain contaminants are present in the reactants. Accurate and reliable quality control can be used to guard against these eventualities.

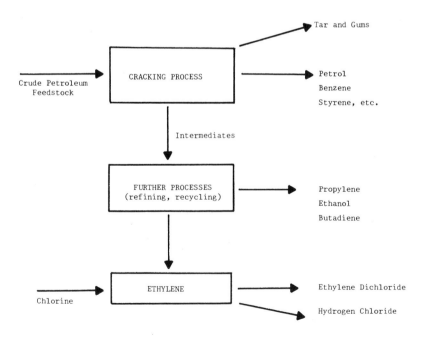

Fig. 1.3 Simplified material flow in a petrochemical plant.

Successful sale of chemicals produced in a manufacturing process will
undoubtedly depend upon their level of purity. Usually, marketable
products will need to meet the stringent specifications embodied in
quality assurance analysis. The requirements are agreed upon by both
producer and consumer at the time a business negotiation is entered into.
Subsequently, when goods change hands, the exchange of material is usually
accompanied by documentary evidence that reflects the standard of the
product. Typically this takes the form of a Certificate of Quality, an
example of which, is presented in Fig. 1.4 (Bar66).

To: B.E.Control.
      Technical Superintendent

                    CERTIFICATE OF QUALITY - BUTADIENE

                    (Day Tanks and Hortonspheres)

      The sample from tank .... at .... hrs ...../...../.....

has been analysed and $\dfrac{\text{conforms}}{\text{does not conform}}$ to Specification.

Laboratory Sample No:

Butadiene (by G.C.)	% mol.	993	98.3% (MIN)
Butenes	% mol.	0.3	0.5% (MAX)
Butanes	% mol.	0.1	0.5% (MAX)
C3s	% mol.	0.10	0.15% (MAX)
C5s	% mol.	0.12	0.15% (MAX)
Acetylenes, as vinyl acetylene,	% wt/wt.	0.008	0.02 (MAX)
T.B.C.	ppm	35	AS REQUIRED
Residue,	% wt/wt	0.002	0.06% (MAX)
Peroxides, as H2O2, ppm		4	6 ppm (MAX)
Dimer, % wt/wt.		0.06	0.1% (MAX)
Sulphur, as H2S	% wt/wt.	0.002	0.005% (MAX)
Carbonyl, as acetaldehyde, % wt/wt.		0.005	0.01% (MAX)
Appearance at -20°C		COLOURLESS CLEAR AND FREE FROM ENTRAINED MATTER	
Acetonitrile, ppm		11	20 ppm (MAX)
Oxygen in vapour space	%	<0.05%	0.3% v/v (MAX)

Fig. 1.4 Analytical report showing the purity of a manufactured product.

The effort involved in producing a report of this type is substantial. It
may involve several different analysts each conducting a particular facet
of the overall analysis. The results of individual experiments have to be
recorded and the activity of the laboratory coordinated in order to ensure
the smooth flow of the sample from input tray to output station. The
results of all analyses will need to be entered into the laboratory log
book and a copy of the certificate of quality retained for possible future
reference.

Final product analysis of the type outlined above represents only one
aspect of quality control. Analysis of by-products and waste materials

are equally important. Often these can be used as indicators of the efficiency of conversion of a process, of the likely contamination to be expected in a product, and, of the possible pollution problems that may have to be solved in order to avoid adverse effects upon the environment. Thus, continual monitoring of the locality in which a manufacturing plant is sited may be necessary in order to ensure that the quality of the environment is not suffering as a consequence of the inadvertent release of pollutants.

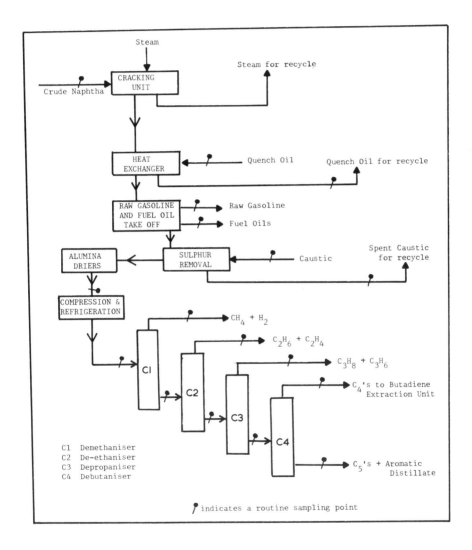

Fig. 1.5 Location of analytical sampling points in a cracking plant.

Observations made on intermediate products and processes are as important as raw material and final product analysis. Collectively, the former are most often referred to as process control analyses. The results that they yield are used to control the plant operating conditions. In a large petrochemical plant there will be a significant number of locations at which chemical samples are taken, analyses performed, results reported and appropriate control action taken. Figure 1.5 illustrates a typical distribution of analytical sampling points for a simple cracking plant.

Wherever possible the materials flowing through a section of the plant will be sampled and analysed at three different stages: (A) prior to entering a processing unit, (B) during the actual process itself, and (C) after leaving the processor. This generalised sampling scheme is illustrated in the diagram shown in Fig. 1.6. Samples are taken at routine sampling points for the purpose of "keeping watch" on the progress of the overall manufacturing process. Analyses are performed at regular intervals and the results compared with those expected for normal running conditions.

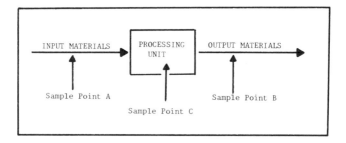

Fig. 1.6 Generalised sampling scheme.

About fifteen years ago most of the analytical procedures involved the use of manual techniques (Bar66). These were conducted in a process control laboratory sited in close proximity to the manufacturing plant. Today, computer techniques are used extensively to complement the activity of the control lab. However, in some cases, as a consequence of the availability of highly reliable, accurate, and precise instruments that are capable of operating independently of human operators, many manufacturing plants are becoming totally automatic (Che80). Despite this progress, conventional analytical techniques are still employed in situations where the introduction of a computer is not cost effective or where there exists technical and social barriers to the introduction of automation.

Analytical Methods

Most organisations involved in chemical manufacture will have a process control lab that is stocked with a range of analytical methods appropriate to the products and processes that they handle. As an example, consider the scheme outlined in Fig. 1.3. Typical of the analytical methods likely to be employed to support such a regime might be:

(a) freezing point measurements,
(b) measurement of refractive index,
(c) titrimetric methods,
(d) gas chromatography, and,
(e) colorimetric techniques.

Crystallisation or freezing point (FP) methods are very important since they can be used as measures of purity by reference to appropriate calibration graphs. Pure materials have a well defined freezing point; benzene freezes at $5.51^{\circ}C$ and styrene freezes at $-30.6^{\circ}C$. The presence of impurities such as polymer or inhibitor can cause the crystallising point of a sample to fall below that of the pure material. The observed depression of freezing point can be used to estimate the purity of the sample of material under test. Consider the case of benzene purity analysis; the crystallising point is determined on a sample saturated with water and a correction is applied to obtain the value for the anhydrous material. A precision thermometer (range $3.9-6.2^{\circ}C$) is used and it is read to the nearest $0.005^{\circ}C$. The recorded crystallising point (wet sample) is corrected by adding $0.09^{\circ}C$ to obtain the dry crystallising point. The repeatability (by the same operator) should be within $0.01^{\circ}C$ and the reproducibility (by two separate analysts) should be within $0.035^{\circ}C$. The purity of the sample is determined by reference to a calibration graph.

In the manufacture of styrene, after the alkylation step, ethyl-benzene is fed into processing units called 'dehydrogenators'. These are responsible for removing hydrogen from the molecule, thereby producing unsaturation. Since dehydrogenation is never 100% complete, the effluent from the processor consists of a mixture of styrene and ethyl-benzene - called dehydrogenated mixture or DM. Measurement of refractive index can be used to ascertain the composition of the mixture.

The refractive index of a material at a given temperature is critically dependent upon its purity. Consequently, refractive index measurements are often used as a criterion of purity of a material - for example, 'the refractive index of Final Product Styrene at $30^{\circ}C$ must be no less than 1.5408'. In addition to being a purity criterion, refractometry can be used as a method of analysis to estimate the amount of material present in a mixture. Calibration graphs similar to that shown in Fig. 1.7 are used for this purpose. They can be constructed using an Abbe refractometer that has its sample holder suitably thermostated to ensure that all readings are taken at the same temperature. Once a calibration graph has been constructed for dehydrogenated mixture, the effective use of refractive index as a tool for its analysis is then a simple matter.

For a variety of reasons, titrimetric techniques (manual, mechanised, or automated) are a common means of quantitative analysis. Acidimetry and alkalimetry are both widely used. As an example, consider the use of an acidimetric method for the estimation of NaOH and Na_2S in a mixture of the two. This determination is often required when caustic soda is used to remove hydrogen sulphide from a gas stream.

Most crude petroleum oils contain sulphur compounds. After cracking, the sulphur usually appears as H_2S. It is removed from the system by washing (or 'scrubbing') the cracked gases with caustic soda. The following equation represents the reaction involved:

$$NaOH_{aq} \longrightarrow 2Na^+ + H_2O + S^{--}$$

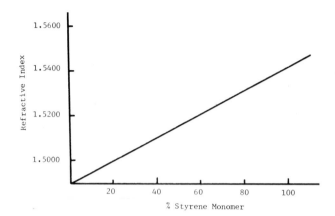

Fig. 1.7   Calibration Graph for the determination of
Styrene Monomer using an Abbe refractometer.   Graph of
refractive index versus Styrene content at 30°C.

Analysis of the caustic soda solution that is cycled through the H_2S
extractor has to be performed at frequent intervals.   These analyses are
designed to

   (1)   ensure that the NaOH content does not fall below some
       preset limit, and
   (2)   check that the level of Na_2S does not become too high.

The results of these analyses are used to determine whether the
circulating caustic soda needs to be exchanged for a fresh charge.   The
analysis can be easily effected using a double indicator technique.

Suppose a sample of the caustic soda solution is titrated with standard
hydrochloric acid (a) to the phenolphthalein end-point and then (b) to the
methyl orange end-point.   If P ml is the titre for (a) and M ml is the
titre for (b) using standard HCl of normality N and a caustic sample size
of V ml, it can be shown (Bar66) that the concentrations of sodium
sulphide and sodium hydroxide are given by the expressions:

     % Na_2S w/v   = $\dfrac{7.8N(M-P)}{V}$, and,

     % NaOH w/v   = $\dfrac{4N(2P-M)}{V}$

Thus, if a 5 ml sample of spent caustic requires titres of 14.9 and 17.5
ml of 1.026N standard HCl to reach the (a) and (b) end-points, then the
concentrations of Na_2S and NaOH would be 4.16% and 10.10%, respectively.

Gas chromatographic methods of analysis are extensively used for assaying
gas and liquid samples from a wide variety of sources.   Details of this

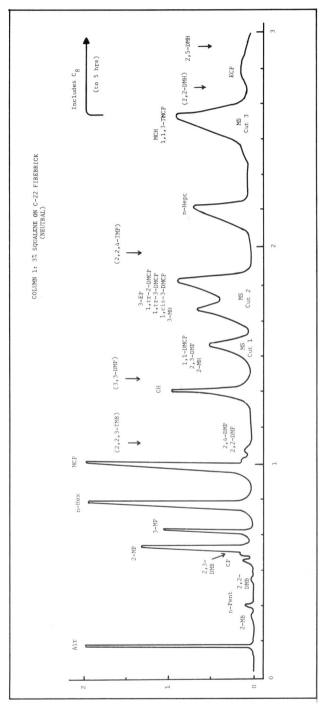

Fig. 1.8 Chromatogram of $C_5 - C_7$ saturated hydrocarbons.

technique will be presented later. All forms of gas chromatography are used - ranging from simple Janek chromatographs through to sophisticated automatic microcomputer based systems. The results of gas analysis conducted by this method are usually presented graphically on a strip-chart recorder. The concentration-time graph so produced is referred to as a chromatogram. A typical example is shown in Fig. 1.8 (Egg58). Ideally, each peak on the graph represents a single component of the gas mixture presented for analysis. The area under the peaks (or, if they are narrow, their height) may be used to estimate the amount of each component present in the mixture. There are various ways of performing the necessary calculations that convert the graphical data into the final analytical results. Appropriate measurements may be taken manually from the chromatograms and processed using a pocket calculator or desk-top computer. Alternatively, some form of automatic 'computing integrator' can be fitted to the chromatograph so that it provides a direct read-out of concentration for each of the components.

The basis behind colorimetric methods of analysis is the accurate measurement of the colour of a solution containing some derivative of the material being sought. The first step is thus to convert the compound of interest into a coloured complex which, if possible shows some sharp absorption in the visible region. Then, after determining the absorbance or optical density of the solution containing the complex, it is possible to determine the concentration of the compound by use of the Beer-Lambert law. Selection of an appropriate wavelength at which to conduct the analysis is facilitated by inspection of the absorption spectrum of the species of interest. An example is presented in Fig. 1.9. (Bar66).

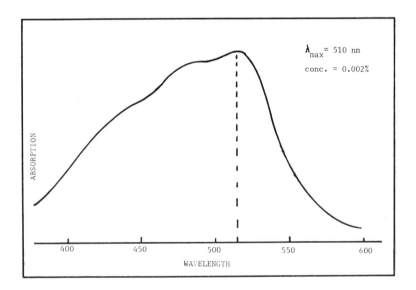

Fig. 1.9 Absorption spectrum for Fe^{++}/o-phenanthroline complex.

The colorimetric determination of $FeCl_3$, in reactor solution (Bar66) may be cited as an example of the techniques involved. Ferric chloride is a catalyst in the manufacture of ethylene dichloride (see Fig. 1.3),

$$Cl_2 + C_2H_4 \xrightarrow{\quad FeCl_3 \quad} C_2H_4Cl_2$$

The catalyst is present to the extent of about 0.6% w/v of reactor solution. In the interests of efficiency its level has to be continually monitored. The method is quite straightforward. A weighed sample of reactor solution is extracted with aqueous HCl (1+3). The extracts are placed in a volumetric flask and a little hydroxylamine hydrochloride added (in order to reduce the ferric iron to its ferrous state. This is followed by the addition of concentrated ammonia (with shaking) until the pH of the solution is between 5 and 9. After standing for 10 minutes the optical density of the solution is read against a reagent blank using 4 cm glass cells and a spectrophotometer set at the absorption maximum (510 nm) for the Fe^{++}/o-phenanthroline complex (see Fig. 1.9). The optical density value is then converted into a concentration value by an appropriate calibration chart.

Conclusion

A brief description has been given of the role of analysis in petrochemical manufacture. Its major function is as a tool for process control and quality assurance of raw materials and final products. Details of some of the more widely used manual analytical techniques are presented in order to indicate the intricacies of the tasks involved.

Example 2   -   The Clinical Laboratory

Application Description

There are a variety of ways in which analytical chemistry plays a vital role in safe-guarding the health of society. Its role in medical science may be illustrated by reference to the type of activities that take place in a chemical laboratory attached to a hosital or health centre. Figure 1.10 shows how such a laboratory is likely to be used as a tool to support a medical practitioner's decision making activity. Such a facility may be called upon as an aid to diagnosis or as a means of following the effects of some course of treatment that the doctor has prescribed for a particular patient. Samples for analysis are obtained from the patient either during a consultation with the doctor or as a result of visiting the local health centre or hospital. Ultimately, these reach the laboratory where they are analysed using a variety of techniques - involving both manual and automated procedures. The results of the analyses are returned to the doctor and copies are archived for future reference.

If the laboratory is part of a large hospital then it is likely that many analyses will be performed - typically in excess of 1,000,000 per annum (Sci80). These may range from routine serum electrolyte analyses and simple blood sugar tests for diabetics to faecal fat analyses and breath tests (using C-14 tracer techniques) in order to study gastro-intestinal diseases. In a situation of this kind several types of problem are likely to be encountered. Amongst the principal ones are:

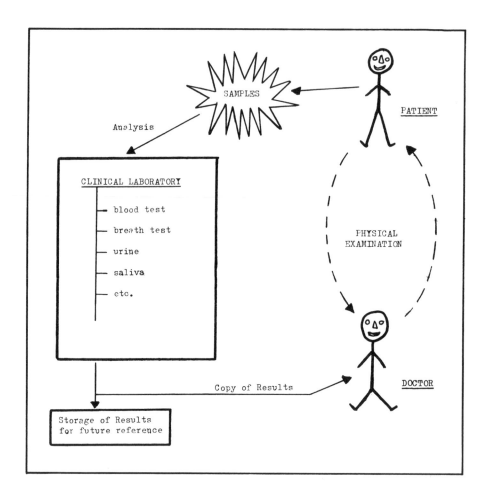

Fig. 1.10   Analytical chemistry as a health screening tool.

(a) achieving the desired analytical throughput to meet the demands
     of routine screening and research work,
(b) catering for the wide range of analytical techniques that are
     necessary to diagnose the many different human ailments, and
(c) arranging for the simultaneous availability of results to
     several doctors located in different parts of the hospital.

In order to solve problems of this type the clinical laboratory has turned
to the use of automated equipment and computer systems where these have
been able to offer plausible solutions.

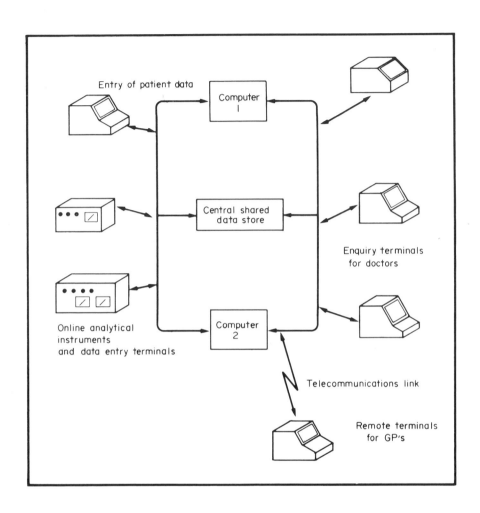

Fig. 1.11   Integration of analytical instrumentation
and computer systems to produce medical information
systems for hospitals.

Figure 1.11 shows the type of configuration likely to be employed (Sci80,
Spa80). Two interlinked computers share a large central store containing
details of hospital patients (typically, 50,000). Laboratory data produced by clinical analysis is entered into the system manually via data
entry terminals and automatically from on-line instruments. These may be
located anywhere throughout the hospital campus. The computer system
enables analytical results to be made available (simultaneously) to
medical staff via data interrogation terminals. Like the data entry
facilities, these are distributed at convenient points within the

hospital. Similarly, using appropriate telecommunication links, doctors who are not resident on the hospital campus can gain access to the stored results via data interrogation units located within their local surgeries.

Analytical Methods

Many of the analytical procedures used in the clinical laboratory are not unlike those used in an industrial process control lab. Titrimetry, colorimetry, gas chromatography, liquid, plate and column chromatography are amongst the many standard procedures employed (Mei77). However, because of the nature of the samples and chemistry involved, a variety of other techniques are also routinely used. Common amongst these are:

(a)   reaction rate analysis,
(b)   electroanalytical techniques,
(c)   atomic absorption analysis,
(d)   radioimmunoassay and scintillation counting, and
(e)   continuous flow methods.

Reaction rate analysis involves initiating a chemical reaction and then observing its progress as a function of time. The technique is often used for assaying dehydrogenases in body fluids (Kin65). An important example of this class of compound is α-hydroxy-butyrate-dehydrogenase (HBD) whose level in blood serum is often used to diagnose myocardial infarcation. It may be determined by a method due to Rosalki (Ros60, Ros63) using a special type of spectrophotometer called a reaction rate analyser. The principle of the method depends upon the fact that HBD catalyses the reaction

$$\text{substrate} \quad + \quad \text{NAD} \;\; \underset{\longrightarrow}{\overset{\longleftarrow}{\rule{3cm}{0pt}}} \;\; \text{product} \quad + \quad \text{NADH}_2$$

in which NAD denotes β-nicotinamide-adenine-dinucleotide. The rate of reaction can be monitored at 340 nm thereby providing a measure of enzyme activity and, hence, concentration of HBD. The instrument that is used will permit batches of up to 16 samples to be automatically processed in sequence. Results may either be presented graphically for manual interpretation or passed to a computer (via a suitable electrical interface) for automatic processing.

Electroanalytical methods are a popular means of observing body functions either in vivo or via samples presented to the laboratory for analysis. Because of the nature of the signals they produce these methods are often ideally suited for computerisation. The most widely used techniques are based upon some form of electrode - probably, the most common and simple being pH determination using a pH meter. Special instruments based upon the use of combinations of electrodes are often used for the determination of the pH of blood and for blood gas analysis - pCO_2 and pO_2.

Carbon dioxide is a natural product of cell metabolism. Disturbances in its partial pressure indicate disorders in acid-base balance and possible difficulty with respiratory exchange of carbon dioxide between the lungs and the blood. The partial pressure of oxygen in the blood or plasma reflects the extent of oxygen exchange between the lungs and blood, and in some cases, the ability of blood to adequately perfuse body tissue with oxygen (Gam67).

What is analysis?

Various types of electrodes (Sev68, Cla56) and techniques (Ham69, Ada67, Mat59, Bat64) are available for the measurement of pH, pCO_2 and pO_2. Particularly important is the method of sampling. Because whole blood and plasma tend to equalise oxygen and carbon dioxide partial pressures with their surroundings considerable care has to be taken. Consequently, special sample collection and storage techniques are required. Also, since the performance of the electrodes is dependent upon their proper calibration and maintenance, they have to be continually checked (every few hours) using calibration gases of appropriate concentrations. The automation of these calibration procedures can be achieved through the application of appropriately designed computer systems.

Body fluids are often conveniently analysed by means of atomic absorption (Wel76) and emission spectroscopy (Str73). The method is quite straight-forward. After suitable pre-treatment, the sample to be analysed is introduced in the form of a solution into a flame. This disperses it and produces a cloud of neutral atoms. In emission spectroscopy, the energy of the flame is used to excite the atoms. When they return to the neutral state they emit light at a wavelength characteristic of the species. The intensity of the light is directly proportional to the concentration of the atom in the flame. Absorption analysis is similar. Radiation from a suitable source is passed through the flame and the neutral atoms decrease the intensity of the beam as they absorb. The amount by which the radiation level is depleted depends upon (1) the absorption coefficient of the element at the wavelength at which the measurement is made, (2) the length of the light path, and, (3) the concentration of the element being determined. By keeping the first two factors constant, the amount of light absorbed can be measured and used as an indication of the concentration of the element.

Absorption and emission are complementary techniques. The analyst selects the one which has the greatest sensitivity and which produces the greatest precision and accuracy for a given application. Flame emission may be less specific than atomic absorption and is subject to greater spectral interference. Many elements can be determined by either method (Al, Ba, Ca), others are best analysed by absorption (for example, Be, Bi, Au, Zn) while the remainder are best handled by flame emission (U, Ru, Na, Th, etc).

When blood serum is analysed its lithium content is measured by emission and its copper concentration determined by absorption. Prior to analysis the performance of the machine is optimised for the wavelength being used and is then calibrated using solutions of known concentration. Optimisation of the instrument may be performed manually or automatically by means of a built-in microcomputer system.

Methods of analysis based upon the use of radiochemicals are widely employed in the clinical laboratory. The use of C-14 measurements (Str80) has already been mentioned. Perhaps more commonplace are methods based upon the use of radioimmunoassay (Mil81). These provide a sensitive method for determining the concentration of an antigenic substance in a sample by comparing its inhibiting effect on the binding of a radioactive labelled antigen to a limited amount of specific antibody with the inhibiting effect of known standards. Usually, the technique depends upon measuring the radioactivity level of a sample containing I-125 or Co-57.

Iodine occurs naturally in many molecules of biological significance. Furthermore, it can easily be introduced (using iodination reactions) into many compounds of clinical importance. Consequently, I-125 activity can be used as the basis for many important methods of assaying iodine containing compounds. One of significant importance is thyroxine whose molecule contains four iodine atoms. This substance is produced by the thyroid gland and its level in the body has a marked effect on behaviour. Too low a level causes lethargy and obesity while too high a concentration produces hyper-activity and extreme leanness. Its level can be measured by radioimmunoassay using a gamma counter. The method involves comparing the radioactivity count produced by an appropriately pre-treated serum sample with the counts produced from samples of known concentration. Various data reduction techniques may be used to correlate gamma activity with thyroxine concentration - log-logit with linear regression, semi-log with linear regression, and so on. The data reduction technique used will depend upon the exact details of the assay involved. Invariably, desk-top computers are used to help with the data processing activities that are involved.

Liquid scintillation counting based upon the use of tritium is another often used method of radio-analysis (Str73). It is frequently employed in those situations where iodine cannot be introduced into a molecule and to which the previous method of analysis therefore cannot be employed. The method is used for assaying many steroids and hormones. Its application to the measurement of urinary free cortisol by competitive protein binding has been described by Beardwell (Bea68). The basic principles of the method are similar to those described previously for thyroxine.

Continuous flow analysis uses similar principles to those involved in reaction rate measurement. However, instead of employing replication of discrete samples as a means of automating operation, this technique uses a continuously flowing solvent/reagent stream (All74, Ama72). The basic arrangement of the equipment is illustrated in Fig. 1.12.

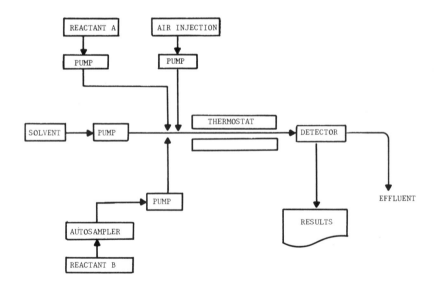

Fig. 1.12 Basic principle of an air-segmented
single channel continuous flow analyser.

What is analysis?

Successful application of the method depends upon the fact that when two or more reagents are mixed together they produce a species that is capable of measurement by either spectroscopic or electrochemical means. An autosampler is used to introduce a sequence of samples into the flow stream. Sample mixing is prevented by means of suitably chosen time delays between sample injection and by use of air-segmentation - a technique whereby tiny air bubbles are introduced into the flowing stream of liquid. Materials are pumped through the system by means of peristaltic pumps.

The basic process implemented in the above single channel analyser could be automated in a second dimension by increasing the number of channels available for sample analysis. Automation of this type is employed in the Technicon SMAC computer controlled biochemical analyser (Tec76). This is a sophisticated instrument capable of performing up to twenty simultaneous analyses in parallel. Typically, the machine is used to analyse serum for all (or a pre-defined subset) of the following materials:

Sodium	Urea	Calcium	Total Protein
Potassium	Creatinine	Phosophorus	Albumin
Bicarbonate	Glucose	Cholesterol	Bilirubin
Chloride	Urate	Triglycerides	Alkaline Phosphatase
SGOT	LDH	CPK	Globulin

Detailed flow diagrams indicating the way in which the Technicon system operates will be presented later (see chapter 3). Usually the detectors used within the analyser are spectroscopic and operate at wavelengths of 410-660 nm for visible and 340 nm for ultraviolet determinations. Sometimes, as in the case of K^+ and Na^+ measurements, ion selective electrodes are used to measure the concentration of a species. The overall activity of the instrument is controlled by its own built-in computer. This is used to monitor the status of the machine, record results and process them, and also keep a check on sample integrity. Above all, the system offers a useful solution to the high through-put requirements of the clinical laboratory.

Conclusion

The clinical chemistry laboratory uses many different analytical methods for processing specimens that originate from a wide variety of sources and which may differ considerably in type. Usually, because of the larger number of samples that need to be processed, techniques that are rapid to perform and which lend themselves to automation are particularly useful. Often these methods involve the use of some form of instrumentation. In large hospitals there is a need to use on-line instrumentation that is capable of entering results directly into a computer system.

Example 3 - Food Chemistry

Application Description

If one interprets the term 'food' in the most general way possible, then, without any doubt, the food industry is, collectively, one of the largest and most important of all the modern manufacturing endeavours. It acts as a hub around which much other scientific, technological and business activity takes place. Indeed, based upon this industry, there exists a

complex spectrum of ancillary professions ranging from waste product
disposal through to refrigeration and catering.

Essentially, there are three basic steps involved in the development of a
food industry:

1. Securing of Raw Materials
   - from agricultural or dairy products, animal breeding
   or the fishing industry,
2. Formulating and Implementing the Production Processes
   - such as preparation, cooking, preserving, packaging,
   distribution and so on, and
3. Developing and Supporting Consumer Outlets
   - for the wide variety of goods that are produced.

Foods for human consumption vary in scope from simple beverages and
confectionery through to complex and sophisticated gastronomic
delicacies. For animals the range is equally diverse. It includes a
multitude of pet-foods and a variety of feed-stocks for farm use and for
the support of wild-life in adverse conditions. Some of these latter
varieties of food may be based upon the recycling of human waste or the
growth of special purpose crops. Increasingly, however, there is growing
interest in the use of synthetic foods generated from petrochemicals and
other raw materials as a means of producing dietary material for animals
(ICI81).

Because the health and well-being of society are so dependent upon the
food industry, the supporting role that analytical science has to play is
one of significant importance. Within the overall context of the chemical
analysis of foods one is likely to encounter many different aspects. Some
of these are indicated by the various relationships shown in the simple
sketch presented in Fig. 1.13.

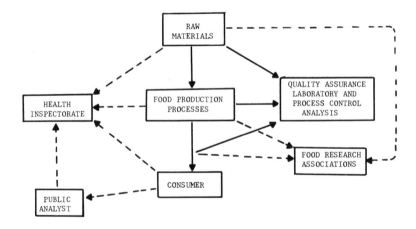

Fig. 1.13 The role of analytical science in food chemistry.

This illustrates the conversion of raw materials into consumable products
via appropriate manufacturing processes. Throughout these, there is

likely to be a considerable demand for intra-organisational process control and quality assurance analysis. However, as a result of the complex nature of many of the problems that the food industry faces, many specialised research centres have developed on the basis of their being able to offer expert advice on particular specialised analytical techniques. Similarly, because of the implications for public (and animal) health several other organisations have substantial interest in food analysis. Of these, some of the more important ones are the public analyst and the numerous other bodies that are sponsored by the various government health departments.

Analytical Methods

Two important factors dominate food analysis: the wide range of products involved and their highly varied nature. Food has to be analysed with many different objectives in view - for example, its appearance, its aroma, its quality (calorific value, nutritional content, carbohydrate level and so on), and its safety - freedom from contamination and the absence of substances likely to harm the consumer. In order to cope with the sample diversity and wide range of requirements a significant repertoire of analytical techniques (both classical and modern) are required. These will be designed to enable the various physical, chemical and microbiological specifications of a given material, product or intermediate to be checked both accurately and reliably.

The more popular techniques of food analysis (Fol77, Yer79) are based upon methods that involve the use of:

- TLC, column and high pressure liquid chromatography,
- gas chromatography,
- atomic absorption and emission spectroscopy,
- UV, Visible and IR spectroscopy,
- mass spectrometry,
- low resolution NMR spectrometry,
- electroanalytical methods (electrophoresis, titrimetric, etc),
- luminescence analysis (phosphorescence and fluorescence),
- X-ray fluorescence, and,
- continuous flow analysis.

In order to gain some understanding of how these tools are used it is necessary to make a more detailed examination of some of the situations where analysis is employed within the food industry. Using the model depicted in Fig. 1.13 as a framework, six examples have been chosen for further study. These, together with an indication of their relevance, are listed below:

(a) crop nutrient analysis - raw material quality,
(b) food sterilisation - long term storage of materials,
(c) food packaging - product protection,
(d) biochemical assay - consumer health protection,
(e) protein assay - quality assurance control,
(f) odour analysis - environmental protection.

Soil and crop nutrient analysis is important for a variety of reasons (Int80). Since the soil is indirectly responsible for a large proportion of the raw material input to the food industry, maintaining it in a condition that sustains healthy crops (and livestock) is a prime objective.

An important factor that effects the quality of the soil is the level of
trace elements such as iron, cobalt, copper, magnesium, manganese,
molybdenum and zinc. The levels of the micronutrients can significantly
influence the health of both crops and livestock - animals begin to suffer
from Co deficiency if they are fed on pastures produced in soils
containing less than 3 ppm of cobalt. Because of effects of this type it
is important for both the farmer and commercial grower to be aware of the
levels of nutrients in the soils they use and in the tissues of the plants
they grow.

A chemical test of the soil can generally reveal what is available in the
soil and what is missing. Tissue analysis can be used as an indication of
what the plant itself is actually short of, or, of those nutrients which
are present in adequate amounts. Based upon the results of this type of
analysis the farmer/grower can treat the soil so as to produce the best
yields of healthy crops. The analysis for micronutrients is not difficult
- atomic absorption and emission can be used. However, the sampling and
recording of results is a more involved process since, usually, for a
given area of pastureland many samples will need to be taken over what
could be a substantial period of time. Results (in terms of crop
improvement and nutrient levels) will need to be correlated with any soil
treatment that the farmer employs. To aid the scientist in the book-
keeping and result processing aspects of this type of work a computer is
invariably used.

Automated food processing techniques are capable of yielding products at
rates which can greatly exceed consumer demand. Consequently, some form
of shelf storage may be necessary prior to distribution or consumption.
For this reason preservatives and sterilants are often used to extend the
storage life of foods or food products by retarding or preventing changes
in flavour, colour, odour, nutritive value, texture or consistency of
appearance (Gam73). A variety of gaseous chemicals (ethylene oxide,
methyl bromide, proplyene oxide and ozone) have been used for this
purpose. Because of the possible health risks government agencies lay
down strict regulations relating to the addition and use of chemicals in
food processing and so it is important to be able to monitor the residual
sterilant in a food product. Furthermore, to ensure appropriate levels of
protection it is necessary to check that sterilisation units maintain
adequate levels of sterilant.

As an example, consider the use of ethylene oxide. This has been used for
many years in the treatment of foods for the prevention of moulds and
fungal growth. Of the different methods available for determining its
concentration, gas chromatography is probably the most suitable for
process control application. Phillips (Phi80) has described a novel
system of automated analysis that provides a reliable means of determining
vessel gas concentrations in order to verify that the sterilisation
process meets established specifications. The method is based upon the
use of a computerised GC system and an automatically actuated 8-way
switching valve and sample unit to which are attached six different sample
probes. These are distributed at appropriate positions within the
sterilisation chamber in order to obtain representative samples. The
computer program that controls the system is responsible for initiating
the analyses, analysing the six GC chromatograms that are produced and
then reporting the results.

Another important factor that strongly influences the shelf life of a food
product is its packaging. There are a variety of ways in which the effect

of different approaches to packaging can be studied. One interesting method based upon mathematical modelling has been described by Karel (Kar73). In this approach modelling techniques are used to study the kinetics of food deterioration and to predict the required packaging protection for a given required storage life.

The model used to represent food-packaging interactions with an environment is shown pictorially in Fig. 1.14. The properties of food that determine its quality (F) depend upon the initial condition of the food (F_0) and upon reactions which change these properties with time (t); these reactions depend upon the internal environment of the food (I). Typical factors that influence deteriorative mechanisms (and hence shelf life) are likely to be oxygen partial pressure (PO2), relative humdity (RH) within the package, temperature (T) and so on. These factors can be combined into a mathematical equation of the form:

$$\frac{dD}{dt} = f(RH,PO2,T) \qquad \dots\dots\dots\dots\dots\dots\dots\dots\dots\dots\dots\dots\dots \qquad (A)$$

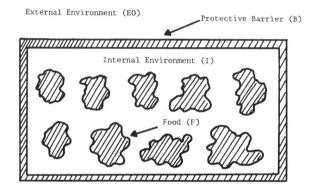

Fig. 1.14 Simple model for the effect of packaging on food.

Similarly, the internal environment (I) depends upon the condition of the food (F), packaging properties (B) and on external environmental parameters. Thus, if Q is the water activity in the food at any time, then the following equation can be formulated:

$$Q = f(Q_0,t,RH,k_1 \dots k_n,T \dots) \qquad \dots\dots\dots\dots \qquad (B)$$

where Q_0 is the initial activity and k_1 through k_n are constants characterising the sorptive and diffusional properties of food and of the package.

The various equations of types (A) and (B) shown above can be combined with values obtained form analytical measurements and the resulting equations solved with the aid of a digital computer to yield solutions that enable storage life or required package properties to be predicted. The validity and usefulness of the technique are illustrated by the correlation between the predicted and experimentally observed results depicted in Fig. 1.15. These show the predicted and actual increase in moisture content of dehydrated cabbage stored in various packaging materials.

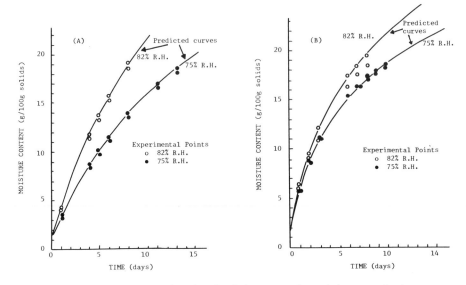

Fig. 1.15 Predicted and actual increase in moisture content
of dehydrated cabbage stored in (A) polyamide pouches and in
(B) polyethylene-coated cellophane; RH: relative humidity.

Because of its influence on the health of the consumer food must be
manufactured under scrupulous conditions of hygiene. This is necessary in
order to avoid the possiblity of contamination by bacteria. As a means of
ensuring the safety of a food product a variety of biochemical methods are
used to detect bacteria and measure their concentration. Unfortunately,
quality control bacteriology in the food industry has often posed a number
of problems: high material costs; difficult technical and labour problems;
and, lengthy analytical times imposed by conventional incubation periods.
Taken together, these factors have often imposed severe limits on the
number of analyses that could be made.

Based upon these observations it is easy to see that bacteriology is an
area where instrumentation and automation of analysis is needed in order
to increase the analytical throughput that can be achieved using this type
of quality control. Some of the early approaches to solving the problems
associated with automation and instrumentation have been described by
Sharpe (Sha73). He describes the use of bioluminescent reactions as a
basis for quantitative instrumentation and a variety of mechanised
equipment to aid the overall assaying process and thereby improve the
analytical throughput of this type of biochemical analysis. To this end,
nowadays, a variety of computerised aids are used - microcomputer
controlled pipettes, syringes, diluters, dispensers and a wide range of
instrumental techniques (Ric78, Che81).

Customer protection in terms of value/price is another important reason
for quality control analysis in the food industry. Protein assay is a
typical example of this type of activity. The protein content of a food
substance can be determined using a pulsed nuclear magnetic resonance
spectrometer in conjunction with a relaxation reagent (Col80). A

relaxation reagent is a solution containing a paramagnetic material (for example, copper) which controls the magnetic relaxation rates of the hydrogen nuclei in the water of the reagent. Addition of a substance which complexes or binds the paramagnetic ions will cause measurable changes in the magnetic relaxation rates. Measurement of these changes enables the quantity of substance to be inferred - after prior calibration. This technique forms the basis of a fast automated instrumental technique for protein assay (New80).

The instrument is calibrated using a conventional Kjeldahl total nitrogen analysis (Vog61). Once calibrated an individual assay takes only 20 seconds; samples may be batched and run through the instrument automatically. The system provides a microprocessor controlled calibration and data reduction facility that permits direct protein content read-out. These results may be reported either as a concentration in solution or as a percentage weight. The instrument contains an on-board printer to provide hard copy of the experimental results that are produced and a decimal keypad to permit digital data entry. In addition there is a built-in oscilloscope (to examine NMR traces) and a digital display screen for the presentation of results and messages to the operator.

This instrument provides an excellent illustration of the way in which computer technology, electronics and analytical chemistry combine to produce an analytical tool that is of substantial utility for solving real world problems. A further example of how these different subject areas unite in this way is illustrated by instrumental approaches to the problem of odour analysis.

Odour analysis and control is an important aspect of food manufacturing processes. Uncontrolled odours can influence employees involved in plant operation and local inhabitants of the environment in which the factory may be sited. For a variety of reasons the analysis of odours represents an interesting analytical problem. Bedborough (Bed80) has described the use of human sensory measurement of odour based upon the use of panellists. Both Bailey (Bai80) and McGill (McG77) have described instrumental methods of approaching the problem. Unfortunately, instrumental methods cannot be used as a definite measurement of quality of odour. Its main applicability is in identifying those chemical compounds contained in the gas mixture that are likely to be responsible for the odour.

Methods of odour analysis are most often based around the use of gas chromatographic separation followed by subsequent identification using a variety of techniques. These include the application of special gas chromatographic detectors (for example, microwave plasma detector, flame ionisation detector and so on) and mass spectrometry. Indeed, coupling the gas chromatograph to the mass spectrometer (called GC/MS analysis) provides a powerful analytical tool if a dedicated computer is available for data acquisition and processing. The addition of an infrared spectrometer to the system further enhances its power and capability for identification (Wil81). Systems of this type provide an easy means of identifying the many components of complex mixtures since it is possible to compare the mass spectrum (and infrared absorption pattern) of each component with those contained in a computer based library. Using the computer it is possible to perform sophisticated pattern matching operations thereby enabling many different kinds of identification procedures to be implemented. Tools of this type will be discussed in more detail later.

Conclusion

Because of the extremely varied nature of food, its analysis can often require the application of a large number of techniques rang'ng from conventional wet chemistry through to sophisticated instrumental and biochemical methods of assay. Often, techniques of analysis are required which can be made automatic and, hence, be included in on-line process control instrumentation. Where this is not feasible, standard manual methods have to be employed. Sampling is also an important consideration in food analysis. This often requires the development of special techniques and procedures in order to gain a representative value for the overall composition of the food. Also, because food can rapidly deteriorate, the time span between sampling and analysis may also be an important factor that is likely to influence the results.

WHAT IS ANALYSIS? - CONCLUSION

From what has been said in this chapter it is easy to see that the fundamental role of the analyst within Society is that of an observer. He (or she) uses the scientific method to observe (and report on) a multitude of different types of process in which the nature of chemical substances are transformed in some way. The art, skills and techniques used in fulfilling this role are often collectively referred to as <u>analytical science.</u>

In this introduction an attempt has been made to describe the nature of analytical science by examining some of the types of activity that takes place in particular application areas: petrochemicals, health chemistry and the food industry. The examples cited have been sufficient only to describe just a few of the methods and techniques that are involved in the scientific analysis of matter in its very many varied forms. The diversity of skills needed by the analyst can be matched only by the wide range of problems he/she is likely to be confronted with, for example,

    What was used to poison the victim?

    What is the nicotine content of this cigarette smoke?

    Has this athlete been taking amphetamines?

    What is the composition of this motor vehicle's exhaust?

    Does this sample of milk contain lead?

    How much sulphur is in this coal sample?

    Is this well water too high in iron?

    What is the composition of this slag?
    etc

A deliberate attempt has been made, through the nature of the examples that have been chosen, to demonstrate the importance of analytical science as it effects people in their everyday lives. Only a very small number of case studies have been presented - many more could have been prepared. Virtually every aspect of modern life is influenced by the activities of the analyst:

Public Health   Paper Manufacture
Perfumery   Clothing/Textile Industry
Printed Circuit Board Manufacture   Drug Screening and Testing
Brewing   Food Production
Semiconductor Industry   Confectionery
Explosives Industry   Pollution Monitoring
Dying Industry   Petroleum Industry
Paint Manufacture   Nuclear Fuel Industry
Detergents   Ceramics
Legal Work   Space Research
Marine Research   Pesticides
Clinical Chemistry   Rubber Production
Fertilizers   Cosmetics
Geological Research   Surface studies
etc

In fewer than 100 years analytical chemistry has progressed from a mystic somewhat unreliable art (Cam79) to a precise reliable science that affects a large number of other areas of human activity. Today, as is evidenced by the above list, analysis makes an extremely important contribution to the functioning of a considerable number of organisations many of which exhibit a large multi-national structure. Bearing this in mind the role of the analyst should not be considered in isolation from those other activities that take place within an organisation. In most industrial, commercial and public service environments the analyst is a part of a sophisticated team. This team is responsible for the production of information that ultimately effects both its parent organisation and Society in many significant ways.

Instrumentation of various types is now to be found in most analytical laboratories. In many situations the conventional techniques of analysis are slowly changing. Increasingly, the analyst is turning to the use of computers as an aid in many of the tasks involved in analysis. Computers appear either in the form of powerful desk-top calculators or as an integral part of the apparatus used in many routine analyses. In this book an attempt is made to look at some of the ways in which the computer and its related technology may help the analyst conduct his/her job more easily, more efficiently and with less effort than was previously necessary.

Although a multitude of instrumental methods now exist, there is still a considerable demand for many of the classical analytical techniques. Despite the trend towards the use of (computer based) instrumentation, very often, classical methods are still used in a large number of laboratories. Modern instruments are very expensive. Consequently, in many situations the use of an instrumental method (if one exists) may not be justifiable and so some form of manual technique has to be employed. Even where instrumental methods are extensively used, it is not unusual to find a wide variety of classical methods used as 'back-up' procedures to cover for the possibility of an instrument not being available because of some form of malfunction. There are many other reasons why the standard methods of classical analysis are still extensively used in a significant number of laboratories. However, while respecting their importance, little else will be said about them. Instead, selected areas of computer application within analytical chemistry will be the subject of subsequent chapters of this book.

REFERENCES

Ada67   Adams, A.P., Morgan-Hughes, J.O. and Sykes, M.K., pH and Blood
             Gas Analysis: Methods of Measurement and Sources of Error Using
             Electrode Systems, Anaesthesia, Volume 22, 575, 1967.

All74   Allen, S.E., Grimshaw, H.M., Parkinson, J.A. and Quarmby, C.,
             Chemical Analysis of Ecological Materials, Blackwell
             Scientific Publications, ISBN: 0-632-00321-9, 1974.

Ama72   Amador, E. and Urban, J., Simplified Serum Phosphorus Analysis
             by Continuous Flow Ultraviolet Spectrophotometer, Clinical
             Chemistry, Volume 18, No. 7, 601-604, 1972.

Bai80   Bailey, J.C., Instrumental Analysis of Odours, Chapter 5, 31-42,
             in Odour Control - A Concise Guide, Edited by Valentin, F.H.H.
             and North, A.A., Warren Spring Laboratory, ISBN: 0-85624-2144,
             1980.

Bar66   Barker, P.G., Some Aspects of Petrochemistry, Bound Version of
             the Presidential Address to the University of Swansea Student
             Chemical Society, 1966.

Bat64   Bates, R.G., Determination of pH: Theory and Practice, John
             Wiley, New York, 1964.

Bea68   Beardwell, C.G., Burke, C.W. and Cope, C.L., Urinary Free
             Cortisol Measured by Competitive Protein Binding, J.
             Endocrinology, Volume 42, 79-89, 1968.

Bed80   Bedborough, D.R., Sensory Measurement of Odours, chapter 4,
             17-30, in Odour Control - A Concise Guide, Edited by Valentin,
             F.H.H. and North, A.A., Warren Spring Laboratory, ISBN:
             0-85624-2144, 1980.

Bei72   Beishon, J. and Peters, G. (Eds), Systems Behaviour, Open
             University Press, SBN: 06-318011-1, 1972.

Ber68   von Bertalanffy, L., General Systems Theory: Foundations,
             Developments, Applications, Penguin Books, ISBN:
             0-14-060-004-3, 1968.

Cam79   Cambell, W.A., Fact and Fantasy in Chemical Analysis,
             Endeavour(New Series), Volume 3, No. 1, 38-41, Pergamon Press
             Ltd., 1979.

Che72   Checkland, P.B., A Systems Map of the Universe, Chapter 3,
             50-55, in Systems Behaviour, edited by Beishon, J. and Peters,
             G., Open University Press, SBN: 06-318011-1, 1972.

Che80   Imperial Chemical Industries Ltd., The "Olefines 6" Plant at
             Teesside, Chemistry in Britain, 299, Volume 16, No. 6, June
             1980.

Che81   Anon., Counting Bugs - AC Banishes Tedium, Chemistry in
             Britain, Volume 17, No. 3, page 97, March 1981.

Che81   Anon., Counting Bugs - AC Banishes Tedium, Chemistry in
         Britain, Volume 17, No. 3, page 97, March 1981.

Cla56   Clark, L.C., (Jnr), Monitor and Control of Blood and Tissue
         Oxygen Tensions, Trans. Am. Soc. Artif. Intern. Organs, Volume
         2, 41, 1956.

Col80   Coles, B.A., Protein Determination by Nuclear Magnetic
         Resonance, Journal of the American Oil Chemists' Society,
         Volume 57, No. 7, 202-204, July 1980.

Egg58   Eggertsen, F.T. and Groennings, S., Determination of Five to
         Seven Carbon Saturates by Gas Chromatography, Analytical
         Chemistry, 20-25, Volume 30, No. 1, January 1958.

Eme69   Emery, F.E., Systems Thinking, Penguin Books, ISBN:
         14-080071-9, 1969.

Fol77   Foltz, A.A., Yeransian, J.A. and Sloman, K.G., Food -
         Applications Reviews, Analytical Chemistry, Volume 49, No. 5,
         194R-220R, April 1977.

Gam67   Gambino, S.R., Blood pH, pCO_2, Oxygen Saturation and pO_2,
         American Society for Clinical Pathologists Commission on
         Continuing Education, Chicago, Illinois, USA., 1967.

Gam73   Gammon, R. and Kereluk, K., Gaseous Sterilisation of Foods,
         91-99, in Engineering of Food Preservation and Biochemical
         Processes, Ed: King, C.J., Volume 69, No. 132, American
         Institute of Chemical Engineers, 1973.

Ham69   Hamilton, L.H., Respiratory and Blood Gas Analysis, Prog. Clin.
         Path., Volume 2, 284, 1969.

ICI81   Imperial Chemical Industries Ltd., ICI's Bugs Come On-Line,
         Chemistry in Britain, Volume 17, No. 2, p.48, February 1981.

Int80   Interlates Limited, Gladden Place, Skelmersdale, UK., Chelated
         Trace Elements, Fourth Edition, July 1980.

Kar73   Karel, M., Quantitative Analysis of Food Packaging and Storage
         Stability Problems, 107-113, in Engineering of Food
         Preservation and Biochemical Processes, Ed: King, C.J., Volume
         69, No. 132, American Institute of Chemical Engineers, 1973.

Kin65   King, J., Practical Clinical Enzymology, Van Nostrand, 1965.

Mat59   Mattock, G., Electrochemical Aspects of Blood pH Measurement,
         page 19, in Symposium on pH and Blood Gas Measurement,
         Churchill, London, 1959.

McG77   McGill, J.R. and Kowalski, B.R., Intrinsic Dimensionality of
         Smell, Analytical Chemistry, Volume 49, No. 4, 596-602, April
         1977.

Mei77   Meites, S., (Ed.), Pediatric Clinical Chemistry - A Survey of
         Normals, Methods and Instrumentation with Commentry, American
         Association for Clinical Chemistry, ISBN: 0-915274-03-5, 1977.

Mil81   Miller, J.N., Antibodies - Ideal Reagents for Analysts,
         Chemistry in Britain, Volume 17, No. 2, 62-67, February 1981.

New80   Newport Instruments, Blakelands North, Milton Keynes, Bucks.,
         MK14 5AW, UK., The Newport P-100 Protein Analszer, product
         description, August 1980.

Phi80   Phillips, S.J., Automated Determination of Ethylene Oxide,
         Spectra Physics Chromatography Review, Volume 6, No. 1, 10-11,
         Autumn 1980.

Ric78   Richards, J.C.S., Jason, A.C., Hobbs, G., Gibson, D.M. and
         Christie, R.H., Electronic Measurement of Bacterial Growth, J.
         Phys. E: Sci. Instrum., Volume 11, 560-568, 1978.

Ros60   Rosalki, S.B. and Wilkinson, J.H., Reduction of α-keto-butyrate
         by Human Serum, Nature (Lond.), Volume 188, 1110-1111, 1960.

Ros63   Rosalki, S.B., α-Hydroxy-butyrate dehydrogenase Activity of
         Tissue Homogenates, Clin. Chem. Acta, Volume 8, 415-417, 1963

Sci80   Perkin Elmer Computers Speed Lab Data to Doctors, Scientific
         American, page 19, Volume 242, No. 5, May 1980.

Sev68   Severinghaus, J.W. and Bradley, A.F., Electrodes for Blood pO_2
         and pCO_2 Determination, J. App. Physiol., Volume 13, 515,
         1968.

Sha73   Sharpe, A.N., Automation and Instrumentation Developments for
         the Bacteriology Laboratory, 197-232, in Sampling -
         Microbiological Monitoring of Environments, Society for Applied
         Bacteriology Technical Series No. 7, Eds: Board, R.G. and
         Lovelock, D.W., Academic Press, ISBN: 0-12-108250-4, 1973.

Spa80   Spaziano, R., A Hospital's CARES, Datamation, 156-166, Volume
         6, No. 6, June 1980.

Str73   Strobel, H.A., Chemical Instrumentation: A Systematic Approach
         to Instrumental Analysis, Addison Wesley, ISBN: 0-201-07301-3,
         1973.

Str80   Strange, R.C., Reid, J., Holton, D., Jewell, N.P. and
         Percy-Robb, I.W., The Glyceryl C-14 Tripalmitate Breath Test: A
         Reassessment, Clin. Chem. Acta, Volume 103, No. 3, 317-323,
         ISSN: 0-09-8981, 1980.

Tec76   Technicon Instruments Corporation, 511 Benedict Avenue,
         Tarrytown, New York, USA., The Technicon SMAC High Speed
         Computer Controlled Boichemical Analyzer, product description,
         1976.

Val80   Valentin, F.H.H. and North, A.A., Odour Control - A Concise
         Guide, Warren Spring Laboratory, Department of Industry,
         Gunnels Wood Road, Stevenage, Hertfordshire, SG1 2BX, ISBN:
         0-85624-2144, 1980.

Vog61   Vogel, A.I., A Text Book of Quantitative Inorganic Analysis
         Including Instrumental Analysis, (3rd Edition), Longmans, 1961.

Wei75   Weinberg, G.M., An Introduction to General Systems Thinking,
            John Wiley, ISBN: 0-471-92563-2, 1975.

Wel76   Welz, B., Atomic Absorption Spectroscopy, Verlag Chemie,
            Weinheim, ISBN: 3-527-25680-6, 1976.

Wil81   Wilkins, C.L., Giss, G.N., Brissey, G.M. and Steiner, S., Direct
            Linked Gas Chromatography - Fourier Transform Infrared - Mass
            Spectrometer System, Analytical Chemistry, Volume 53, No. 1,
            113-117, January 1981.

Yer79   Yeransian, J.A., Sloman, K.G. and Foltz, A.K., Food -
            Applications Reviews, Analytical Chemistry, Volume 51, No. 5,
            105R-134R, April 1979.

2

Analytical Techniques

INTRODUCTION

As in most other areas of scientific problem solving, there is a pattern in handling problems in analytical chemistry. This pattern consists of the general ideas and thought processes that analytical chemists employ as they initiate and pursue the steps that lead to the solution of the particular problems in which they are involved. Some of these steps will be outlined in this chapter. Particular emphasis will be given to those areas where computer systems are being employed as a tool to aid the analytical task. In addition, some possible future uses of computers in these areas will be outlined.

Analytical chemistry, as a science, has been practised for almost 100 years and so the steps that constitute the general problem solving pattern referred to above are fairly well defined. Accordingly, each of these is described in one of five sections of this chapter. Obviously, because there is such a wealth of recorded literature and only a limited amount of space available in which to describe it, a high degree of selectivity has had to be exercised. Unfortunately, there is only sufficient space available to devote a small number of pages to each. Hopefully, these will be sufficient to provide some background material appropriately illustrated with suitable examples and references to the literature.

The stages of the overall analytical process with which this chapter is concerned are as follows:

(a) Sampling
(b) Separating
(c) Finding an Analytical Method
(d) Performing the Analysis
(e) Reporting the Results

Obviously, this list does not indicate any of the complex interactions that exist between the various steps. Furthermore, there is a considerable amount of approximation involved in this simple representation. Despite this, it is not totally unrealistic to assume that most of the analyst's time is distributed over one or more of these basic activities. The order in which they are presented does not necessarily reflect the order in which each stage is carried out. Sometimes the analyst is presented with a sample in a bottle and asked to report on its composition. On the other occasions finding a suitable analytical method is the first step and then appropriate sampling and separating techniques are designed. Of course, finding an analytical technique may involve the chemist in a considerable amount of innovative design work based upon a host of considerations that need to be taken into account for the particular situation involved. Although each of the above steps is almost always involved in an analysis, unfortunately, there is no definite algorithmic approach that can be applied to each and every situation.

The goal of every analytical problem is to get the most conclusive answer in the shortest possible length of time. As will be described later, a computer can often help minimise the amount of time taken to reach a solution. It can do this in a variety of ways - perhaps, by helping the analyst handle the literature of analytical chemistry, or, through the use of an 'expert system', enable possible solutions to a problem to be outlined. The computer may act as a powerful 'electronic notebook' enabling results to be recorded automatically during an analysis; later it might act as a sophisticated multi-colour display medium through which the results and conclusions may be presented to those involved in their inspection. Alternatively, the computer may be used as a powerful simulation tool to enable the design and evaluation of analytical methods thereby minimising the amount of costly experimentation that needs to be performed. Many analyses will need to be performed under strictly controlled conditions. Here again, the speed and accuracy of the computer can be used to check that important parameters do not change during the course of an analysis thereby adversely affecting the final results. Then, of course, there is the question of automation. Once an analytical technique has been perfected there may be a need to automate it - either to enable it to be applied to the automatic analysis of large batches of samples (as in health screening) or for use in an automatic control loop for application within some area of manufacturing industry. The computer is a useful tool to aid the analyst in the design and implementation of an automated technique.

There is another task, not involved in the original list, but in which many analysts become involved - training activities. This term is used to describe those mechanisms by which expertise and knowledge within various areas of analytical chemistry is passed on to those others who may benefit from it. This might involve providing newcomers with the background material necessary to understand older/current analytical techniques. Alternatively, its purpose might be to familiarise established practitioners with new methods of analysis and current trends in analytical chemistry. Computer based training methods are now widely employed for each of these purposes. The interested reader will find many examples of this technique in the literature.

SAMPLING

The Need for Sampling

Sampling is probably one of the most difficult aspects of the whole analytical process. For a variety of reasons, both theoretical and practical, this stage in the overall analytical scheme needs to be given considerable attention. In order to illustrate some of the difficulties to be overcome, consider the following analytical situations:

   (a) it is required to determine the compostion of an ingot of
        steel measuring 10' x 10' x 10',
   (b) the tar content of a particular brand of cigarette is to be
        estimated using a 'smoking machine',
   (c) the compostion of the gas flowing in a pipe is to be
        monitored using a gas-liquid chromatograph,
   (d) the composition of a coke + sand + sulphur mixture in a
        stock pile is to be measured, and,
   (e) the average value for the sodium content of blood serum for
        coloured males in the age range 30-50 is to be ascertained.

Before any reasonable progress with these tasks can be made some 'common sense' observations need to be applied. These relate to the amount of material that will need to be subjected to the analytical process. In cases (a) and (d), analysis of the entire bulk of material available would be impractical and too costly - if a destructive analytical technique was used. Similarly, in case (b), smoking every cigarette would not be feasible nor would the analysis of every coloured male in case (e). In case (c) the limitations of the instrument (dictated by its inherent dead-time) do not permit its use for continuous monitoring of the gas mixture.

Each of the above examples illustrates the need for the formulation of suitable sampling strategies that would enable the desired properties of the material/population to be inferred from an analysis of a representative portion of the whole. These concepts are summarised in Fig. 2.1 which depicts the extreme 'end values' of the two <u>sampling variables</u> outlined previously.

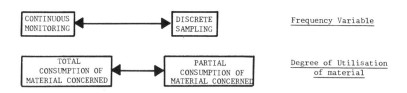

Fig. 2.1 Important sampling factors.

Continuous monitoring refers to situations wherein some variable of the system is constantly recorded. Thus, the measurement of the pH of a river using an ion selective electrode with subsequent display of the results on a chart recorder is an example of continuous monitoring. Similarly, the measurement of the temperature of a furnace with a thermocouple could constitute a second example of this mode of sampling. Two further examples are: the weighing of each and every steel ingot produced in a casting shop and the routine recording of X-ray pictures of the whole of the population of a city for health screening purpose. Often, however, continuous monitoring is not a practical proposition either because of instrument dead-time or because of the large amounts of information that is produced. Each of these difficulties could be overcome: in the first case by choosing an instrument that has minimal (or zero) dead-time and then using several of these in parallel with each other and operating at suitably staggered intervals; the second problem could be obviated by designing appropriate information filters such as 'reporting by exception'. Obviously, the use of several instruments in tandem represents only an approximation to continuous monitoring - it is really an example of discrete sampling at high sampling periodicity. Such approaches as this, and continuous monitoring generally, can be expensive. The cost problem is one which is more easily overcome.

Discrete sampling refers to the situation where a limited selection of the members of a possibly large population are choosen for analysis based upon some suitable plan. For example, 'analyse three cigarettes from every one-hundredth packet coming off the production line'; 'measure the pH of the river every twenty minutes on Mondays, Wednesdays and Fridays'; 'extract a flake of metal from each side of the ingot and obtain samples for analysis by drilling holes to a depth of 2 feet into each of the faces of the ingot at given points'. Many more examples could be cited. When discrete sampling is used the analyst has the problem of deciding upon the correct sampling periodicity (either with respect to time or any other system variable) for the situation under consideration. This will vary from application to application. Of course, when computers are involved there are additional factors to be taken into account. In many applications of laboratory or process control computers the data acquisition rate is another particularly important factor relevant to a discussion of sampling. Thus consideration needs to be given both to: (1) the time intervals at which measurements are made, and, (2) the data acquistion rate at the chosen sampling time. Obviously, (2) will determine a limit for (1). A discussion of data acquisition rate will be presented in chapter 5.

Returning to the second of the above 'sampling variables' - the degree of utilisation of material. In many situations it would be too costly, impracticable or too time consuming to consider the total consumption of the material of interest - see example (a) above. Consequently, some suitable and representative portion of it has to be obtained and subjected to the analytical scheme. The process of sampling is used to achieve this goal - obtaining a suitable fraction of material whose analysis will reflect the composition of the whole. Sampling techniques thus offer alternative procedures to continuous monitoring or total utilisation of materials. There is considerable scope for ingenuity and innovation in the design of suitable sampling systems.

There is a vast body of knowledge available on the theorectical, practical and procedural considerations of sampling and the results that can be inferred by adopting particular types of strategy or plans. Many analytical chemistry text books devote chapters to this topic (Sko63) and there are many specialist texts dealing with particular specific areas, for example, the sampling of cokes, oils, effluents, soils and so on. The statistical and mathematical aspects of sampling are covered in numerous books. Those by Wetherhill (Wet77), Paradine (Par60) and Chatfield (Cha78) each contain readable descriptions of the statistical considerations that need to be taken into account when designing various sampling methods - for example: 'design a single sampling scheme such that lots containing 2% defective have 95% chance of acceptance while lots containing 8% defective have 5% chance of acceptance'. Various standard methods of sampling are often published by many of the recognised standards institutions such as the British Standards Institution (BSI72,

BSI75) or the American Society for Testing and Materials (AST51).
Descriptions of recent advances in various aspects of sampling appear in
the literature of analytical chemistry (Kra81, Cau80), statistics and
computational science. In the context of this type of work the computer
is often helpful in designing sampling strategies and providing the
computational facilities to enable them to be theoretically evaluated
before they are put into practice.

The Pragmatics of Sampling

This term refers to the practical aspects associated with physically
obtaining a sample ready for analysis. When thinking about this problem a
large number of factors need to be taken into consideration. Of
particular importance are the material's

origin	-	human body, the moon, a chemical plant, etc,
state	-	gas, liquid, solid, heterogeneous mixture, etc,
stability	-	whether the sample decomposes in any way,
reactivity	-	whether the sample will be affected by the material from which the sampling system is constructed,
size	-	how much sample will be available for analysis,and,
environment	-	will the sample be easy to obtain or will hazardous conditions be encountered?

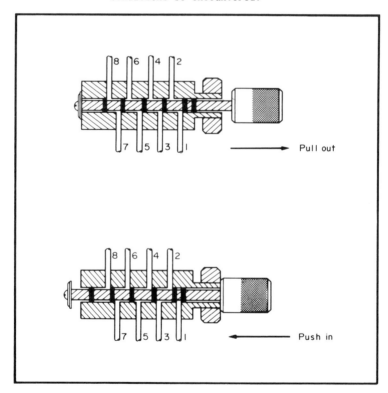

Fig. 2.2 Typical eight port cylindrical switching valve.

In addition, the mode of sampling will need to be considered, that is, whether continuous or discrete sampling is required. In the latter case a suitable sampling frequency must be decided upon. This will need to be within the limitations imposed by the sampling device used - automated syringe, sampling valve, dropper or whatever. Sampling could involve the design of quite complex stream splitting equipment (or flow diversion techniques) in order to guide the sample (be it liquid, gaseous or particulate) into a suitable sampling vial, flow cell or around an appropriately designed probe or transducer. As an illustration, consider the gas/liquid sampling valve shown in the diagrams contained in Fig. 2.2 and Fig. 2.3.

Figure 2.2 shows a cylindrical eight port change-over valve for switching gas and liquid streams in both laboratory and instrumentation applications. Figure 2.3 illustrates how a similar eight port valve may be inter-connected to enable the construction of a gas chromatographic sampling valve. Each of the examples shown is intended for manual operation on a 'push-pull' basis. Similar valves are available wherein

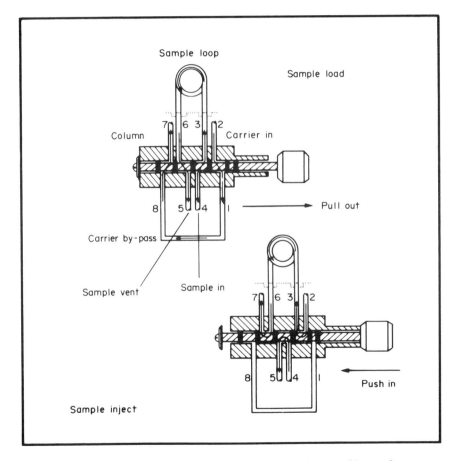

Fig. 2.3   Construction of a gas chromatography sampling valve.

the manual operation is replaced by electromagnetic action through the use
of a solenoid. This enables such valves to be operated automatically
under computer control. They are therefore often used for on-line process
control applications in situations where gas mixture compositions need to
be recorded. Suitable pressure transduction equipment is required in
those situations where there are significant pressure differentials
between the system being sampled and the gas chromatographic equipment. An
analogous arrangement could be used for the sampling of liquids - particularly where an analytical unit based upon the use of liquid chromatography, refractive index, polarimetry, etc is to be employed. However,
where sensitive gas chromatographic methods are to be used the system
would need to be modified. Because the volume of liquid sample required
is much smaller than that required for gas analysis a suitable sample
dilution or micro-sampling technique might need to be devised. Various
types of system exist, the exact design details depend upon both the type
of sample being monitored and the analytical method involved. Many
interesting examples are to be found in the literature. Bodmer and
Rushton (Bod78) describe the use of on-line monitoring of water purity in
thermal power stations. Here automatic analysers continuously monitor
water and steam purities. A range of on-line instruments is used to
enable the continuous recording of conductivity, pH, and the content of
ammonia, hydrazine, phosphate, oxygen, hydrogen, sodium, chloride and
silica. The description of a sampling system for steam operated power
stations presented by Svoboda and Schmid (Svo78) contains an interesting
discussion of techniques useful for interfacing to equipment used for
chemical analysis. Similarly, Schlatter and Heinicke (Sch78) describe a
sampling technique for solids that are to be subjected to analysis via an
X-ray spectrometer. Each of these involve the use of an on-line computer
system.

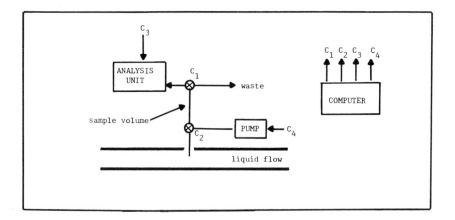

Fig. 2.4 Computer control of an on-line sampling system.

A simple scheme for discrete sampling from a liquid stream is shown
schematically in Fig. 2.4. In this sketch C_i values denote control
signals from the controlling computer to the instrumentation equipment.
Suppose t_i denotes sequentially increasing values of the time variable.
At time t_1 signal lines C_1 and C_2 are activated thereby enabling the

sampling valve to become 'purged' with sample. A little later, at time
t_2, the same lines are used to isolate the sample volume containing the
material to be analysed. The combined effect of signals on lines C_1
C_2, C_3, and C_4 together at time t_3 activates the analytical unit
and pump, the latter serving to inject the sample into the analysis unit.
Subsequent control signals on the appropriate lines return the system to
its original state so that on receiving a signal from the analytical unit
the computer can initiate the cycle again – depending upon how the system
has been programmed to operate. Similar types of sampling and sample
handling techniques can be designed for most on-line instruments such as
mass spectrometers, photoionisation and infrared gas analysers and so on.

Where continuous monitoring of a system is required a suitable type of
flow cell needs to be devised (capable of handling gas, liquid or
particulate materials) or a direct insertion probe designed. Direct
insertion probes are either responsive to some physical/chemical property
of the substance being monitored or enable continuous 'leaking' of the
sample from the system into the sample handling section of the analytical
unit. The most appropriate type of sampling arrangement will depend upon
the actual application involved.

The principle of the flow-through cell is illustrated by the diagram shown
in Fig. 2.5. This is typical of those used for on-line infrared gas
analysers. In use, such a system would require suitable stream splitting
piping in order to direct a portion of the flowing material into the cell
for analysis.

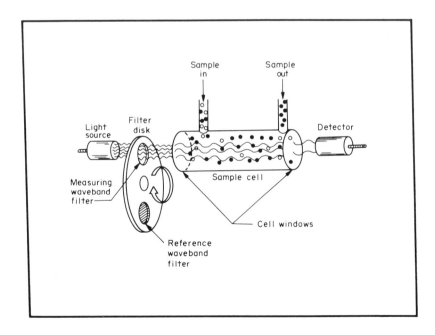

Fig. 2.5   Principle of the flow-through sample cell.

Sometimes the analyser unit is built into the actual flow lines of the system as an integral unit as shown in Fig. 2.6. This avoids the necessity of having stream splitters or of having to pump/suck the sample into the flow cell. Obviously, in situations such as these, it is important to pay particular attention to factors which influence the reliability and deterioration of the analyser. In the example illustrated in Fig. 2.7, in addition to observing the absorption signals of material(s) in the flow stream, the computer also monitors both the source and detector electronics for possible sources of malfunction. Subsequent to the analysis of the signals produced by the analyser, the computer relays results to a display in the plant operations control room and also makes appropriate control adjustments to plant equipment.

Perhaps the most well known type of direct insertion probe is that based upon the use of an ion selective electrode. A typical example of the kind used in an industrial environment is illustrated in Fig. 2.8.

A considerable number of electrodes is now available (Ori78, Pye77) which permit the determination of a wide range of species both in aqueous and non-aqueous environments. These electrodes can easily be incorporated into laboratory and plant equipment to provide analytical arrangements that are ideally suited to on-line computer control and monitoring. A typical example of a flow-through electrode unit has been described by Forney et al (For75) while their use for automation of analysis has been outlined in a paper by Rigdon and his collaborators (Rig78).

The subject of 'sampling' is one which covers a broad spectrum of activity. In the few pages that have been devoted to it in this chapter only a limited amount has been covered. However, some of the major factors involved have been mentioned. Particularly important in the context of computers are those methods of sampling which facilitate (a) the automation of an analytical technique for use in the laboratory, and (b) enable the on-line analysis of plant or laboratory processes to be achieved automatically without the need for human intervention. In this section greater emphasis has been given to (b) than to (a). This imbalance will be compensated for later in the section devoted to laboratory automation (chapter 8).

The Use of Robots and Remote Sampling Equipment

Very often an analyst may be responsible for the collection and subsequent analysis of a series of samples relevant to some investigation in which he/she is involved. Unfortunately, many situations arise where it is not feasible for the analyst (or an appointed agent) to directly obtain the collection of samples for analysis. The most common examples of cases of this type relate to the collection of specimens from environments which, because of their hazardous nature, might endanger human life. Similarly, cases exist in which conditions are too difficult or where there are limitations imposed by the awkward and uninviting physical location of the materials to be sampled. The following list presents some examples of practical situations in which one or more of these conditions prevail:

    (a)   retrieving samples from the moon or a planet,
    (b)   obtaining samples from the sea bed,
    (c)   collecting samples from the atmosphere or outer space,
    (d)   taking samples in an area contaminated with radioactivity,

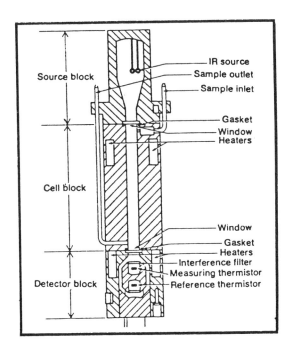

Fig. 2.6 Special purpose flow-through cell.

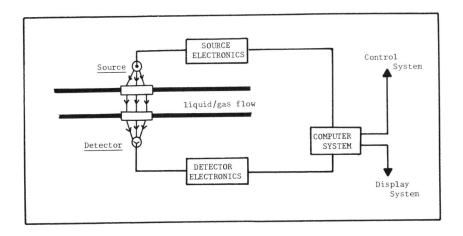

Fig. 2.7 Computer monitored on-line analyser for continuous sampling.

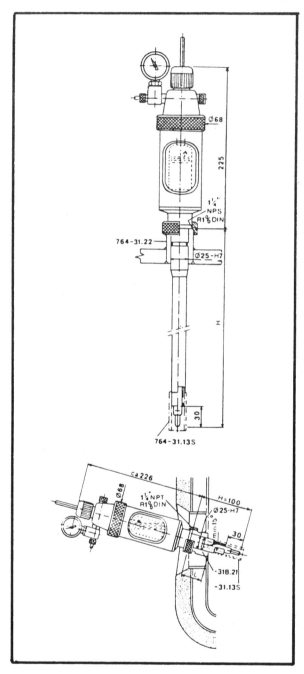

Fig. 2.8   Industrial insertion probe for continuous measurement of pH.

(e) performing experiments on and analysis of space materials
     in a sealed space laboratory,
(f) collecting samples at/near volcanoes.

Obviously, cases such as these necessitate some alternative means of
sample collection. In most of the examples cited there is a need to be
able to revert to some remote means of sample gathering that can be
controlled by a human at a location that may be some distance from where
the samples are to be collected. Quite often this will entail the use of
mechanical samplers having the ability to access the location to be
sampled and then return the samples to the laboratory. These might entail
the use of simple robotic equipment or less sophisticated devices such as
master-slave manipulator systems in which the slave arm at the remote site
is mechanically or electrically coupled to a geometrically identical (or
similar) master arm. As the master arm is moved so the slave arm follows
its motion.

In a recent review paper Bejczy (Bej80) has used the term 'teleoperation'
to describe manipulative activities performed by mechanical devices at a
remote site under remote control. Usually, the remotely performed
mechanical actions are associated with the normal work functions of the
human hand. Through teleoperation the manipulative capabilities of the
arm and hand can be extended to remote physically difficult or dangerous
environments. The first system of this type was devised about 35 years
ago to allow an operator to handle radioactive materials from a workroom
separated from the radioactive environment by a concrete wall. The
operator observed the work scene through viewing ports in the wall. Since
these early systems the development of teleoperator devices has progressed
substantially.

Nowadays a teleoperator is regarded as a robotic device having video
and/or other sensors, manipulator arms and some capability for mobility.
It is controlled remotely over a telecommunication channel by a human
operator. The human operator can be a direct in-the-loop controller who
observes a video display of the teleoperator and continuously controls the
position of the teleoperator vehicle, its arm(s) or its sensor(s)
orientation. Alternatively, the teleoperator may employ an onboard
computer capable of executing control functions automatically through
local force or proximity sensing. In this type of situation the remote
human operator shares and trades control with the computer.

The logical organisation of a simple robotic sampling device is
illustrated in Fig. 2.9. There are essentially five parts that enable the
device to interact with its environment (that is, its sample space). The
sampling mechanism will consist of facilities to enable the extraction of
the sample - a mechanical grab, scoop, pipette system or specialised
probe. This is responsible for obtaining the sample and transferring it
to an appropriate storage device (ampoule, specimen tube, gas bottle, etc)
within the storage subsystem. A record is kept in the computer system of
the sampling conditions (or other parameters such as geographical
location) and the location of the corresponding sample within the sample
storage subsystem. The computer also maintains local control of the
equipment and is responsible for communication with the remote control
station via the telecommunications link (TL). The sample dispatch
subsystem is responsible for ensuring the return of samples back to the
control centre. The magnitude of the sampling space will depend upon two
factors: vehicular mobility and architectural geometry. The mobility of
the device refers to whether or not it is fitted with a traction facility

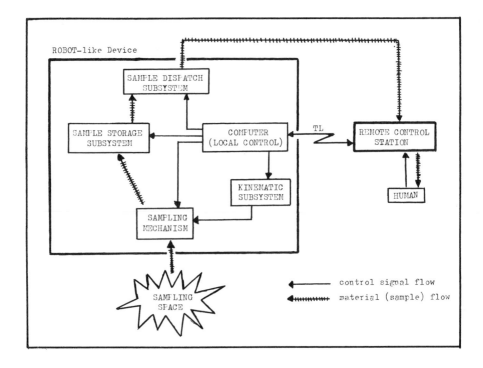

Fig. 2.9 Basic principles involved in the use of robots for sampling.

(wheels, caterpillar tracks, air-cushion, etc) to enable it to move over a geometrically plane surface. The alternative is an immovable device that is designed to operate from a single fixed geographical position. Obviously, the former is more useful but is also more expensive. The architectural geometry of the device refers to the different possible movement capabilities of the limb or probe used for sampling. Included in this design aspect are the three dimensional movement capability via the use of telescopic projection devices, rotational ability about various axes, the number of pivoting joints, and so on. A simple example of an industrial robot incorporating some of these principles has been described by Murphy (Mur80).

Bejczy (Bej80) has described a more sophisticated design of teleoperation device based upon the research work that has been conducted at the Jet Propulsion Laboratory at the California Institute of Technology. This work involves sophisticated types of human-machine interaction based upon the use of video cameras, to enable the operator to see what the robot is doing, and a voice input/output system as a means of providing a facile control capability. A diagram showing the basic arrangement of the system is shown in Fig. 2.10. Possible applications for such robots in space research have been described by Mason (Mas73), Merritt (Mer76) and Arnold (Arn80).

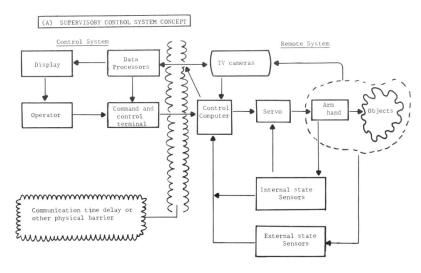

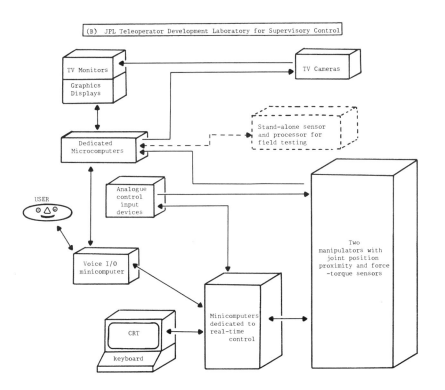

Fig. 2.10   Advanced robot based teleoperation system.

SEPARATING

Separation is the general name associated with those processes that are
designed to enable simple (or complex) combinations of entities to be
resolved into the constituent elementary components. Like sampling, the
term separation (in the context of analytical chemistry) is one which is
associated with a diverse spectrum of activity ranging from simple
filtration or solvent extraction through to sophisticated highly automated
separatory techniques. The purpose of this section is to explain some of
the ways in which this term is used and at the same time provide some
background material of relevance to the topics to be covered in the next
chapter on instrumention.

The conventional analytical chemistry meaning of the term refers to those
industrial or laboratory processes for physically/chemically separating
chemical species for the purposes of quantitative identification or
qualitative estimation. Over the years a considerable body of theory and
practical methodology on this topic has accumulated. A comprehensive
description of physical and chemical methods of separation is contained in
the textbook by Berg (Ber63) while other books devote chapters to a more
general coverage of the subject (Bar65). Chapters covering particular
aspects of separation methods in analytical chemistry are dealt with in
considerable detail in Kolthoff and Elving's 'Treatise on Analytical
Chemistry' (Irv61, Her61, Ros61).

Often, separation implies the differential movement of the species
comprising a 'mixture'. This may be brought about by differences in their
physical properties by the application of an appropriate force such as
pressure, electric potential, a magnetic field, gravitational field,
centrifugal force or as a result of changes in temperature. The efficiency
of the physical separation will often depend on the magnitude of the
differences in the physical properties of the species to be resolved
(solubility - as in the separation of a NaCl/sand mixture, volatility,
molecular size, ability to diffuse, molecular polarity, ion mobility,
etc). Some of these physical properties form the basis of a large number
of instrumental methods of chemical analysis such as gas chromatography,
dialysis (as in the Technicon SMAC chemical analyser mentioned in chapter
1), electrophoresis, ultracentrifugation and so on.

The chemical separation process depends upon chemically converting bound
(or free) entities into forms that are more easily determined by other
means. Thus, the controlled oxidation of a hydrocarbon (either
eudiometrically or thermally) in order to measure the carbon and hydrogen
content (as CO_2 and H_2O, respectively) is a typical case in point.
Alternatively, where the species to be separated are not bound together
chemically, the selective reactivity of one or other of the compounds with
a particular reagent might be used to produce a mixture of materials more
amenable to analysis. Typical examples of this approach include selective
precipitation, mixed titrations and the use of complexing reagents.

The analyst can combine any number of physical and chemical separating
methods (provided their efficiency is acceptable) in order to form a
suitable sequence of processes that ultimately enables the composition of
the sample to be determined. However, to do this 'at the drawing board'
as opposed to 'at the bench' one needs to have available a considerable
amount of both physical and chemical data in order to design an
appropriate separation strategy. Obviously, there is here an ideal
application for the analyst to utilise the storage and logical capability

of a computer along similar lines to those employed in the LHASA program (Wip74, Cor71, Cor72). This program was written in order to help organic chemists design synthetic reaction pathways - aided by the storage and computational facilities of a powerful computer system. LHASA is an acronym for Logic and Heuristics Applied to Synthetic Analysis and is the name assigned to the sophisticated program responsible for providing the facilities that are available. It uses artificial intelligence techniques and provides the means by which its users may utilise graphical input and output devices in order to communicate with it. LHASA was produced as a result of experiences with an earlier program called OCSS (Organic Chemical Simulation and Synthesis) which incorporated a 'logic centred' approach to the design of organic syntheses. Essentially, the method involves starting from a 'target molecule' and stepping backwards through possible sets of precursor molecules. Each of these is itself considered to be a target molecule and is analysed similarly thereby generating a tree of synthetic intermediates. Each precursor is in some way simpler than the target from which it was derived or leads to further precursors which are simpler. The analysis terminates when precursors are produced which are considered to be relatively simple or readily available. Prior to the availability of the LHASA program, the large number of intermediates which would be involved in a comprehensive synthesis tree for a complicated molecule could only be generated by the expenditure of significant amount of time and effort by a knowledgeable chemist.

Naturally, there is no reason why techniques similar to those outlined above could not be used for finding possible ways of performing an analytical separation of a specified mixture of components. In an ideal situation it would be beneficial if an analyst could be provided with a computer based facility that enabled the following types of question to be posed:

How might a mixture of phospholipids be separated?

How should one separate component Z from a mixture containing W, X, Y and Z?

How can the instrumental parameters for separation method xyz be optimised in order to produce the best possible separation?

There are no intrinsic reasons why questions such as these should not be capable of solution by computer provided appropriate models of the systems involved could be formulated and all the necessary parametric data stored within a suitably structured data base (see chapter 10). Consider what might be involved in attempting to answer the first question. Perhaps a high speed search through a data base containing phospholipid data might cause the reply: 'try the method of abc reported in journal pqr'. In the case of the second question, a careful analysis of the properties of each of the components combined with other built-in knowledge (see EXPERT SYSTEMS in the next section) might lead the machine to make suggestions about how best to proceed with the separation task. Unfortunately, the construction of computer programs to enable these systems to be constructed is no simple matter. While simulations for particular types of separations do exist (particularly those relevant to industrial manufacturing processes) there is no generally applicable mathematical model of separation (or supporting data base) available.

As an illustration of the principles involved, consider the application of
these methods to the separation of a mixture of components using liquid
(column) chromatography (Bar65, Ros61). A simplified sketch of the basic
arrangement of the equipment is shown in Fig. 2.11. A computer is
included since much of the instrumentation currently available utilises
some form of programmable logic or facilities for storing sequences of
control instructions.

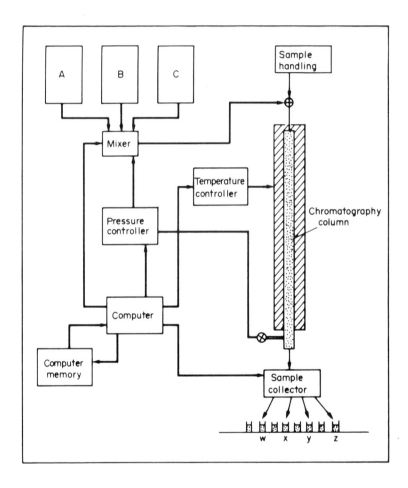

Fig. 2.11   Basic components of a computer controlled liquid chromatograph.

In this diagram, A, B and C represent three different column eluents
having differing eluting properties. These can be mixed in any proportion
in order to produce a mixture that varies in eluting power as a result of
changes in concentration, ionic strength, pH, polarity and so on – this is
the basic principle of gradient elution. It is the function of the mixer

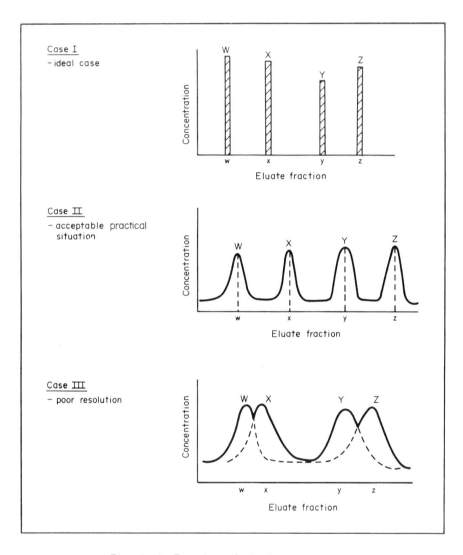

Fig. 2.12   Examples of elution chromatograms.

unit to control the concentration (or any of the other properties listed above) of the mobile liquid phase moving onto the column. The pressure controller enables high pressures to be applied to the front of the column or the application of reduced pressures at the tail. Similarly, the temperature controller enables the column to be operated isothermally or, alternatively, it may be temperature programmed in a variety of ways. The possibilities include variation of column temperature with time or the creation of various types of temperature gradient along the column.

Column parameters such as length, composition, affinity, degree of
deactivation can each be varied. Control of many of the essential
parameters (temperature, pressure, eluent composition, sample collection
scheme) is effected by means of the computer system. This has a memory
(or internal storage facility) which contains a user specified sequence of
stored control instructions. These specify the series of states through
which the instrument is to pass during a particular interval of time.

Suppose a mixture of four components (W, X, Y and Z) are to be separated
using such a system. The most acceptable (ideal) type of elution
chromatogram that could be achieved is one which showed that each
component of the original mixture had been isolated in an individual
sample collection vessel. That is, component W was contained in vessel w,
component X in vessel x, and so on. This situation is illustrated in the
elution chromatogram (case I) shown in Fig. 2.12. The situation wherein
each component is distributed over a narrow range of sample vessels (case
II) would be acceptable even though it might be less convenient. In this
case, component W would be contained in a small sequence of vessels $w \pm \delta w$,
component X would be contained in the sequence $x \pm \delta x$, and so on.

Case III illustrates a situation in which there is considerable interference between the eluent fractions. In most practical analytical
applications this would not be useful. However, as will be discussed
later, there are situations in which this type of eluent chromatogram
could be used - in conjunction with mathematical curve fitting and
simulation techniques. But, for the majority of cases of this type, the
analyst is likely to spend considerable time performing practical experiments with the system in an attempt to obtain the optimal instrumental
conditions necessary to achieve the best possible separation. Much of
this experimental effort could be avoided if a suitable computer model of
the system was available. This could enable the analyst to simulate
conventional experimental operations. By facilitating simulative
investigation of the effects of eluent composition, column temperature,
pressure, composition and so on (in rapid succession and in any combination), it would be possible to determine whether it is indeed feasible
to convert a profile similar to that shown in case III to one of the forms
depicted in case II or case I. Having isolated the appropriate conditions, it would be a simple matter to set up the 'real' equipment in the
laboratory to effect the results predicted through simulation experiments.

While the necessary computer models can be formulated for particular types
of separation (similar to that outlined above), no sufficiently powerful
mathematical models of the overall activity of separation exist to enable
a general purpose computer simulation model to be constructed. Consequently, the goals of this approach, although laudable, are at present
not a practical reality. Thus, the analysts' equivalent of the organic
chemists' LHASA still awaits creation. However, despite the little
progress that has been made in this context, considerable advances have
taken place in the area of automatic parameter optimisation. This has
been achieved in many of the more sophisticated analytical instruments
through the use of built-in microcomputers (see chapters 3 and 4).
Technological advances of this type make it possible to overcome some of
the problems of poorly adjusted instrumental conditions. This is an
important advance since incorrect settings on instruments can often be
partly responsible for the type of phenomena described in the previous
example - poor resolution of signals.

Separation - A Simplified Classification

In a book of this type it would be both impossible and unnecessary to investigate all the different aspects of separation relevant to a study of analytical chemistry. The important topics are covered in the more specialised texts and reference books cited in the bibliography at the end of this chapter. It remains only to provide a frame of reference for the discussion that will take place in subsequent sections of this book. Consequently, an indication of the more important areas will be needed. These are depicted in the hierarchical classification tree shown in Fig. 2.13.

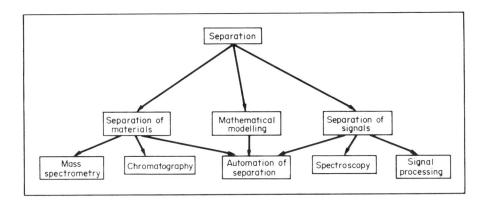

Fig. 2.13   Semantics of the term 'separation'.

Based upon the approach to be used in this book, Fig. 2.13 is an attempt to sub-divide the concept of separation into three major areas: mathematical modelling, separation of materials and separation of signals. The basic ideas underlying the concepts of mathematical modelling have been introduced in an informal way by the example that was described earlier in this section. Further details of these techniques will be presented in chapter 9. Of the two remaining categories, separation of material will serve to introduce the techniques of chromatography and mass spectrometry - as described in the next chapter. These represent two of the most important and versatile methods of instrumental analysis presently available. The final category, separation of signals, will be used as a means of describing the many broad sub-divisions of spectroscopy (infrared, ultraviolet, visible, NMR, ESR and so on) and, in a later chapter, the principle of signal processing.

FINDING AN ANALYTICAL METHOD

In many 'routine' process control laboratories in which analysis is used to monitor and report on the quality of materials being handled (final products, intermediates, raw materials) there will usually be a set of well defined 'standard' analytical methods in use. For example: standard methods of water analysis, recommended methods of hydrocarbon analysis, routine methods of serum analysis and many others. Within such a

laboratory any departure from the use of its routine techniques of
analysis will usually only be required when some unusual or non-standard
analytical situation arises or when a new process (manufacturing,
screening, sampling, etc) is being introduced. Consider the situation in
which a chemical plant is producing a product X from raw materials P, Q
and R in a reactor held at temperature T and at a pH Z the contents of
which are stirred at a rate K. As a result of a critical variation in pH
some solid substance M is precipitated. The process control laboratory is
required to characterise M. Similarly, in the case involving the
introduction of a new manufacturing process, the control engineer
responsible for it may require the analytical laboratory to provide
suitable methods of analysis to enable the new process to be monitored.
Characterisation of the material M probably borders on the edge of a
'research' investigation and may require facilities not available in the
routine quality control laboratory. Similar comments could apply in the
case of the new manufacturing process. Analytical techniques appropriate
for use in quality control will need to be devised (or 'researched') and
tested for their suitability. Previously, suitable techniques may never
have existed. In this case, those new methods that are formulated will
ultimately become part of the laboratory's routine repertoire when the
process is commissioned.

Each of the above situations have much in common with the type of problems
that are often encountered by those organisations that undertake research
investigations within analytical chemistry - universities, consultancies,
research associations and various industrial/government laboratories.
Their common denominator is the need to be able to formulate techniques of
analysis that are applicable to the solution of particular analytical
problems. In this context, the research may involve investigating current
(and older) methods of analysis with a view to determining how they might
apply to new situations or to problems in which they have not previously
been tried. Alternatively, a considerable amount of research effort may
be channelled into attempting to formulate entirely new methods of
analysis, evaluating them and carefully documenting their advantages and
shortcomings. Nowadays, this may involve devising sophisticated equipment
utilising entirely new technological advances or it may involve modifying
existing equipment to meet the needs of a current situation.

Many analysts, no matter what their background, have at one time or
another been placed in the situation of having to find an analytical
method to solve a problem in which they have been involved. Obviously,
the analyst's most important asset in this context is experience and
intuition. These will ultimately lead to one of any number of thought
patterns, one possibility for which is depicted in the simple decision
graph shown in the diagram presented in Fig. 2.14. As this graph reveals,
there are at least seven major decisions to be made.

Once the analytical problem has been explicitly defined it should be an
easy matter for an experienced analyst to determine whether or not a
standard method can be modified to meet the requirements of the problem in
hand. If no solution is found at this stage, discussion with an immediate
colleague may help, in which case the cycle is re-iterated - the colleague
may have suggested a new standard method to try. Each of these stages (1,
2 and 3) are fairly routine in-house affairs. However, the subsequent
steps (4, 5 and 6) often necessitate the analyst searching outside of
his/her own laboratory in order to find a solution to the problem.

In the first instance the obvious primary sources of information are the

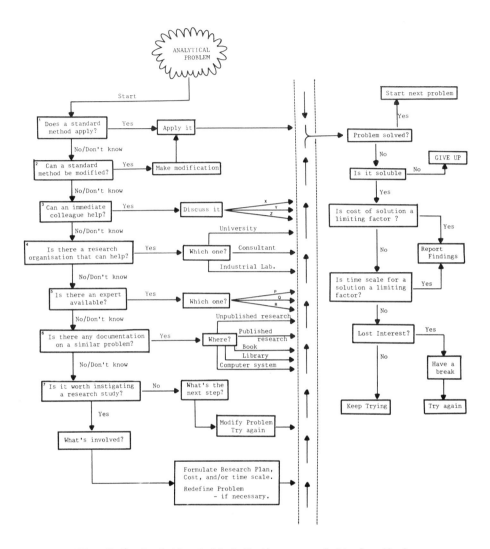

Fig. 2.14 Logistics behind finding an analytical method.

various forms of technical documentation and the literature of analytical chemistry - and related fields. Technical documentation will include a wide range of resources of both a formal and informal nature. Typical examples include unpublished/published research reports, private communications, manufacturers' literature and product briefs, professional bulletins, newsletters and many others. The literature of analytical chemistry will also provide numerous possibilities. Amongst these are included collected works (Wil59, Kol80, Sne66), survey publications (Sig68, Wes73), specialised publications and text books of various types (Lai75, Mei58, Mei63). In addition there will be various reports from learned societies, journals, abstracts, patents, theses and conference proceedings. The analyst will thus have a considerable volume of literature to search through in order to determine if the required information is available. Sometimes this responsibility is delegated to a colleague who specialises in the handling of scientific literature - such a person is often referred to as an 'information scientist'. However, many people are unable to afford the services of such a specialist. In this case the analyst must perform the literature search unaided. Manual searching of the literature (perhaps aided by special publications like Chemical Abstracts, Current Contents or the Science Citation Index) is a slow, tedious and often error prone process. Sometimes the 'lone analyst' can only hope to cover a representative sample of the total literature available - perhaps just the few publications that relate to one specialist field - gas chromatography, mass spectrometry, inorganic analysis, radio-analysis or whatever.

Because of the vast amount of scientific literature that is published each year it becomes increasingly difficult for research workers or research groups to make anything like an exhaustive literature search of current material. A consideration of retrospective information creates an even more significant problem. Consequently, many organisations are tending to use computer systems to perform automatic searches of the literature on behalf of the analyst. The various information services that are available and the techniques that are used for searching the analytical chemistry literature will be discussed in more detail later (chapter 11).

One of the most powerful methods that the analyst has available to help overcome the literature problem is probably the technique of information indexing (Bec67). As reports and journals (typically: Analytical Chemistry, Talanta, The Analyst, Analytical Abstracts, and so on) are scanned a note would be made of those articles or papers that are of particular personal interest. A suitable indexing system could then be built to enable details of these reports (analytical methods or techniques) to be retrieved at some future date. By building up a series of indexes to the literature some of the problems of searching for a suitable analytical method may be obviated. There is a variety of ways in which an accumulation of knowledge may be indexed. Some of these are shown schematically in Fig. 2.15.

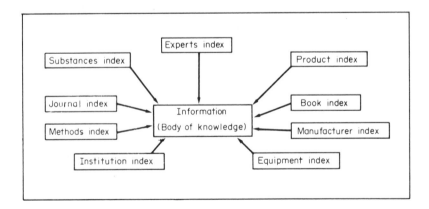

Fig. 2.15 Some common techniques used in the indexing of information.

Common methods of implementing indexing schemes involve the use of conventional index cards, computers or combinations of each of these. The most familiar is probably the manual index based upon the use of collections of cards that are appropriately labelled and structured. Usually the structures set up will be hierarchically organised classifications with appropriate cross-referencing if this is deemed to be necessary. The basic card records will contain whatever bibliographic details the analyst feels it is worthwhile to enter on them. Manual indexing is not without its problems - particularly when updating and retrieval operations involve the use of large indexes. Of course, any manual system based on cards can be replaced by an equivalent computer based system. Consequently, many people prefer and are moving towards the use of computer based indexes. Two of these, which are convenient and easy to use, are the KWIC and KWOC indexes (Hea78). KWIC is an acronym for <u>K</u>ey <u>W</u>ord <u>i</u>n <u>C</u>ontext while KWOC stands for <u>K</u>ey <u>W</u>ord <u>O</u>ut of <u>C</u>ontext. Of these, KWIC is probably the more useful of the two. The appearance of an index of this type is shown in Fig. 11.4 (chapter 11). Many computer systems provide standard software packages (this term is explained in chapter 4) for producing KWIC indexes. Typical of these is that available from the IBM Corporation (IBM62) called KWIC/360. This is simple to use and is an effective way of indexing large volumes of documentation. Further details of this will be found in the reference cited and a description of a simple system based upon the KWIC principle will be outlined in chapter 11.

Having searched through any personal indexes and those available through one or more local libraries, the analyst may still not have found the method of analysis being sought. At this stage a decision may be made to solicit the help of some external organisation - a university, a consultancy or some acknowledged expert in a particular field. The approach that is adopted will often depend upon each individual analyst's experience and knowledge of who to contact in order to discuss a particular type of problem. Over the years, most analysts will build up a circle of contacts via both formal and informal means. These contacts often form the basis of extremely useful information networks (the grape-vine). From time to time, many of the Learned Societies and research organisations publish lists of staff (and members in the case of research associations or collectives) along with their areas of specialism. Each of these sources of information can be useful in helping to track down an appropriate expert or a suitable research organisation.

Even if a suitable consultant is located, the analyst's problem does not necessarily end there. It might happen that the consultant is only able to confirm the analyst's own impression that no simple readily available solution to the problem exists. The evidence may now indicate that some basic research investigation is needed in order to formulate the exact solution to the problem in hand. This could involve the research division of the analyst's parent company or, alternatively, hiring the skills of some external research organisation. The path adopted will obviously depend upon both financial constraints and the urgency with which the problem is to be solved. Finding an analytical method can thus involve quite a complex chain of activity. In many cases there is no easy solution to the problems that arise.

The 'Expert Analyst' -
Application of Artificial Intelligence Techniques

Arising from the extensive work done at the artificial intelligence
laboratory at the Massachusetts Institute of Technology on 'Expert
Systems' there is currently much interest in this topic within many areas
of problem solving that require the application of sophisticated skills.
An expert system (or knowledge based system) incorporates knowledge of a
specific subject area from one or more experts, thereby enabling it to
advise its users on that subject and answer queries about why it gave a
particular piece of advice.   Such systems constitute a new form of problem
solver based on representing substantial bodies of human knowledge in a
computer's memory system.   Expert systems are currently operating
successfully in the areas of oil prospecting, medical diagnosis and
chemical analysis.   Many large chemical producing companies and research
organisations have shown much interest in these recent developments.   A
detailed description of the potential role of expert systems has been
given by Michie (Mic79).

Buchanan (Buc79) has proposed that the logical structure depicted in Fig.
2.16 be used to represent an expert system.   For most purposes the expert
may be thought of as a human being.   There is no reason however, why the
source of the expert's knowledge (scientific observations, empirical data,
textbooks, papers, etc) should not be used.   Accumulating knowledge of a
domain in a form that a program can interpret and use is essential for the
program to assist with problems of reasoning.   The transfer agent is
responsible for transferring knowledge from the expert to the program.
Three general methods are used to add scientific knowledge to programs:
hand-coding, automatic dialogue and automatic rule formation.   Each of the
methods is concerned with giving the computer program the same kind of
knowledge that an expert in a subject domain uses for problem solving.
The verifier provides a mechanism to enable the program's level of
expertise to be determined.   In additon, it enables the detection of
errors in the knowledge base by expert users of the system.   Clancy
(Cla79) has described the implementation of an expert system called
GUIDON.   In this system the expertise to be taught is provided by a rule
based consultation program the design of which is based on natural
language studies of discourse in artifical intelligence.   Domain relations
and facts take the form of rules about what to do in a given
circumstance.   Further details of GUIDON and related systems will be found
in the reference cited.

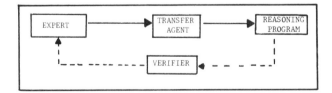

Fig. 2.16 Basic structure and information flow in an expert system.

CONDUCTING THE ANALYSIS

The exact mechanism involved in conducting an analysis will depend upon a large number of factors: the type of estimation involved (qualitative, quantitative or both), its time scale, the number of times it is likely to be performed, the accuracy, sensitivity and selectivity required, the type of equipment available and the nature of the information that is to be derived from it. This list represents only a few of the many considerations that need to be taken into account when formulating a strategy for conducting an analysis. Given that a computer of some description is available within the laboratory (or at the site at which the analysis is to be conducted), there are several ways in which an analysis could proceed:

(A) the analyst conducts the experiment unaided by computer,
(B) the analyst and a computer perform the experiment together, or
(C) the computer conducts the analysis automatically.

As far as this book is concerned, examples of case A represent the conventional approach to analysis and so this category will not be discussed further. It is assumed that this mode of analysis will be familiar to most readers.

Analyst - Computer 'Symbiosis' -
An Example of Human-Machine Interaction

The second category, B, is more interesting since in this situation the analyst employs some of the useful facilities of the computer in order to make the analytical task a little easier. Four important contributions that the computer can make in this context are: (1) storage of experimental data during the analysis, (2) monitoring and controlling experimental conditions and reporting/applying appropriate 'correction factors' where these are thought necessary, (3) the automatic optimisation of instrumental parameters, and, (4) the computational processing and archival of results. It is worth discussing each of these in turn.

The first situation listed above is depicted in the sketches shown in Fig. 2.17. Three basic possibilities can arise. In case I, referred to as manual data acquisition, the analyst plays an active role in conducting the experiment. This role involves controlling the equipment being used and, at the same time, entering the experimental data produced by the analytical equipment into the computer using an appropriate input channel. In this situation the computer performs the role of a simple 'electronic notebook'. The observations that are made are thus recorded electronically and later processed to yield the required analytical result - either qualitative or quantitative (or both) evaluations of composition. The computer input channel referred to above usually takes the form of a simple numeric keypad or alphanumeric keyboard. This enables the analyst to 'type in' the data values as they are observed. More recent approaches to this type of data entry technique involve the use of touch sensitive screens and pressure sensitive writing pads. The latter type of device enables the analyst to record the experimental results on an ordinary piece of paper which is placed over a pressure sensitive surface. Then, as the results are written down so the computer 'recognises' the values and stores them in its memory. A typical example of such a device will be described in Chapter 8 (laboratory automation). This manual data collection approach can become very tedious if a large

number of experimental readings are to be recorded. Furthermore, for fast
reaction systems it would not be a viable proposition. Consequently, Case
II, automated data acquisition, is more often used.

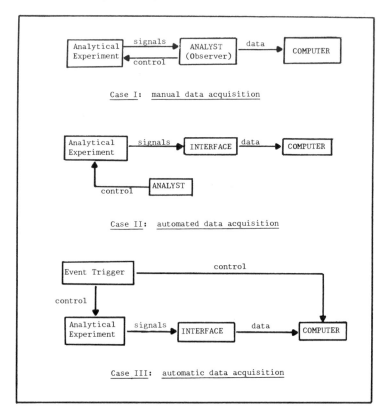

Fig. 2.17   The role of the computer as a data acquisition tool.

In this approach to data acquisition the analytical results are
automatically recorded by the computer without any need for intervention
by the analyst who simply controls the experiment. Use of this method may
require the design of a special interface to enable the data signals
produced by the analytical equipment to be converted to a form that is
acceptable to the computer. Early approaches to this technique were based
upon the use of paper tape punches and readers: a paper tape produced by
the analytical equipment was later read into the computer by the paper
tape reader (see Fig. 2.18).

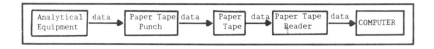

Fig. 2.18   Use of paper tape for transferring data to a computer.

More recent systems, however, use electronic interfaces that enable the
analytical equipment and computer to be directly interconnected - either
permanently or just for the duration of the analysis. In automated data
acquisition an attempt is made to replace human manipulative effort
(recording of results) by mechanical or instrumental devices. This
differs from the third category (case III of Fig. 2.17) which is referred
to by the title of automatic data acquisition. In an automatic device,
specific operations are performed at particular points in time without the
need for human intervention. As can be seen in the diagram (Fig. 2.17)
the data collection activity is initiated by some 'event trigger'
associated with the analysis. This trigger could be an electronic timer,
a level sensor, a pH sensor and so on, depending upon the exact specifications of how and when the analysis is to be performed. An introduction to
the techniques of digital data collection will be found in an article by
Shepherd and Vincent (She73). A more detailed discussion of this topic
will be presented in chapter 5.

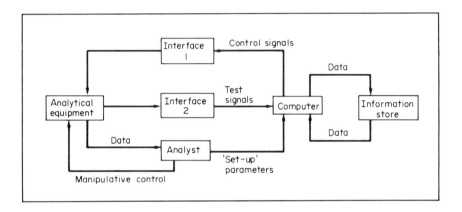

Fig. 2.19 Computer interfaces to enable instrument monitoring and control.

The second area in which the computer may assist the analyst is in the
monitoring and control of experimental conditions. As before, interfaces
will be required in order to enable the computer system and analytical
equipment to be connected together. In the diagram shown in Fig. 2.19 two
types of interfaces are depicted. Type 1 enables the computer to set
various experimental/instrumental conditions (temperature of a water bath,
speed of rotation of a mixer, fluid flow rates, pH, pressure, etc) through
the use of appropriate control signals that are applied to various parts
of the equipment. Type 2 interfaces allow the computer to receive various
test signals from appropriate sensing probes attached to different parts
of the equipment. These permit the computer system to measure the
experimental conditions under which an analysis is being performed and
then change them as and when it becomes necessary. An arrangement similar
to that shown in Fig. 2.19 enables the analyst to conduct the analysis in
either of two ways depending upon how the analytical conditions are
specified - directly or indirectly. When the direct approach is used the
analyst activates the computer system which then requests him/her to enter
the details of the experimental conditions that are to be used for the
analysis. The computer then sets these up and informs the analyst when

the equipment is ready for use. This approach is time consuming if the analysis is to be performed on a routine basis and if the list of experimental conditions is lengthy. In this type of situation, the alternative approach, indirect specification, can be more useful. When this method is used the list of experimental conditions is entered into the computer store, assigned a unique reference number and then referred to on future occasions by that reference number. Thus, each time the analysis is conducted its reference number is entered into the computer which retrieves the previously entered details and then sets up the analytical conditions accordingly. If the computer has a suitable storage system this can be used to store tables of correction factors that are to be applied to experimental results - perhaps to correct for the effects of changes in ambient temperature, atmospheric pressure, humidity or any other factor which may influence the analysis and which cannot be directly controlled. This mode of equipment control can of course be combined with any of the data acquisition strategies outlined previously. Such a combined approach enables quite effective analytical equipment to be set up.

Related to the monitoring and control applications of the computer outlined above is the related area of parameter optimisation. As was mentioned during the discussion of separation, this is particularly important in the area of instrumental analysis. In a complex instrument there will be a large number of parameters the values of which directly influence the instrument's performance - particularly, sensitivity, resolving power and selectivity. In view of this it is important to know that the machine is performing to the best of its ability. A computer can be used to tune the various instrument settings in order to produce the most favourable operating conditions. A similar procedure can also be used in order to check the performance of the individual components of the instrument and diagnose error conditions or mal-functioning parts - using a technique called 'signature analysis'. As far as the analyst is concerned these procedures manifest themselves in the form of switch settings on the operating console of the instrument. Having selected the analysis to be performed, the machine is switched into 'check mode' followed by 'optimise' for the analysis that has been selected. In some instruments the switching between check and optimise modes is performed automatically prior to conducting the analysis.

The last of the four general application aids provided by the computer lies in the area of computation and archival of results. Having conducted the analysis and collected appropriate data the analyst will usually wish to 'process' the experimental values in various ways in order to produce meaningful information. This processing may involve some simple computation; alternatively, it may require the application of more sophisticated techniques involving standard numerical or statistical procedures. Details of these procedures can be stored in the computer and the data processed accordingly. Once the results have been obtained they may need to be stored away (or archived) for future reference. It may be that the results need to be capable of recall for periods of up to several years - depending upon the application - for legislative purposes. Here again, the computer facilitates this type of retrieval operation. These applications will be discussed in greater depth later.

Automating the analysis

The third of the categories of analysis listed at the beginning of this section (category C) referred to the situation in which the computer conducts the analysis automatically. In this context there are three basic ways in which the analysis could be performed in an automatic fashion:

(1)   the computer conducts the analysis after a cue from the analyst,
(2)   as for case (1) but without the need for the analyst's cue, and,
(3)   totally automatic operation of the equipment.

There are many variants of the above three approaches. Obviously, the degree to which an analysis may be automated (or made automatic) will depend upon the basic nature of the instrumentation available. Some types of equipment are particularly useful in this context while others are less so. Much depends upon an individual instrument's design and its operating characteristics. A general scheme for the application of automation to analysis is depicted in Fig. 2.20.

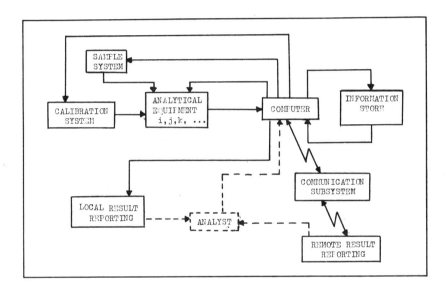

Fig. 2.20   The role of the analyst in automated analysis.

Referring to the previous categories of automation, in case (1), the analyst simply presents the sample to the instrument or initiates sampling by pressing the appropriate button. From then on the instrument performs all aspects of the analysis without the need for any further attention. At the end of the analysis the analyst is presented with the results of the experiment. Typical examples of this approach include automated titrimetry and infrared spectroscopy. In the first example the equipment automatically performs the titration and subsequent computation of results using a program stored in computer memory. Similarly, in the second situation (IR analysis), the system takes an absorption spectrum of the sample; it then performs a pattern matching exercise that involves comparing observed absorptions with tables of standard values stored within computer memory. The results presented to the analyst consist of the probable composition of the sample.

Method 2 (of the previous list) is essentially similar to method 1 except that some agent other than the analyst triggers the operation of the equipment. In other words, the equipment depends upon some change in its environment in order to initiate the analysis. In category (3), there are again several possibilities. Two important classes are:

(a) repetitive operation of an instrument as in batched analysis or on-line process control, and,
(b) the use of 'intelligent' machines for analysis.

Case (a) is fairly straightforward, it is based upon the repetitive application of methods (1) and (2) described previously. Case (b) is more interesting since in this situation the computer controls an array of instruments (i, j, k, ...) and selects the most appropriate combination for a particular application. This procedure represents a heuristic approach to analysis rather than an algorithmic one as might be the case in methods (1) and (2). The overall system 'sees' as its goal the complete (or pre-specified) analysis of a sample - both qualitative and quantitative. As progress is made towards this goal the machine automatically determines what analyses need to be performed in order to achieve its objective. The system 'calls' for human assistance only when this is required. The underlying logistics of this type of system is depicted in Fig. 2.21.

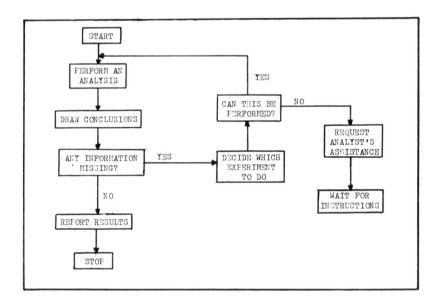

Fig. 2.21 Logistics underlying totally automatic analysis.

As individual instruments become more 'intelligent' and become easier to inter-connect, so, such systems will become increasingly sophisticated. Some examples of simpler systems of this type (such as gas chromatography/ mass spectrometry and gas chromatography/infrared spectroscopy) will be described in the next chapter.

Some introductory examples of analyses performed under computer control
using methods (1) and (2) are described in Perone's book (Per75) and some
applications of totally automatic analysis applied to on-line process
control (method 3a above)have been outlined by Siggia (Sig59), Sandford
(San77), Hall (Hal79) and Jutila (Jut79). A more detailed description of
automated batch analysis will be presented later in chapter 8.

REPORTING THE RESULTS

When the analysis has been performed the often arduous task of reporting
the results of the investigation must commence. There is a variety of
ways in which computer technology is able to help the analyst in this
respect. Much will depend upon what is to be done with the results and
the way in which they are to be reported. Result reporting may be totally
dependent upon computer equipment that is an integral part of the
analytical apparatus. Alternatively, there may be heavy dependence upon
the use of some computer system that is totally independent of the
experimental equipment. There will be various levels of result
reporting and types of result to be communicated. Futhermore, result
reporting may be initiated in a variety of ways depending upon the
sophistication of the equipment and the requirements of the application.
The simplest types of reporting chain are summarised in Fig. 2.22.

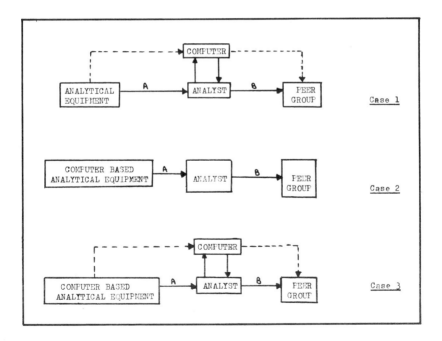

Fig. 2.22 Mechanisms for computer based result reporting.

In these diagrams the solid arrows denote the most probable directions of
result flow while the broken arrows indicate other possibilities that are
likely. Case 1 covers the simplest type of situation wherein the analyst
utilises some computer system to aid the process of result presentation.
This computer may be used to perform particular types of computation on
the data or to convert them into some other form such as graphic or
pictorial images. In case 2 the analyst relies solely on the computer
facilities built into the analytical equipment while case 3 represents a
combination of cases 1 and 2. As indicated in these diagrams there are
essentially two broad types of result, A and B, involved in the overall
result reporting process. Experimental data or results (A) produced by
the equipment will be displayed or reported to the analyst. These are
then used to make decisions about the various phenomena being studied.
These decisions may then form the basis of further results that are
generated (B) and which are passed on to a peer group. Typical
illustrations of results that fall into these two types are shown in the
list presented in Table 2.1. They are classified into three general
classes.

TABLE 2.1   Classes of Experimental Results

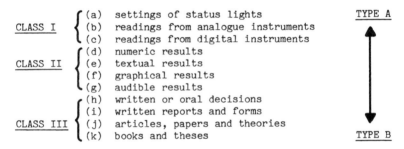

CLASS I	(a)	settings of status lights	TYPE A
	(b)	readings from analogue instruments	
	(c)	readings from digital instruments	
CLASS II	(d)	numeric results	
	(e)	textual results	
	(f)	graphical results	
	(g)	audible results	
CLASS III	(h)	written or oral decisions	
	(i)	written reports and forms	
	(j)	articles, papers and theories	
	(k)	books and theses	TYPE B

The results in class I are sometimes referred to as basic results since
they are the fundamental values that are obtained directly from the
analytical equipment. These may be manipulated in various ways, either by
the analyst or the computer system to produce the results in class II
which are referred to as derived results. These, in turn, may be used to
produce the 'high level' results that constitute the members of class III.

In the simple models of result reporting that were presented in Fig. 2.22
no facility was provided that would allow the analyst or the peer group to
'interrogate' the equipment in order to determine the progress or outcome
of a particular analysis. Usually, in computer based systems there will
be storage media available that will enable the results of analysis to be
stored until they are no longer required - these may provide either on-
line or off-line storage facilities. Such additional storage is capable
of forming the basis of various types of interrogation facility as is
illustrated in Fig. 2.23.

The additional facilities provided by a system such as this now permits
automatic dispatch of results to those interested in receiving them. It
is thus possible to formulate requests such as:

'Analyse sample X and report the results directly to Dr. Y'

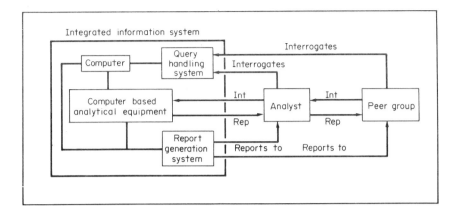

Fig. 2.23   Totally integrated approach to result presentation.

Alternatively, it now becomes feasible to enable those who are authorised
to directly interrogate the system using requests such as:

'Is the Sodium content of solution XYZ available?'

provided that appropriate data querying and retrieval systems exist within
the information system. Obviously, the facilities provided by systems
such as this will depend upon their purpose (process control, medical
applications or whatever). The application area for which these real time
information systems are designed will dictate the overall system design
and response criteria that are to be met. Some examples of systems
employing these principles of result reporting were briefly outlined in
chapter 1.

Basic Display Hardware Associated with Result Reporting

The term 'display hardware' is commonly used in the context of computer
based instrumentation. It is used to describe the various types of
physical devices that are employed to (a) display the results of analyses,
and, (b) show the internal status of an instrument. In all cases the
ultimate objective is to display both kinds of information in the most
desirable and cost effective way. A wide range of result reporting
hardware is available ranging from coloured lights to sophisticated
animation based upon the use of colour graphics systems and from simple
audible alarms to devices based upon the use of voice synthesis
equipment. Some of the more commonly encountered devices are listed in
Table 2.2.

The primitive devices (bulbs, neons, alarms, etc.) are often controlled by
a small computer - typically a microprocessor system. These are also used
in some of the more sophisticated types of display. In computer based
instrumentation laboratories one of the earliest types of printing,
computer compatible display device was probably the teletype. The ASR33
(Cam79) is undoubtedly one of the most well known. Nowadays there is a

wide variety of such devices available (WHI80a). Between them they show a
considerable range in the facilities they offer: speed, communications
adaptors, type face, character set, line width, print mechanism, portability, facilities for reading/punching paper tape and so on. For many
applications, particularly in the more sophisticated types of instrumentation, the visual display unit (VDU) and other sorts of graphics
devices are replacing the conventional printing terminal. However, where
printed forms have to be completed some form of print facility needs to be
retained. This can be achieved through the use of a simple printer
capable of handling special pre-printed stationery designed for the
particular application concerned.

TABLE 2.2 Devices Used for Displaying Analytical Results

    Indicator lights such as bulbs, neons and light emitting
        diodes (LEDs) controlled by a computer system.
    LED bar displays.
    Legendable indicators (rear illuminated).
    Audible tones and alarms.
    Voice synthesis equipment.
    Calibrated meters and displays of various forms.
    Digital displays based on liquid crystals.
    Oscilloscopes.
    Single line alphanumeric displays.
    Teletype-like devices.
    Video monitors.
    Visual Display Units (VDU).
    Graphics Display devices
        (monochrome/colour, storage tube/refreshed).
    Graph Plotters (drum, flat-bed, multiple pens/colours, etc).
    Chart recorders (strip chart and circular chart).
    Printers of various types
        (thermal, impact, laser based, etc).

A visual display unit consists of a cathode ray tube (CRT) screen similar
to that used in a domestic television receiver. This is indirectly
connected to a keyboard in such a way that information typed at the
keyboard appears on the VDU screen. The size of the screen varies from
model to model, 24 lines each containing 80 characters (that is, a total
capacity of 1960 characters) is not untypical. However, smaller screen
sizes are often encountered. A cursor (WHI78) is used to indicate the
position on the screen at which information or data is entered. Like
teletypes, most VDU devices are totally character orientated, that is,
they display lines of characters and are not capable of producing high
quality graphics or pictorial representations of data. There are many
different types of VDU commercially available (Hal77, Mer79, WHI80b).
They vary both in price and sophistication. At the lower end of the
spectrum are the simple 'dumb' teletype devices such as VDUs which offer
few facilities. At the opposite end of the range are the 'intelligent
terminals' that contain their own built-in microprocessor and provide many
sophisticated facilities such as multiple character sets, screen editing
and screen partitioning to enable the display of multiple images (see Fig.
2.24).

For many types of application the VDU offers several advantages over
printing terminals and printers. Often they are capable of providing more

useful display formats and the results that they display can be
instantaneously updated. They are silent in operation, cheaper to run and
contain no moving parts that are likely to break down. Application areas
such as process control are ideal - provided no continuous record of
results is required. Where the input of results is not a system design
requirment the keyboard can be dispensed with. In situations such as this
a video display monitor can often be used as a substitute for the VDU
provided suitable character generation electronics are provided. A more
detailed description of visual display units and their applications has
been presented by Grover et al (Gro76).

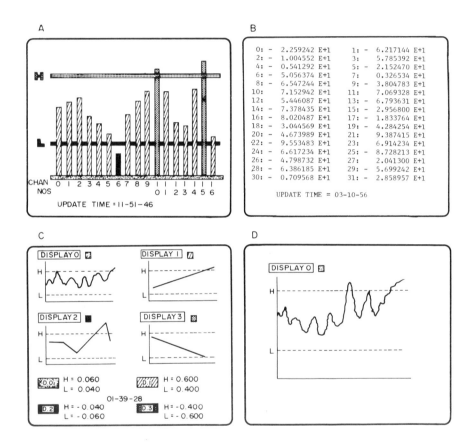

Fig. 2.24 Typical result reporting techniques
based upon the use of a graphics display unit.

A more sophisticated, and often more useful, type of display device is the
interactive graphics terminal (Gil78, New73, Wal76, Par69). Although more
expensive than the VDU, a graphics terminal often offers greater
capability in situations where it is required to present results in the
form of high quality pictures or diagrams. By means of appropriate
programming techniques (see for example Jon80) they can be used to display
several sets of independent results simultaneously. In addition, they

have the ability to display both textual and graphic information in any combination, colour and degree of animation. Some examples of the ways in which such a terminal might be used are illustrated in Fig. 2.24 which shows just four of the many different result reporting possibilities.

There are three principle types of graphics terminal available:

    (1)   vector refresh displays,
    (2)   storage tubes, and,
    (3)   raster refresh displays.

Vector refresh displays are expensive because in addition to the CRT and keyboard, they embody a small computer system that is used to display and manipulate the information on the screen. This type of terminal permits pictures to be easily modified and so is ideally suited to situations where scenes change dynamically as in the examples presented in Fig. 2.24. One major disadvantage of this type of terminal is the flicker that can arise when an attempt is made to display too much information on the screen.

Graphics terminals based upon the use of storage tubes do not require an ancillary computer to store the picture. Instead, the information used to generate the image is stored inside the CRT tube itself on a special electrostatic grid. Unlike the refresh graphics devices, storage tubes lack the capability of selective erasure, and so, are not suitable for dynamic computer graphics. They are, however, well suited for applications that require the display of large quantities of data since they do not suffer from screen flicker.

Raster refresh displays are similar to vector refresh displays in having an image display processor that is responsible for refreshing the screen. The screen of a raster display is divided up into a number of cells called pixels. Associated with each pixel is a small portion of computer storage that is used to store all the display details associated with that pixel - colour, brightness level, and so on. A microcomputer within the terminal interprets the stored data and controls the display hardware in such a way as to produce the appropriate graphical images. Raster refresh graphics are becoming increasingly popular as a consequence of the falling cost of computer memory.

In addition to the interactive graphics devices and VDU units that have been briefly outlined above there is a wide variety of other results reporting hardware available for the presentation of results. Particularly important are those that are used for producing printed versions of the information displayed on graphics screens. This category of device includes graph plotters and various types of chart recorders all of which are used for a variety of purposes within the laboratory.

Software Considerations for Results Reporting

The term software is generally used to describe the sequences of mathematical operations that are stored (in the form of programs) in the control store of a computer - a more detailed description of these concepts will be presented in chapter 4. A program is thus simply a specification of the sequence of operations that have to be performed in order to transform data from one form to another. In this context, software of various degrees of sophistication will usually be available.

Its exact nature and mode of operation will depend upon the format of the data, how it is to be presented and the hardware that is to be used to display the results. Some possible modes of result reporting are illustrated in Fig. 2.25.

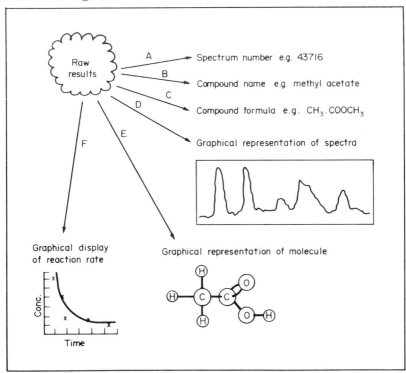

Fig. 2.25   Software techniques for result presentation.

This diagram depicts the types of transformation to which raw data produced directly by the analyst (or indirectly via a computer based instrument) may be subjected. Each of the transformations labelled A through F will require particular types of computer software. Graphical presentation of results will often necessitate the use of some special graphics package such as GINO-F (GIN80), ORTEP (Joh70), CONSYS, GHOST, PLUTO (NUM79), and so on. Many of these are available at most large computing installations. The results of processing raw data can be presented on appropriate CRT units or hardcopy devices such as the graph plotters, printers and chart recorders that were mentioned in the last section. Some of the transformations (data retrieval, smoothing, curve fitting, plotting, etc) depicted in Fig. 2.25 will be discussed in the chapter devoted to data processing technqiues - chapter 9.

There are many occasions when the analyst will have to produce written reports on the work he/she has undertaken or prepare papers for publication in learned journals. This is yet another activity in which the computer can be of assistance to the analyst - through the use of word processing techniques (Mea70). Word processing is the general term used to describe the application of computers to the preparation of all types of printed documents ranging from routine correspondence through to

multi-volume books. The basic principle is as follows: the first draft
of a document is entered into the computer's storage system. The stored
document can subsequently be modified in various ways (spelling corrections made, paragraph ordering modified, sections of text deleted, etc)
through the use of a VDU or teletype. After amendment, a copy of the
updated document can then be printed on a printer after it has been
reformatted using a special text processing package. The process of
amendment and reprinting updated versions of a document can continue
indefinitely since the master copy is always held in an easily modifiable
form within the computer's memory system. The use of word processing is
increasing in a large number of offices and laboratories. There are many
advantages to this type of document preparation facility - particularly,
the fact that it forms the basis of the electronic journal principle which
will be described in chapters 11 and 12.

CONCLUSION

The five principle areas of activity of the analyst have been outlined.
Some indication of the role that the computer and computer based
technology might play within these areas has been given.

Four of the most important areas for future development in which computer
technology will probably have its greatest influence are in:

(a)   the application of robotics,
(b)   the development of expert systems,
(c)   the production of intelligent instruments, and,
(d)   widespread automation of analysis.

Considerable developments have taken place in areas (c) and (d) during the
last few years. There is currently much active research taking place in
each of the other areas. How future developments in these areas are
likely to influence future approaches to analytical chemistry remains to
be seen.

REFERENCES

Arn80   Arnold, J.R., The Frontier in Space, American Scientist,
           Volume 68, No. 3., 299-304, May-June 1980.

AST51   Symposium on Bulk Sampling, Proceedings of the 54th Annual
           meeting of the American Society for Testing Materials, Atlantic
           City, New Jersey, June 1951, Special Publication No. 114.

Bar65   Barnard, J.A. and Chayen, R., Modern Methods of Chemical
           Analysis, (Chapter 6), McGraw-Hill, 1965.

Bec67   Becker, J. and Hayes, R.M., Information Storage and Retrieval:
           Tools, Elements, Theories, John Wiley, 1967.

Bej80   Bejczy, A.K., Sensors, Control and Man-machine Interface for
           Advanced Teleoperation, SCIENCE, Volume 208, No. 4450,
           1327-1335, June 1980.

Ber63   Berg, E.W., Physical and Chemical Methods of Separation,
           McGraw-Hill, 1963.

Bod78   Bodmer, M. and Rushton, C.A., On-line monitoring of Water Purity
           in Thermal Power Stations, Brown Boveri Review. No. 8,
           546-554, 1978.

BSI72   British Standard: 6001, Sampling Procedures and Tables for
           Inspection by Attributes, British Standards Institution,
           London, 1972.

BSI75   British Standard: 6002, Draft British Standard Specification
           Sampling Procedures and Charts for Inspection by Variables,
           British Standards Institution, London, 1975.

Buc79   Buchanan, B., Issues of Representation in Conveying the Scope
           and Limitations of Intelligent Assistant Programs, Chapter 20,
           407-425, in Machine Intelligence, Volume 9, Eds: Hayes, J.E.,
           Michie, D. and Mikulich, L.I., John Wiley, ISBN: 0-470-26714-3,
           1979.

Cam79   Camp, R.C., Smay, T.A. and Triska, C.J., Microprocessor Systems
           Engineering, Matrix Publishers Inc., Oregon, USA, ISBN:
           0-916460-26-6, 1979.

Cau80   Caulcutt, R., Acceptance Sampling, Analytical Proceedings of
           the Chemical Society, Volume 17, No. 5, 166-172, May 1980.

Cla79   Clancy, W.J., Tutoring Rules for Guiding a Case Method Dialogue,
           International Journal of Man-Machine Studies, Volume 11, No.
           1, 25-49, January 1979.

Cha78   Chatfield, C., Statistics for Technology - A course in Applied
           Statistics, Chapman and Hall, ISBN: 0-412-15750-0, 1978.

Cor71   Corey, E.J., Computer Assisted Analysis of Complex Synthetic
           Problems, Quarterly Reviews, XXV, 455-482, 1971.

Cor72   Corey, E.J., Wipke, W.T., Cramer III, R.D. and Howe, W.J.,
           Computer Assisted Synthetic Analysis. Facile Man-Machine
           Communication of Chemical Structure by Interactive Computer
           Graphics, J. Amer. Chem. Soc., Volume 94, No. 2., 421-430,
           1972.

For75   Forney, L.J. and McCay, J.F., A Flow through Electrode Unit for
           Measurement of Particulate Atmospheric Nitrate, The Analyst,
           Volume 100, 157-167, 1975.

Gil78   Giloi, W.K., Interactive Computer Graphics - Data Structures,
           Algorithms, Languages, Prentice-Hall Inc., ISBN: 0-13-469189-X,
           1978.

GIN80   GINO-F User Manual, Issue 2, Computer Aided Design Centre,
           Cambridge, UK., 1980.

Gro76   Grover, D., (Ed.), Visual Display Units and their Application,
           IPC Business Press Ltd., ISBN: 0-902852-65-5, 1976.

Hal77   Hall, J., CRT Terminals - windows into your process, Instru-
            ments and Control Systems, Volume 50, No. 11, 27-33, November
            1977.

Hal79   Hall, J., On-line Analysers tackle New Demands, Instruments and
            Control Systems, Volume 52, No. 8, 33-38, August 1979.

Hea78   Heaps, H.S., Information Retrieval - Computational and Theore-
            tical Aspects, Academic Press, Inc., ISBN: 0-12-335750-0, 1978.

Her61   Hermann, J.A. and Suttle, J.F., Precipitation and Crystallisa-
            tion, Chapter 32, 1367-1409, in Treatise on Analytical
            Chemistry, Eds: Kolthoff, I.M. and Elving, P.J., Part I, Volume
            3, John Wiley, 1961.

IBM62   IBM Corporation, Keyword in Context (KWIC) Indexing, Form:
            GE20-8091-0, 1962.

Irv61   Irving, H. and Williams, R.J.P., Liquid-Liquid Extraction,
            Chapter 31, 1309-1365, in Treatise on Analytical Chemistry,
            Eds: Kolthoff, I.M. and Elving, P.J., Part I, Volume 3, John
            Wiley, 1961.

Joh70   Johnson, C.K., ORTEP: A Fortran Thermal-Ellipsoid Plot Program
            for Crystal Structure Illustrations, Oak Ridge National
            Laboratory, Oak Ridge, Tennessee, USA, 1970.

Jon80   Jones, P.S., Development of an Intelligent Graphics Terminal,
            M.Sc. Thesis, University of Durham, 1980.

Jut79   Jutila, J., Multicomponent On-stream Analysers for Process
            Monitoring and Control, InTech, Volume 26, No. 7, 38-44, July
            1979.

Kol80   Kolthoff, I.M. and Elving, P.J., (Eds), Treatise on Analytical
            Chemistry, John Wiley, Parts I, II and III, 1959-1980.

Kra81   Kratochvil, B. and Taylor, J.K., Sampling for Chemical Analysis,
            Analytical Chemistry, Volume 53, No. 8, 924A-938A, July 1981.

Lai75   Laitenen, H.A., and Harris, W.E., Chemical Analysis (2nd
            edition), McGraw-Hill, ISBN: 0-07-036086-3, 1975.

Mas73   Mason, B., Chemistry of the Moon's Surface, Chemistry in
            Britain, Volume 9, No. 11, 456-461, October 1973.

Mea70   Meadow, C.T., Man-Machine Communication, Chapter 9, 244-278,
            (Editing Text), Wiley-Interscience, ISBN: 471-59001-0, 1970.

Mei58   Meites, L. and Thomas, H.C., Advanced Analytical Chemistry,
            McGraw-Hill, ISBN: 0-07-041335-5, 1958.

Mei63   Meites, L. (Ed.), Handbook of Analytical Chemistry,
            McGraw-Hill, ISBN; 0-07-041336-3, 1963.

Mer76   Merritt, R., Mission to Mars: the Search for Life, Instrument
            Technology, Volume 23, No. 6, 27-34, June 1976.

Mer79   Merritt, R., CRT Terminals - an Update, Instrument and Control
        Systems, Volume 52, No. 5, 20-28, May 1979.

Mic79   Michie, D., Expert Systems in the Micro-Electronic Age,
        Edinburgh University Press, ISBN: 0-85224-381-2, 1979.

Mur80   Murphy, E., Slow March for Britian's New Robot, New Scientist,
        Volume 87, No. 1213, 455, August 1980.

New73   Newmann, W.M. and Sproull, R.F., Principles of Interactive
        Computer Graphics, McGraw-Hill, ISBN: 0-07-046337-9, 1973.

NUM79   Hall, N.F., Graphical Facilities at NUMAC, NUMAC Documentation
        Group, University of Newcastle upon Tyne, UK., December 1979.

Ori78   Orion Research Limited, Analytical Methods Guide, December
        1978.

Par70   Paradine, C.G. and Rivett, B.H.P., Statistical Methods for
        Technologists, English Universities Press, ISBN: 0-340-04901-4,
        1960.

Par69   Parslow, R.D., Prowse, R.W. and Green, R.E., Computer Graphics
        - Techniques and Applications, Plenum Publishing Company, ISBN:
        0-306-20016-3, 1969.

Per75   Wilkins, C.L., Perone, S.P., Klopfenstein, C.E., Williams, R.C.
        and Jones, D.E., Digital Electronics and Laboratory Computer
        Experiments, Plenum Press, ISBN: 0-306-30822-3, 1975.

Pye77   Pye Unicam Limited, Pye Ingold pH and Redox Electrodes, March
        1977.

Ros61   Rosenthal, I., Weiss, A.R. and Usdin, V.R., Chromatography:
        General Principles, Chapter 33, 1411-1467, in Treatise on
        Analytical Chemistry, Eds: Kolthoff, I.M. and Elving, P.J.,
        Part I, Volume 3, John Wiley, 1961.

Rig78   Rigdon, L.P., Moodley, G.J. and Frazer, J.W., Determination of
        Residual Chlorine in Water with Computer Aided Automation and a
        Residual Chlorine Electrode, Analytical Chemistry, Volume 50,
        No. 3, 465-469, 1978.

San77   Sandford, J., On-stream Analysers become Faster and More
        Reliable, Instruments and Control Systems, Volume 50, No. 3,
        March 1977.

Sch78   Schlatter, H.G. and Heinicke, G., Plant Monitoring and Control
        Systems in the Cement Industry, Brown Boveri Review, No. 7,
        433-442, 1978.

She73   Shepherd, T.M. and Vincent, C.A., Digital Data in Chemistry,
        Chemistry in Britain, Volume 9, No. 2, 66-70, February 1973.

Sig59   Siggia, S., Continuous Analysis of Chemical Process Systems,
        John Wiley, 1959.

Sig68   Siggia, S., Survey of Analytical Chemistry, McGraw-Hill, 1968.

Sko63   Skoog, D.A. and West, T.M., Fundamentals of Analytical
           Chemistry, Holt, Rinehart and Winston, Inc., 1963.

Sne66   Snell, F.D. and Hilton, C.L., Encyclopedia of Industrial
           Chemical Analysis, Volumes 1 through 20, John Wiley, 1966-1974.

Svo78   Svoboda, R. and Schmid, P., Sampling System for Steam Operated
           Power Stations, Brown Boveri Review, No. 3, 179-188, 1978.

Wal76   Walker, B.S., Gurd, J.R. and Drawneek, E.A., Interactive
           Computer Graphics, Edward Arnold Ltd., ISBN: 0-7131-2505-5,
           1976.

Wes73a   West, T.S., MTP International Review of Science, Volume 12:
           Analytical Chemistry - Part I, Butterworth, ISBN:0-408-70273-7,
           1973.

Wes73b   West, T.S., MTP International Review of Science, Volume 13:
           Analytical Chemistry - Part 2, Butterworth, ISBN:0-408-70274-5,
           1973.

Wet77   Wetherill, G.B., Sampling Inspection and Quality Control (2nd
           Ed.), Chapman and Hall Science Paperbacks, ISBN: 0-412-14960-5,
           1977.

WHI78   Parlez-Vous Terminal?, WHICH COMPUTER?, Volume 2, Issue 8,
           ISSN: 01-40-3435, 55-57, August 1978.

WHI80a   Survey of Printing Terminal Suppliers, WHICH COMPUTER?, Volume
           4, ISSN: 01-40-3435, 69-79, April 1980.

WHI80b   Guide to VDU Suppliers, WHICH COMPUTER?, Volume 4, Issue 3,
           ISSN: 01-40-3435, 27-40, March 1980.

Wil59   Svehla, G., (Ed.), Wilson and Wilson's Comprehensive Analytical
           Chemistry, Elsevier Scientific Publishing Company, 1959-1977.

Wip74   Wipke, W.T. and Dyott, T.M, Simulation and Evaluation of
           Chemical Synthesis. Computer Representation and Manipulation of
           Stereochemistry, J. Amer. Chem. Soc., Volume 96, No. 15,
           4825-4834, 1974.

3

Instrumentation

INTRODUCTION

An instrument is a device that facilitates the activities of measurement and control. These activities may take place either in a laboratory or in some real-time process control environment. Within the laboratory, the analytical chemist usually has available a wide range of instruments to facilitate both the qualitative and quantitative analysis of matter in its various forms. Instruments vary in complexity, versatility, reliability, sophistication and cost. At one end of the scale there are simple devices such as the pipette, burette, stop-watch and the Lovibond Nessleriser (Vog61). The other more sophisticated end of the spectrum is represented by instruments such as the infrared spectrometer (Nak62), gas chromatograph (Amb71), mass spectrometer (Hil66) and the computer. Often the analyst will choose the tool according to the nature of the problem to be solved. The practice of using instruments to perform chemical analysis is often referred to as instrumental analysis. There are many excellent text books devoted to this area of study (Bau78, Ewi75, Str73, Rob70, Wil65 and others). The reader should consult one of these if a detailed understanding of the principles of instrumentation is required. Each of these books will also contain an appropriate taxonomy (particularly Bau78 and Str73) of the various methods and techniques available, for example,

> spectrometric/optical methods,
> electroanalytical methods,
> chromatographic techniques,
> radio-isotope analysis,
> thermal methods of analysis,
> kinetic methods,
>   o
>   o

Of course, in addition to helping the analyst identify the presence of a species and estimate it quantitatively, a variety of instruments will also exist to support many of the ancillary operations of sampling, separating, and sample pre-processing prior to performing the actual analysis. This class of equipment includes devices such as balances for weighing,

77

pipettes (automatic) for measuring and diluting samples, syringes and
valves for stream injection, automatic grading and separating tools such
as centrifuges and counter-current machines. Many more examples could be
cited. This book, however, is concerned with only a small number of
instruments that

(1) lend themselves to automatic operation via computer control,
(2) require the use of a computer because the system complexity
     deems this necessary,
(3) produce results that require some form of computer analysis
     or processing,
(4) incorporate computer technology in order to produce an
     instrument that gives better performance and reliability,
(5) employ computer technology to produce an instrument that is
     ergonomically better than its precursor, or
(6) are required for special purpose applications in areas such
     as space or planetary research.

Obviously instrumentation cannot be divorced from physics and
electronics. Pure and applied research in each of these areas provides
the knowledge and results which form the basis for the design and
construction of virtually all modern instruments. There are many
authoritative text books devoted to this area (Gri62). Unfortunately,
rapid technological advances in instrumentation often render much of the
published practical information out of date within a few years of its
publication. However, most of the theory of operation and many of the
basic principles remain valid. In order to examine current trends in this
area or to seek out a new instrumental technique it is a simple matter for
the researcher to locate up-to-date text books and appropriate research/
engineering journals such as:

        Reviews of Scientific Instruments
        - produced by the American Physics Society
        Scientific Instruments
        - issued by the Institute of Physics, UK
        ISA Journal
        - issued by the Instrument Society of America
        Instruments and Control Systems
        Instrument Technology
        InTech, etc

and appropriate journals in the area of electronics. Particularly
important is the area of digital electronics relevant to computers and
computer-instrument interfacing.

Instrument design and fabrication is very often a multi-disciplinary
activity involving an analyst, physicist, electronics expert and a design
engineer. Increasingly, as computers become components of measuring
instruments, there is a need to expand this team to include the computer
specialist. The degree of involvement of the latter will depend upon the
part the computer is to play in the functioning of the instrument. Often
the computer scientist will be involved in the design and construction of
special purpose software and firmware (these terms will be explained in
the next chapter) that control the way in which the instrument interacts
with its environment - both the process to be measured/controlled and the
operator responsible for the instrument. The computer expert may be
required to design special pieces of hardware (computer logic circuits,
interfaces, etc) to enable the results of an observation to be recorded

locally within the instrument's built-in memory or else enable the data to be transmitted over thousands of miles to some remote storage facility. Some of these possibilities will be discussed in later sections of this chapter and throughout the remaining chapters of this book.

The diverse nature of chemical instrumentation is indicated in the tree diagram shown in Fig. 3.1. This summarises most of the different types of instrumental technique commonly encountered either in the laboratory or in a process control environment. Further details on any of these techniques of analysis and their underlying theory will be found in most of the text books that were cited at the beginning of this section. In subsequent parts of this chapter particular types of instrument have been selected for the purpose of illustration. Much of what is said about one type will apply to many of the others. Before discussing some of these, such as the gas chromatograph and mass spectrometer, a simple model of an instrument will be described.

BASIC STRUCTURE OF AN INSTRUMENT

The diagram shown in Fig. 3.2 illustrates the relationship between the various basic functional units from which an analytical instrument is likely to be constructed. In this diagram rectangles that have been drawn with solid lines denote necessary parts of the instrument while those rectangles composed of broken lines represent optional components. The latter type are units which are not necessarily to be found in all analytical instrumentation. The direction of material or signal flow (that is, sample absorptions/emissions, etc) is denoted by heavy solid lines.

Common to all instruments will be sample handling, sample disposal and detection sub-systems. In addition, there will be either a display or a control system to enable the instrument to communicate with its outside environment. Sometimes both of these are present. The control sub-system enables the analytical instrument to be interfaced into a suitably designed control loop - perhaps involving a computer - in order to monitor some external process. In addition, it also performs the task of controlling the internal status of the instrument according to the values of various operational parameters set by the operator during the course of an analysis (temperature, liquid/gas flow rates, pH, detector selectivity, etc) - this too may involve the use of an internal or external computer system. Internal instrument control is an extremely important consideration when there is a large number of parameters likely to affect its performance. The display system enables the activity/status of the instrument to be made available both to engineers/technicians and those others who use it. In most cases it is the display system that enables the user to obtain the results of an analytical measurement - optical density, pH, $\%CO_2$, particle count, gas mixture composition or whatever else happens to be of interest.

As can be seen from the different types of rectangle shown in Fig. 3.2, many of the basic units are not likely to be present in every instrument. The units that an instrument contains will depend upon its basic design and the type of analysis for which it is to be used. Thus, in many instruments there will not be pre-processing or post-processing units. Similarly, in others, such as polarimeters, hydrometers, refractometers, viscometers, plastometers, etc) there may be no obvious separation system

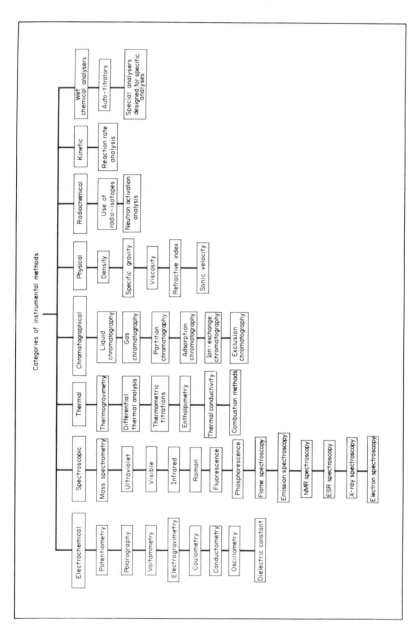

Fig. 3.1 A selection of instrumental methods suitable for computerisation.

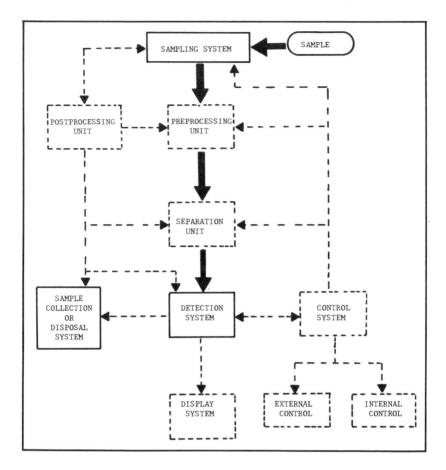

Fig. 3.2   Basic model for a simple analytical instrument.

while in gas/liquid chromatographs, mass spectrometers and most of the
spectroscopic tools some form of separation unit will be immediately
apparent.

Obviously, instruments will vary in complexity. Each instrument should
therefore be considered in its own right when any detailed discussion is
necessary. The simple model presented here is used only to tie together
areas of commonality for the purpose of subsequent discussion. The
following sections present some general points relevant to each of the
different units illustrated in the diagram.

(1) **The Sampling System**

The sampling system is the means by which the sample to be analysed is
introduced into the instrument for analysis or the mechanism via which a
part of the instrument is brought into contact with the material to be

analysed. Some discussion of the concept of sampling was presented in
Chapter 2 where gas/liquid sampling valves and pH/ion selective electrode
probes were cited as typical examples of particular types of sampling
device. In many instruments the sample handling sub-system can often be
more complex than any of the other parts - particularly where radioactive,
explosive or precious materials are involved. Sometimes, when an
instrument has to handle a wide range of materials (solids, liquids, gases
and mixtures of these) special types of sample inlet systems have to be
devised. Similarly, many situations arise in which purpose orientated
sampling units need to be constructed in order to adapt an instrument for
use in a particular application. The sample handling system will often
need to be appropriately interfaced with various other units such as the
post-processing system (perhaps to enable purging and cleaning after a
corrosive material has been processed) and the control unit (in order to
allow automatic sampling or to implement particular types of sampling
strategy such as stream splitting, automatic dilution, etc). Some of
these activities fall more naturally within the domain of the pre-
processing unit.

(2) **The Pre-Processing Unit**

In certain types of instrument some form of sample conditioning is often
needed in order to modify the physical or chemical characteristics of the
samples to produce states suitable for analysis. Thus, after a sample has
been obtained and introduced into the instrument, it may require some
particular type of physical pre-treatment prior to actually conducting the
analysis. This may be a relatively simple pre-processing operation such
as temperature equilibration, pressure reduction/compression, or volatili-
sation. Sometimes samples may be decomposed in a pyrolysis chamber or
subject to some photolytic or electrical decomposition/ionisation process
prior to the analysis. Alternatively, chemical pre-processing may be
required. This may involve the use of a derivative - as in the case of
the TMS derivatives of erythromycin prior to their analysis by gas
chromatography. Many different types of pre-processing technique have been
used. Unfortunately, problems can arise when multicomponent analyses are
performed because physical or chemical interactions which facilitate the
detection and measurement of one constituent may interfere with the
sensing of another. In view of this, many instrument designers are
attempting to circumvent these problems by means of highly selective
detection techniques and mathematical compensation methods; these permit
samples to be processed with little or no pre-treatment.

(3) **The Separation System**

The versatility of an analytical instrument depends upon the number of
items that it can be used to analyse for - simultaneously, if possible -
and the concentration ranges over which it is able to perform quantitative
analyses. Versatility is thus directly related to the ability of an
instrument to separate the entities that comprise a complex mixture - as
was discussed in chapter 2. Almost all instruments used to analyse multi-
component samples have a means of separating the chemical species in a
mixture so that individual detection and estimation is facilitated. The
classic example of such instruments are those that utilise the principle
of chromatography. The separation system may utilise any of a large
number of physical, chemical, optical, electromagnetic and mathematical
methods in order to achieve and sense the discriminations sought after by

the instrument designer. The various types of instrument that are currently available are usually distinguished by the methods used for performing the separation and detection functions - as was outlined in the last chapter. Thus, in chromatography based instruments the underlying principle involved is separation of materials based upon differences in affinity of those materials for some chromatographic reagent. In mass spectrometry, the ability to separate charged ions using magnetic/electrical forces forms the fundamental separating technique. In most forms of spectroscopy, the ability to separate electromagnetic signals using appropriate filters or other types of monochromator device is the underlying technique. In some cases, where complex spectra are produced, the ability to resolve the signals using mathematical signal processing techniques forms the basis of the method. Indeed, mathematical methods and computer techniques for indirect measurement of process variables and parameters are often used in the process control environment. Digital computers are applicable to many tasks involving indirect measurement which are too complicated for solution with instrumentation alone. Indirect measurement by computer is desirable and usually feasible when sensors for direct measurement are not available (Cly70).

Separation techniques are of extreme importance in instrument design. Usually, any reasonably efficient method of resolving materials/signals can form the basis for the design of an instrumental method of analysis provided a suitable means of detection can be found - either for the separated species or for the signals involved.

(4) The Detection System

The detection system of an instrument is probably its most important part. Without a suitable detector an analytical instrument, used in process control or in the laboratory would not be able to detect the presence of a species or measure its concentration. The pre-requisites of an ideal detection system will depend upon the purpose for which it is required. Linearity of response (with concentration), reproducibility (of measurement), selectivity (of species) and sensitivity are probably the most important attributes to consider.

The simplest type of detector is probably the transducer. A transducer is a device that converts one physical variable into another - usually the target variable is an analogue or digital electrical signal. Transducers are not restricted to electrical signal conversion techniques. However, in the main these predominate as electrical methods are universal and provide a common interconnection method for a manufacturing system or a scientific experiment. Transducers provide convenient signals for measuring a process, for automatically recording these measurements when needed, and finally, for providing a signal that can be used to control. As it is not possible to control without measuring, the transducer provides the fundamental basis for automation. An extensive description of the use of a wide variety of transducers for both measurement and control will be found in the book by Sydenham (Syd75). This book describes a number of measuring instruments and discusses the type of transducer upon which their operation is based.

The type of detection system employed will very often depend upon the nature of the instrument involved and the purpose for which it is to be used. In many analytical instruments the detectors that are commonly available fall into two types: radiation detectors and material detectors.

Radiation detectors form the basis of instruments that depend upon the use
of spectroscopic methods of analysis (Fel80) while material detectors form
the basis of many chromatographic (for example, catharometer, flame
ionisation detector, photo-ionisation detector, β-capture detector, etc)
and electrical (for example, pIon analysers, polarographs, etc) methods of
analysis. Sometimes radiation detectors are used as material detectors -
as in the application of ultraviolet or infrared detection units to liquid
chromatography equipment. In many situations an instrument will contain
several different types of detector in order to increase its versatility.
This is often the case in spectroscopic equipment where several sources/
detectors are provided in order to cover different parts of the spectrum
with optimum sensitivity. Similarly, in gas chromatographic equipment as
many as four (or more) detectors are often provided. These may be
connected in series or in parallel in order to increase the range of
materials to which the instrument is capable of responding. A similar
approach is used when ion selective electrodes are used for analysis. A
detection system may be specifically designed to contain a combination of
particular types of electrodes in order to cater for the (often
individual) application involved.

Particular kinds of detection system will often be employed in instruments
of similar type. Thus, detectors used for infrared spectrometers will not
vary greatly with the type of the spectrometer. Similarly, the type of
detector used for a nuclear magnetic resonance spectrometer will differ
from that used in an infrared machine but will not vary greatly from one
NMR spectrometer to another. The details of particular types of detection
systems will be found in appropriate manufacturers' technical manuals or
in books on instrumental analysis (Bau78, Str73).

(5) **The Display System**

An analytical instrument is essentially a provider of information. The
five most frequently presented types of read-out provided by scientific
instrument display systems are probably:

   (a) quantitative information:
        the value of some variable such as pH or $\%H_2O$,
   (b) qualitative information:
        such as the identity of a compound or the
        functional groups that it contains, the
        direction of change of a variable, etc,
   (c) status information:
        giving some indication of a discrete condition
        such as on/off, high/low, in/out,
   (d) representational information:
        such as a pictorial or graphic presentation of
        concentration/temperature gradients in a reaction
        vessel,
   (e) alphanumeric and symbolic information such as
        labels and instructions on how to use the instrument.

Some discussion on display systems was presented in the previous chapter
under the heading of result reporting. The material presented there
obviously applies in the present context. Since the purpose of an instrument is to provide information, it may utilise any display technology that
meets the requirements of cost effectiveness and which provides a suitable
medium for the presentation of results in an appropriate way. As the cost

of display technology decreases, the following trends seem to be taking
place:

(a)   there is a move towards digital instead of analogue
      readout systems where this is viable,
(b)   the incorporation of automatic scaling, integration
      and printing units in those instruments that require
      the application of peak processing procedures,
(c)   there is a move towards the use of video screens for
      the display of time varying signals (such as reaction
      rate profiles, spectra, chromatograms, thermograms,
      etc) alongside/instead of conventional paper chart
      recorders.

The reasons for some of these trends are easy to see. Digital readout can
be made more accurate and easier for the operator of the instrument
(compare, for example, analogue and digital pH meter readouts). Automatic
appliances such as those listed in (b) save time and are often more
accurate. The use of video technology, mentioned in (c), permits the use
of colour and special techniques such as windowing (selection of particular parts of a display for closer analysis), compaction, expansions,
superimposing and animation (for example, the rotation of images of
molecules, crystals, surfaces, etc) in real time. Obviously, display
systems will incorporate computer technology (processing, communication
and storage facilities) where this is economically and technically viable.
A recent article by Steiner (Ste80) describes many of the trends that are
taking place in the area of instrument display systems.

(6)   **The Control System**

As was indicated in Fig. 3.2, a control system may be needed to manage
both the internal operation of the instrument and some external process
that the instrument itself may be involved in monitoring. The required
control mechanisms may be implemented by means of conventional analogue or
digital controllers using conventional circuit techniques. However, there
is an increasing tendency towards the use of digital computers as control
elements - either built into the instrument or through an external
computer linked into the system via an appropriate instrument/computer
interface. Sometimes these latter types of arrangement are referred to as
computer aided measurement systems.

A general introduction to the basic ideas of computer aided measurement
has been presented by Brignall (Bri75, Bri79). These sources provide many
useful suggestions regarding the use of a scientific instrument as a
computer peripheral. In the context of analytical instruments for chemical
analysis, spectrophotometers, chromatographs and complicated spectrometers
(such as NMR, ESR, MS, ENDOR, X-ray, etc) are the most well known examples
of measuring tools that have required the use of sophisticated control
systems and which themselves have often been used (as process analysers)
in control applications. Consequently, they are probably the most well
researched with regard to the use of computer based control systems. A
detailed description of the way in which built-in computer technology is
being used to improve the performance of spectrophotometers is contained
in a series of articles in Hewlett Packard Journal (HPJ80). The coupling
of an external computer to a gas chromatograph has been described by
Leathard (Lea73) and Perone (Per75) while the potential of chromatographs
having a built-in computer has been outlined by Crockett and Mikkelson
(Cro76). Some of the advantages listed were totally automatic data

analysis, better performance, lower cost and the provision of many new
facilities, some of which have already been realised (Dul77). Typical
applications of the use of external computers for controlling spectrometers have been described by Balestra (Bal79), Mattson (Mat72), Ganjei
(Gan76) and Perry (Per77a). A useful mini-survey of some of the control
applications of digital computers in analytical chemistry instrumentation
has been presented by Perrin (Per77b).

As computer systems become smaller (as a result of developments in the
area of microelectronics) and the cost of developing computer based
control systems decreases, so there is likely to be an increasing tendency
for their incorporation into both sophisticated equipment (as outlined
above) and many of the less sophisticated devices (Wu78). Consequently,
computer controlled pH meters, titrators, dilution devices, homogenisers,
etc are likely to be increasingly commonplace in many laboratories in
future years.

(7) The Result Processing System

There are two main ways in which conditioning and processing of signals/
results may take place: either at the time the measurements are being made
(called real-time processing) or after the measurements have been taken
(often referred to as post-processing). In real-time signal conditioning
(such as in digital filtering) there may be no requirement to store
signals within the instrument and so this type of processing can often be
implemented using either conventional circuitry or computer-based systems.
However, when post-processing is required some form of storage for the
acquired measurement will be needed. Computer systems are especially
useful in this context - particularly microcomputer systems. These can be
pre-programmed with various data processing options which are then hidden
away within the instrument so as to provide a selection of result
processing options at the 'touch' of an appropriate button on the
instrument fascia. Some good examples of computer-based result processing
techniques lie in the areas of Fourier transform methods (Bri75, Per75) -
for converting signals in the time domain to the frequency domain - and in
the computer averaging of transients (the CAT technique) used to produce
time-averaged spectra (All63). Some further examples of result processing
will be discussed in a later chapter.

(8) The Post-Processing System

Once an instrument has been used to perform an analysis there will often
be a variety of 'house-keeping' operations that will need to be performed
before a subsequent analysis may be undertaken. The complexity of the
post-processing operations will depend upon the nature of the instrument
and the type of analysis that it is used to perform. One of the most
well-known examples of this type of operation is that of backflushing in
gas chromatographical equipment (Gib79). Backflushing is used to prevent
unwanted components of a sample progressing through the analytical column
of a chromatograph and thereby reducing its performance or increasing the
dead-time of the instrument. Similar types of operation may be required
in other instruments. In mass spectrometers, 'bake-out' is used to remove
sample 'memory' while in UV or IR machines cell windows may need to be
cleaned/polished after the sampling part of the instrument has been
subjected to the effects of corrosive materials. The post-processing unit
may be a very important system component in a process control environment
in which instruments are expected to operate continuously for long periods
of time without human intervention. Automatic post-processing procedures

may be necessary to ensure that sample flow systems do not clog-up or that optical windows do not become opaque as a result of analysing reactive substances. Ultrasonic, mechanical and liquid purging techniques are often used to overcome some of the problems encountered in this area.

(9) The Sample Collection/Disposal System

Before an analysis can be undertaken many instruments will require the introduction of a sample of the material to be analysed. This is inserted through the sample handling system. After the analysis has been completed the problem of disposing of the spent sample arises. The procedure to be adopted will depend upon many factors such as the method of assay (destructive or non-destructive), the value of the sample, its toxicity/stability, the amount involved in analysis and so on. Depending upon the exact requirements, once the analysis has finished, the sample can be returned to the flow stream (in a process control environment), vented to atmosphere or pumped into some special disposal system. Alternatively, it may be retained for future use by means of some special sample collection equipment attached to the outlet port of the instrument. For example, liquid fractions eluting from a chromatography column may be collected in a fraction collector while components eluting from a gas chromatograph might be trapped out by means of some low temperature collection system. Obviously, the exact procedure to be adopted will depend upon the instrument, the sample and the purpose for which the spent material is required.

BASIC PROPERTIES OF AN INSTRUMENT

In order to evaluate the usefulness of an instrument and to be able to predict its performance it is necessary to formulate certain standard instrumental parameters or characteristics. These specifications enable instruments in a particular class to be compared with each other (for example, mass spectrometer with mass spectrometer) and with members of other classes (for example, infrared spectrometer with polarograph). In this section some of the basic properties of instruments are outlined.

Versatility
This is the most desirable characteristic of an instrument. Ideally, it would be useful to have a single instrument that would analyse for any substance over any concentration range under any conditions with minimal pre-processing effort. Unfortunately, such ideals are unlikely to be realised in any simple instrument. However, some of the more sophisticated types of equipment described later in this chapter do show extreme versatility.

Applicability
This term refers to the application areas of the instrument (that is, the type of analysis that it is able to perform), the analytical problems to which it can be applied and the sorts of materials it can be used for analysing. Applicability thus relates to versatility.

Sensitivity
Sensitivity is specified by the relationship between concentration and instrument output and hence by the slope of the instrument response curve ($S = dR/dC$). Sensitivity also specifies the minimum detectable change in

concentration governed by the signal-to-noise ratio of the instrument. Sensitivity is generally defined as the concentration required to give a signal equal to twice the root mean square of the baseline noise.

Repeatability, Reproducibility or Precision

If repeated measurements are made of a static process by an instrument with sufficient sensitivity there will be a scatter of the values around some mean value. The scatter represents the uncertainty of the measuring process used. The most commonly used method of expressing this scatter is by means of the standard deviation (σ). Thus, if an instrument has a repeatability of $\pm 1\sigma$ then it can be shown statistically that there is a 68% chance of the true value lying between plus and minus one standard deviation of the mean. In practice these limits are not tight enough and so repeatability of $\pm 2\sigma$ and $\pm 3\sigma$ or more are often quoted. For $\pm 2\sigma$ there is a 95% chance of the true value lying between $+2\sigma$ and -2σ, while for $\pm 3\sigma$ there is a 99.7% chance of the value being in the range $+3\sigma$ to -3σ. Repeatability is probably the first requirement of any instrument since without it accuracy has no meaning.

Accuracy

This is the most difficult factor to obtain. An instrument may be precise, always giving the same value, but to be accurate that value must be true to the established standards. There is no way of establishing accuracy without resorting to another measuring device. Often accuracy is added to a precision instrument by means of suitable calibration techniques (see below).

Responsiveness

This parameter refers to the speed with which the instrument is able to respond with respect to changes in the signal it is monitoring. Speed of response is usually defined as the time required for the instrument to reach a specified percentage of the total change observed. Responsiveness may be specified in terms of the instrument's rise time, time constant or response time (Bau78).

Dead-Time

This is defined as the time interval, after alteration of the parameter being measured, during which no change in the parameter is observed at the detector. Dead-time is never entirely absent from an instrument. However, it can be minimised by appropriately positioning transducers, keeping the sample lines short and using high flow rates. The dead-time of an instrument is of considerable importance in automated systems since it reflects the period in which system changes cannot be observed.

Reliability

This refers to the instrument's freedom from unavailability because of breakdown. Often, reliability is expressed in terms of the MTBF - Mean Time Between Failure. The value of the MTBF can be increased by building redundancy into the instrument. Reliability is an important consideration in on-line control situations.

Durability and Robustness

These two factors each influence the reliability of an instrument. Robustness refers to the ability of the instrument to withstand violent impact, shocks or impulses of various sorts, while durability reflects the instrument's capability to resist the effects of environmental changes - corrosion of probes or connecting leads, etc. These factors are particularly important in instruments that are used in non-laboratory environments such as industrial sites or in equipment used for field-work studies.

Resolving Power
This is another term that is often met in the context of instrument parameters. The exact meaning of the term varies with the type of instrument to which it is applied. Generally, however, the resolving power of an instrument refers to its ability to separate. Thus in the case of spectrometers the resolving power measures the ability of the instrument to separate frequency signals. In the case of a mass spectrometer it reflects the ability to separate m/e values, while in the case of a chromatograph the term is often used to describe the instrument's capability of separating the components of the mixture it is used to analyse. In a more general context, resolving power is used to reflect how well an instrument is able to separate signals from noise. Usually, the higher the resolving power of an instrument the greater its cost.

Selectivity
The selectivity of an analytical instrument is its ability to discriminate between the species of interest and possible interferences. Many commercially available instruments are designed for the largest possible number of applications with the result that some loss in selectivity may be encountered.

Range and Span
The range of an instrument is defined as the interval over which a quantity is measured. The term span is used to refer to the width of the range. It is thus the difference between the upper and lower range values. Span and range are related to the instruments gain and bias. A change in span requires a change in the unit response of the instrument, that is, a change in gain. A change in range requires a change in the zero point of the instrument, that is, a change in bias. These are further discussed in Bau78.

Interfaces Available
This term refers to the number of standard electrical (or mechanical) interfaces that are available to enable the instrument to be connected to other devices - either other instruments (as in a GC/MS combination) or to a computer system of some form. Alternatively, it may be required to connect the instrument to some form of communications equipment (to allow remote control or data transmission) or to some form of signature analysis equipment for the purpose of error diagnosis.

Ease of Calibration
As was mentioned earlier, the accuracy of an instrument depends upon how well it is calibrated. Most analytical instruments used for identification purposes or for quantitative measurement need to be calibrated in appropriate ways. Ease of calibration is thus an important factor to consider when investigating the potential of an instrument for a particular application. Automatic calibration is a feature which may need to be considered for instruments that are to be used for on-line process control analysis.

Ability to Identify
Many instruments in analytical chemistry are used primarily because of their ability to identify an unknown material. There are two important factors that contribute to capability in this area:

(a)   the inherent potential of the analytical technique that
      the instrument implements. This will determine the nature
      of the basic data that the instrument provides, and,
(b)   the assistance given to the analyst in order to help
      interpret the basic data presented.

As machines become equipped with internally stored libraries of reference
values and other types of correlation data, so the ability of the
instrument to help identify unknown materials will increase. The storage
capability associated with built-in computer systems will help contribute
to this aspect of the instrument.

Expense
As a general rule the more sophisticated an instrument becomes the greater
its overall cost. Expenses arise from three basic sources: the initial
cost of purchasing (or renting) the instrument, running costs and the cost
of maintenance. Obviously, cost justification is an important factor to
consider when investing in expensive instruments.

Many of the particular values associated with the above basic properties
of instruments are intrinsically fixed because of the methods or modes of
measurement that are employed. However, certain of them can be strongly
influenced by the incorporation of computer based components. For
example, in a large number of instruments the techniques of parameter
optimisation and computer control can be used quite successfully to effect
substantial increases in performance. In subsequent sections of this
chapter some of the ways in which the computers are influencing the basic
design of instruments will be outlined. Some of the above properties will
then be discussed in more detail.

ERGONOMICS OF INSTRUMENT DESIGN

The term ergonomics is used to describe the study of the relationship
between people and the environment in which they work. Ergonomics (or
human factors) applies anatomical, physiological and psychological
knowledge to the problems arising between humans and their working
environment. In the present context this environment consists of an
analytical laboratory and the various instruments that it contains.
Scientific instruments allow the human to observe and measure aspects of
the physical universe beyond the range and precision of the unaided human
senses. To do this, however, the instrument must detect, process and
translate a particular signal into a form that can be easily understood by
a human operator. Display of results - as was discussed earlier in this
chapter (and in chapter 2) - is thus an extremely important
consideration. Other important factors to consider when discussing the
ergonomics of instrument design are: ease of use, comfort of use, safety
and ability to detect possible operator errors that might influence the
results produced. These factors are described in some detail in an
article written by Pearce and Shackel (Pea79).

The way in which computer technology is helping to improve the ease of use
of instruments is reflected in many different ways - both in simple
devices (such as digital pH meters, autotitrators, weighing systems,
automatic dilutors, and so on) and in more complex machines (such as gas
or liquid chromatographs, infrared and chemical analysers, NMR spectrometers and so forth). The introduction of a small computer system (or

microcomputer) into an instrument provides three important facilities: the ability to process numerical data, the capability to remember or store various items of information and the ability to communicate more easily with its operator and with other instruments. As an illustration of the ergonomic advantages of installing a microprocessor in an instrument consider the mode of operation of the digital pH meter shown in Fig. 3.3.

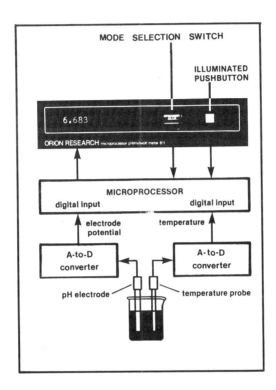

Fig. 3.3   Microprocessor based ion sensitive electrode system; A-to-D: analogue-to-digital conversion of signals.

An instrument of this type is easy to operate since
    (a) it has a digital readout which means it is easy to read, and,
    (b) it has only two controls - a 12-way mode selection switch and
         a simple push-button for calibration.

Because of its ergonomic design this 'clever' instrument can perform many functions which would otherwise have to be performed by its operator, or, in some cases (as in circuit trouble-shooting) an electronics engineer. Amongst its repertoire of ergonomic advantages are its ability to

    (a)   alert the operator if an attempt is used to calibrate
          the instrument using an incorrect buffer solution,
    (b)   automatically detect a defective pH electrode,
    (c)   self check itself when put into 'check mode',

(d)   prevent readings being taken before the electrode has
      achieved a steady state,
(e)   prevent incorrect calibration procedures,
(f)   automatically compensate for temperature changes in
      the buffer solutions used for calibration and samples
      presented for measurement.

A digital pH meter such as this works on a different principle from that
of a conventional pH meter (Vog61) since it calculates pH values instead
of obtaining them from analogue signal manipulation. Because of this it
is able to allow the operator to introduce buffer solutions in any order
during calibration. As the instrument uses two analogue-to-digital
converters (see chapter 5) temperature information is available continuously. This information is used both to calculate corrections for
changes in electrode slope with temperature and to ensure that the meter
is standardised to the correct pH value for a buffer regardless of
temperature. It is able to do this because the internal microprocessor
system is pre-programmed with a table of NBS (National Bureau of Standards
- see Vog61) buffer values.

The microprocessor based pH meter is just one of many new types of
instrument designed in a way that is able to take advantage of the
ergonomic benefits offered by microcomputer technology. Some further
examples are presented in Table 3.1.

TABLE 3.1   Some of the Advantages Gained through the use
            of Microcomputers in Analytical Instruments

Instrument	Advantages of Using a Microcomputer
Autotitrator	addition of memory enables the storage of standard methods of titration, touch sensitive keyboard and simple dialogue facility (Met79).
Specific Ion Meter	digital specific ion meter that automatically calculates calibration curves and stores them internally in its memory (Ori79).
Balances	digital readout, program cards, simplicity of use, attachment of peripherals for printing, self calibration facility (Met79).
Polarographic Analyser	alphanumeric display, touch panel controls, data and calibration curve storage, bi-directional communication interface, can be made fully automatic (Pri80).
Thermoanalyser for TG/DSC	easy operation, modular construction, micro-processor controlled temperature, interface connections for a desk-top computer (TA200).
Liquid Chromatography Solvent Programmer	touch keyboard, self-test diagnostics, storage and editing of programs, automatic control, printer interface, remote inputs (KON200).

Details of other types of microcomputer based instruments could be listed
and their advantage presented. Unfortunately, space limitations do not
permit this and so the reader is referred to the literature for further
details of these: atomic absorption spectrometer (AA875), infrared systems
(MIR80, Per680), UV/Visible spectrometers (HPJ80, DMS90), ultracentrifuge,
gas chromatograph (Pye304) and the process control mass spectrometer
(IQ200). This list is representative rather than comprehensive.

SOME EXAMPLES OF COMPUTER BASED
ANALYTICAL INSTRUMENTATION

In the previous section some illustrations were presented of the way in
which the ergonomics of an instrument could be improved by the addition of
a built-in microcomputer. The considerations presented in Table 3.1
coupled with the possibility of improving the basic characteristics of an
instrument provides considerable motivation for the addition of a computer
system to an instrument. As the size of the computer system grows so does
the number of advantages that can be gained. In this section three
examples of more complex computer based analytical instrumentation will be
described in terms of the simple model presented in Fig. 3.2. A variety
of different approaches may be encountered involving different
combinations of (a) the use of internal microprocessors, (b) the use of
external microcomputers or desk-top computers and, (c) the use of larger
external computer systems known as minicomputers.

(1) The Gas Chromatograph

Chromatography is a separation technique which makes possible the
determination of the composition of complicated mixtures consisting of
substances having similar chemical and/or physical characteristics. The
extremely rapid growth of the chromatographic technique (compared with
other analytical methods which provide separation) is probably due to the
speed at which the separation process can be carried out. This means
short analysis times which is often an important factor in the process
control environment. The speed of analysis which characterises the
chromatographic method has been a prominent feature of gas chromatography
ever since its development in the early 1950's by James and Martin
(Jam52). Since that time there have been many significant developments
that have contributed towards improved separation and speed of analysis -
for example, open tubular columns, temperature programming, stream
splitting and so on.

Andrews and Abbott's book (Abb65) provides an elementary introduction to
gas chromatography. A more comprehensive treatment of the subject will be
found in the books of Purnell (Pur62) and Ambrose (Amb71) while recent
trends in the area are reflected in the review article by Cram and Risby
(Cra78). Introductory discussions on the applications of computers have
been presented by Leathard (Lea73) and Perone (Per75) while references to
the major areas of current development will be found in Cra78. Perone
presents a simple case study of data acquisition which is a good
introduction to the subject while Leathard discusses some of the problems
associated with the mathematical processing of the data that is obtained.
The task of simultaneous data acquisition from multiple chromatographs
(ranging from 1 to 32) and a laboratory data system to handle this type of
situation is described in publication by INSTEM (INS79). More recent

descriptions of chromatogaphical data acquisition and processing have been
described by Lanza (Lan80) and Reese (Ree80).

The fundamental hardware of the chromatograph fits into the scheme
outlined in Fig. 3.2. Various types of sampling facility exist both for
liquids (via a syringe) and gases (via a gas sampling valve - see Fig.
2.3) either of which may be operated under computer control in order to
achieve improved precision. The separation unit consists of one or more
columns which may be packed with various chromatographic reagents. Column
length, temperature, gas flow and the nature of the chromatographic
reagents used all strongly influence the separating power of the instrument. A chromatograph may contain a single column or several columns
interlinked in various ways (both parallel and serial combinations)
depending upon the effects to be achieved. Materials eluting from the
column(s) are detected by one or more of a wide range of detectors
(thermal conductivity, flame ionisation, thermionic, electron capture,
flame photometric, spectroscopic, electrochemical, radiochemical, photo-
ionisation etc). Each of these vary in their sensitivity, selectivity and
response characteristics. There are several books available (Sev75,
Dav74) that describe some of the different types of detectors commonly
employed in gas chromatography.

The essential data produced by the instrument is a plot of detector
response as a function of elapsed time measured from the commencement of
the analysis. The resultant graph is referred to as a chromatogram - a
typical example of an actual chromatogram is shown in Fig. 1.13. Several
chromatograms may be recorded/displayed simultaneously if several
detectors/columns are used concurrently. The procedure adopted will
depend upon the instrumental techniques employed for a particular
analysis. From a chromatrogram it is possible to derive two important
measurements for each eluent: retention data and detector responses. The
former can be used to identify the materials concerned using empirical
calibration techniques, Kovats retention indices (Bau78) or by means of
special detectors based upon the use of infrared or mass spectrometry.
Detector responses may be used to estimate the eluting materials
quantitatively, provided that suitable calibration procedures for the
detector are employed. A useful description of quantitative analysis by
gas chromatography can be found in the book by Novak (Nov75).

The separating ability of the instrument will depend upon a variety of
factors most of which are described in the books mentioned above. Other
relevant texts are those by Baiulescu (Bai75) and Jennings (Jen78). Some
interesting work designed to compare liquid phases used as chromatographic
reagents was undertaken by McReynolds (McR70) who attempted to compare
liquid phases for their ability to separate various classes of compounds
as a means of predicting retention behaviour. The technique is useful for
characterising and checking proper column operation of new columns and for
checking for column changes which occur with use. Over the years various
collections of chromatographic retention data have been accumulated. These
are useful in that they can be used as a guide both to column selection
and compound identification. McReynold's book (McR77) is claimed to be
one of the most comprehensive collections of retention data to be found.
It contains approximately 60,000 entries all of which have been determined
in a single laboratory using a single instrument thus ensuring the
consistency of the data. Another publication containing ASTM data (AST25)
provides a compilation of the gas chromatographic conditions used for the
separation of a wide variety of compounds. The tables contained in this
book are useful since they facilitate the tentative identification of

materials that produce peaks on chromatograms. Data from sources such as these are, in principle, well suited for processing by computer thus enabling the automatic identification of column eluents. Thus, under given chromatographical conditions, the retention data of compounds should be characteristic of their identity and so the results obtained from the automatic measurement of retention data could be used to suggest the likely components of a mixture. There are many problems associated with the technique from the point of view of its use for absolute compound identification. In this context, the addition of a more sophisticated detector such as an infrared or mass spectrometer would undoubtedly provide information which is of greater utility. This will be discussed in more detail later in this chapter.

A review of the role that computerisation is playing in gas chromatography has been presented by Cram and Risby (Cra78) who provide a wide selection of references to many of the important areas of development. A list of some of these is presented below:

- the influences of microcomputer technology,
- the potential role of computer networking and hierarchical systems for control applications,
- data logging,
- data reduction techniques,
- automation of chromatography and the role of chromatography as an automation tool,
- improved instrumental performance,
- design of laboratory data systems,
- application of Fast Fourier Transform techniques to enable the retention times and areas of overlapping peaks to be determined,
- the effects of digital filtering techniques on chromatographical data,
- software development for data processing, data searching from stored libraries and result reporting.

Current trends in the application of computers are devoted to exploiting the improved instrument performance offered by built-in microprocessors and the attachment of desk-top computers of various sorts to enable the processing of GC data (Pye304). In addition, there are significant developments being made in the application of minicomputer systems for improving the performance and capabilities of the gas chromatograph in combination with other instruments such as the mass spectrometer (Ner80).

(2) **The Infrared Absorption Spectrometer**

Infrared absorption spectroscopy is a powerful technique for the analysis of many kinds of gaseous, liquid and solid samples. As an analytical tool the infrared method can often compete with gas chromatography from the point of view of speed of analysis, accuracy and precision. In addition it is able to provide a wide range of information to enable both the qualitative identification and the quantitative determination of a wide variety of materials. An elementary introduction to the theory, instrumentation and interpretation of spectra will be found in the books by Meloan (Mel63), Ewing (Ewi75) and Meehan (Mee61). Current developments are summarised in the American Chemical Society's biennial reviews contained in ANALYTICAL CHEMISTRY. Typical of these is that of McDonald McD78).

A wide variety of infrared spectrometers are available from different
manufacturers. They fall essentially into three broad categories: Fourier
transform, dispersive and non-dispersive. The latter are usually less
expensive since they do not employ a dispersing element to separate
signals. They are often used in process control applications. The
MIRAN-80 (MIR80) is an example of an instrument of this type. Dispersive
instruments - such as the Perkin Elmer Model 683 (Per680) - contain either
a prism or diffraction grating to enable the separation of radiation of
different wavelengths. In a Fourier Transform (FT) spectrometer an
interferometer (Str73) replaces the dispersing element used in a conventional machine. The Nicolet MX-1 FT/IR spectrometer (Nic80) is an
example of an instrument of this type. Each of the spectrometers cited
above contains a built-in computer. They are used to enable improved
performance to be obtained or because they are a necessary system
component as in the case of the FT/IR spectrometer.

The basic arrangement of equipment for an infrared spectrometer fits the
simple model that was presented in Fig. 3.2. Radiation from a suitable
source (Nernst glower, Globar, etc) is passed through the sample which is
held in a suitable sample cell. Absorption of radiation by the molecules
in the sample reduces the intensity of the radiation at particular wavelengths - the reduction in intensity depending upon the amount of material
in the sample beam. Depending upon instrument design, either before or
after passage through the sample the polychromatic radiation from the
source is separated (or resolved) by the dispersing element into mono-
chromatic radiation and its intensity at different wavelengths is measured
by means of a suitable detector such as a bolometer (thermal detector) or
a photomultiplier (photo-detector).

The data collected when a sample's infrared spectrum is taken consists of
the observed detector response as a function of wavelength. Detector
response is converted to absorbance units or percentage transmittance and
is plotted against wavelength expressed in microns (10^{-4} cm) or
wavenumbers (cm^{-1}). The infrared region from 2 to 15 microns (5000 to
666.6 cm^{-1}) is probably the most useful for qualitative analysis. Most
chemical compounds show marked selective absorption in the infrared. For
this reason the near-infrared region is often likened to a fingerprint of
the molecule. It is not uncommon to observe at least 30, or more, easily
resolvable maxima in the 2-15 micron region of the spectrum. Absorptions
at particular wavelength regions are usually characteristic of particular
functional groups contained in a molecule (Fle63). Consequently, the
observed signals may be used as an aid to identifying the compound. The
infrared spectrum of a pure compound provides a sure method of identification provided the analyst has available a suitable library of spectra
of known compounds with which to make comparisons. Many standard
collections of spectra exist (for example, the ASTM Infrared File
containing over 135,000 spectra) some of which are available in computer
readable form. This means that the computer can be used as a tool for the
automatic recognition of compounds based upon mathematical comparison of
observed and standard spectra. Two basic approaches may be used - partial
matching and complete matching. In partial matching, particular signals
in the spectrum of the unknown are compared with standard tables of
characteristic frequencies coded into suitably organised 'look-up' tables
within the search program. When complete matching is required, a standard
procedure is used to condense the observed spectrum. The result is then
used as a search pattern in a library search routine that operates on some
standard collection of condensed reference spectra. In both types of
searching, statistical limits can be set which must occur when pattern

matching is being performed. McD78 cites a selection of references to some of the various computer based techniques (such as the CASE and SPIR programs) that are currently being used or which are under development. Of the spectrometers mentioned earlier, both the Nicolet and Perkin Elmer instruments implement library searching methods. Nicolet employs a technique based upon the use of condensed spectra from Sadtler while Perkin Elmer uses a combination of spectra generated in their own laboratory and spectra obtained from the Coblenz Society.

Quantitative analysis is often used both in laboratory and in process control situations. Both MIR80 and Per680 discuss the accuracy and precision that can be obtained with microcomputer based spectrometers. Meehan (Mee61) amd Ewing (Ewi75) outline some of the problems associated with the quantitative analysis of both pure samples and mixtures. The most interesting cases arise when the materials presented for analysis consist of mixtures. In situations such as this, computer processing may be required to solve the mathematical equations that need to be formulated in order to represent the total absorption in terms of those of the individual components. Examination of the spectra of the corresponding pure compounds from which the mixture is composed enables many of the coefficients for the mathematical expressions to be obtained. If there are n components in the mixture then a set of n simultaneous equations will need to be solved. This will require n experimental measurements to be made at 'optimum wavelengths'. Ideally, these optimum wavelengths are chosen in such a way that there is one wavelength unique to each of the compounds present. This ideal is often not met and appropriate mathematical corrections may need to be applied in order to compensate for this overlap. The application of computer techniques to the problem of multicomponent spectral analysis has been described by Beech (Bee75). An example of the use of a small computer system coupled to an IR spectrophotometer in order to perform spectral subtraction is described in the paper by Lynch and Brady (Lyn78). In this work computer separation of the infrared spectra of complex organic molecules was used as an aid to both qualitative and quantitative processing of data. The literature contains many other examples of the use of the computer for this type of application.

An important area of data analysis in which the computer is of substantial utility is that of Fourier Transform applications. Generally, there are two broad approaches to Fourier Transform spectroscopy:

(a) application of Fourier techniques to data obtained from conventional systems - see, for example, Horlick and Yuen (Hor76), and

(b) the utilisation of special purpose Fourier Transform spectrometers - see, for example, Cournoyer et al (Cou77).

Because of the improved facilities that they offer (Cou77) there is an increasing use of Fourier Transform IR instruments (Str73, Ewi75, Koe75). As was mentioned earlier, the FT/IR spectrometer is based upon the use of an interferometer instead of a monochromator and relies upon the use of a computer for resolution of the signals. A typical arrangement is shown in Fig. 3.4.

The path length of the interferometer can be varied by moving a mirror in a known and systematic fashion - under computer control if need be. The movement of the mirror creates a pattern of bright and dark bands (called interference fringes) to sweep across the detector. With a continuous

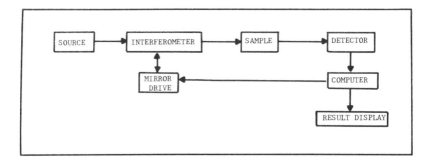

Fig. 3.4   Role of the computer in a Fourier transform instrument.

source very many wavelengths will be present simultaneously and a complex
pattern will be obtained. The record of detector signal as a function of
path difference is referred to as an interferogram. A computer which is
attached to the detector is used to translate the interferogram (a
representation of intensity as a function of distance) into an optical
spectrum (a graph of intensity as a function of wavelength) using Fourier
Transform techniques. One of the major differences between this type of
spectrometer and the conventional type is that all the wavelengths of the
spectrum are presented to the sample simultaneously rather than sequentially. This can provide many advantages. Thus, in many energy-limited
situations (for example, in the measurement of weak sources or where rapid
scanning of spectra is required) greatly improved signal to noise ratios
can be obtained. Some examples of the benefits that can be achieved
through the use of this type of spectroscopy are presented in the papers
by Cournoyer (Cou77) and Vanderwielen (Van79).

The Nicolet Instruments MX-1 IR spectrometer described earlier (Nic80) is
an example of a microcomputer based Fourier Transform instrument that uses
a rapidly scanning Michelson interferometer (Str73, Ewi75). In addition
to applying a Fourier Transform to the interferometer data the microprocessor based data system provides facilities for the storage and recall
of sample, reference and background spectra, subtraction of solvent or
reference spectra for non-chemical separations, and the ability to replot
the spectrum in a variety of different formats without re-running the
analysis.

In addition to its use in conventional spectroscopy the FT/IR spectrometer
has been used as a sophisticated detector in gas chromatographs. Details
of this work are contained in the paper by Coffey (Cof78).

(3) **The Mass Spectrometer**

The mass spectrometer is probably one of the most versatile analytical
tools available to the analyst. It permits the analysis of a wide variety
of substances (liquids, solids and gases) and requires only minute
quantities of sample. Unlike infrared and most forms of gas chromatro-
graphic analysis, the mass spectrometer's major drawback is that the
method of analysis it uses is destructive - that is, it is not possible to
retrieve the sample after it has been processed. However, the many

advantages of this instrument, particularly its sensitivity and the variety of information that it provides, far outweigh the disadvantages.

The instrumentation for mass spectrometry is similar to that shown in Fig. 3.2. Materials for analysis are presented to the system via an appropriate sample handling unit such as a gas or liquid batch inlet, a molecular leak or a direct insertion probe. Once the sample has been introduced into the spectrometer it is ionised using a suitable technique (such as electron impact, chemical ionisation or a spark source, etc). The charged particles produced as a result of ionisation and fragmentation of the initially ionised species (usually the molecular ion) are separated using magnetic and/or electrostatic fields. A detector such as an electron multiplier is used to measure the intensities of the individual ion beams produced by the dispersion and focussing system. The degree to which the different ion beams are separated is referred to as the resolving power of the instrument. This can vary from 500 (low resolution) up to about 150,000 (high resolution). The resolving power of the instrument will determine the type of information that can be obtained from it. Various types of mass spectrometer are available - time of flight, quadruple mass analyser, ion cyclotron resonance spectrometer, etc - depending upon the exact manner in which the ions are produced, separated and detected. An introduction to the basic principles of mass spectrometry will be found in Hil66, Ewi75, Bau78, and Bey60. An introduction to the use of computers in mass spectrometry has been given by Chapman (Cha69) while a review of current trends in most areas of the subject has been prepared by Burlingame et al (Bur78).

The data collected from a mass spectrometer is called a mass spectrum and when presented on a recording device usually takes the form of a bar-graph in which the x-coordinate represents the mass to charge ratio (m/e) corresponding to a particular ion and the y-coordinate represents the intensity of the corresponding peak. Usually the intensities are reported in terms of relative abundance with respect to the base peak - the most intense peak in the spectrum. A typical section of a mass spectrum is presented in Fig. 3.5.

Usually the peaks in the spectrum correspond to positive ions although sometimes negative ions and metastable ions (Bac76) are recorded.

Beneath the mass spectrum shown in Fig. 3.5 a simple list structure is presented. This is used to denote one possible way in which a spectrum might be stored within a computer. The first number of each pair represents the m/e value while the second denotes the relative intensity of the peak.

Mass spectrometry differs from other forms of spectroscopy since the signals that are produced are not generated by energy-state transitions. Instead, the information that is produced is of a chemical nature since it is formed as a direct result of chemical reactions (ionisation and fragmentation). Consequently, in order to be able to understand a mass spectrum the analyst needs to be able to appreciate how the information was derived. The interpretation of mass spectra is a skilled task that requires a considerable degree of expertise and a wide range of knowledge on how charged fragments are likely to decompose in the spectrometer. The information contained in the mass spectrum, sometimes combined with other analytical evidence (such as NMR, IR, GC, etc) usually enables the structure of a molecule to be determined or the composition of a mixture to be computed. Because of the complex rules involved in interpreting

mass spectra and the large amount of data that is produced, computers are
invariably used as an integral part of the spectrometer system.

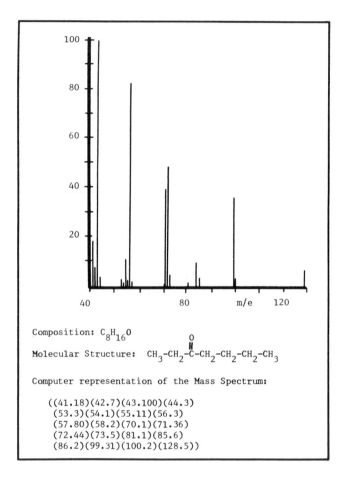

Fig. 3.5   Mass spectrum of 3-octanone.

There has been a substantial amount of effort devoted to solving some of
the problems associated with the automatic analysis of mass spectra by
means of computers.   In the early 1960's a group of researchers at
Stanford University - led by Lederberger (Led64) - pioneered the
application of artificial intelligence techniques to the interpretation of
the chemical data produced by the mass spectrometer.   The result of their
early efforts was a computer program called HEURISTIC DENDRAL.   This
program subdivided the interpretation problem into three broad stages:

(1)   preliminary inference (or planning) in which clues
      from the data are used to infer which classes of
      compounds are suggested or forbidden by the data,

(2) <u>structure generation</u> in which the system enumerates
    all possible explicit structural hypotheses which
    are compatible with the inferences made in the
    previous step, and,
(3) <u>prediction and testing</u>, where the system predicts the
    consequences from each structural hypothesis and compares
    the prediction with the original spectrum in order to
    choose the hypothesis which best explains the original data.

The above three phases of the problem solution are embodied in a computer
program whose flow logic is depicted in the graph shown in Fig. 3.6. The
PREDICTOR module is interesting since it contains an approximate model of
a mass spectrometer which is used to compute the mass spectra based upon
the hypotheses that are made.

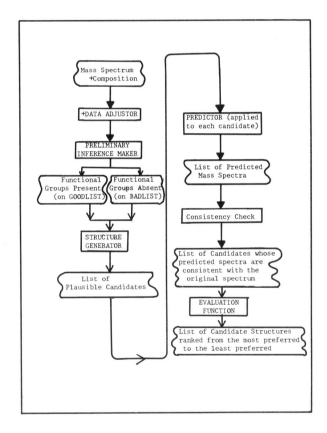

Fig. 3.6 General design of the HEURISTIC DENDRAL program.

When a pure sample is introduced into a mass spectrometer linked to a
system such as this, an ideal program for deducing the structure of the
sample should output exactly one structure as the explanation of the
spectrum. Usually, however, several different best fit structures are

suggested as plausible explanations for the original data. The analyst
then selects the one which is most appropriate. If the program completely
misses the correct structure it is refined by re-programming as a result
of subsequent discussion with an analytical chemist who specialises in the
interpretation of mass spectra.

The history of the development of the HEURISTIC DENDRAL program is
described in a series of thirty or more papers - published in the Journal
of the American Chemical Society and Journal of Organic Chemistry.
Interesting review articles have been written by Buchanan (Buc69a, Buc69b
and Buc79), who wrote in 1969,

> "the HEURISTIC DENDRAL program is expanding from the 'automatic
> mass spectroscopist' to the 'automatic analytical chemist'.
> Other analytical tools such as infrared will be incorporated
> eventually...."

Since the early work an NMR predictor and inference maker have been added
to the system and substantial improvements made to it. The current status
of work on the heuristic programming project at Stanford University is
described in a recent article by Buchanan (Buc79). The DENDRAL package
now consists of three basic programs: CONGEN, the DENDRAL planner and
Meta-DENDRAL. In these programs there is an attempt to exploit to maximum
advantage the best characteristics of both the computer and the human
expert through appropriately programmed human-machine dialogue. This work
is closely related to the description of expert systems presented in
chapter 2.

The work on HEURISTIC DENDRAL represents only one of many different
projects designed to apply computers to the interpretation of analytical
data. Sasaki (Sas68) has described an attempt to feed several different
types of physical data acquired by computers from mass spectrometers, NMR,
IR and UV instruments into a central computer program to effect structural
elucidation based on a combination of all four types of data. Similarly,
McLafferty's STIRS package - Self Training and Interpretive
Retrieval System (McL73) - represents yet another approach to mass
spectral correlation. In the STIRS system the computer selects different
classes of data known to have a high structural significance, such as
characteristic ions, series of ions and masses of neutral fragments lost,
from the unknown mass spectrum, and matches these against the
corresponding data of all the reference spectra contained in a library.
The reference compounds of closest match in each data class are examined
for common structural features; criteria having been determined so that
such features can be identified with approximately 95% reliability.

As in most other analytical chemistry spectral techniques there has been a
considerable amount of research activity devoted towards the use of
computers to solve the following types of problems:

    (a)   compaction of spectra for storage in computer
           systems,
    (b)   retrieval strategies,
    (c)   availability of standard collections of reference
           spectra in computer readable form, and,
    (d)   suitable computer based techniques for handling
           large collections of reference data.

Fast retrieval techniques, minimal storage requirements and the ability to easily disseminate collections of reference spectra between laboratories are some of the most desirable objectives of future developments. Grotch (Gro75) and Gray (Gra76) have discussed the use of binary coded mass spectral data in file search systems for spectrum identification. The main advantages of this technique lie in the resulting economies in storage and reduction in search times that can be achieved. In general, techniques for searching mass spectral files fall broadly into two categories: forward searching and backward or reverse searching. A forward-search method compares an unknown to a library entry while a reverse-search compares a library entry to an unknown. Various methods of comparing an unknown mass spectrum against a reference library have been developed. One method suggested by Biemann (Bie71) divides an unknown mass spectrum into the two largest peaks that occur in every 14 amu region. This condensed mass spectrum is then compared against a library which has been condensed in the same way. Other procedures condense mass spectra by using the six, eight or ten most abundant peaks in the spectrum (Mat76). In all these procedures, comparison is by forward-search. McLafferty (McL74, McL75, McL76) developed a system called Probability Matched Search (PBM) which differs from other search routines in two ways. First, it compresses a spectrum by a process which determines the uniqueness of a mass and its abundance to characterise a compound - common highly abundant masses are given low uniqueness values. Ten peaks are then used in order to perform a compressed library search. The second way in which PBM differs from other methods is that it uses a reverse library search. Abramson (Abr75) has compared the major differences between the two methods and claims that the reverse search has significant advantages: if a sample is pure unknown (that is, the mass spectrum contains no contribution from other sources - see next section) the results from a forward and reverse search should be similar. However, if the unknown is contaminated in any way the reverse search should give superior results.

There are several commonly used library systems for reference spectra. Two of these, the Wiley library and the NIH/EPA MSDS (NBS) library, are compared in Table 3.2.

TABLE 3.2 Some Libraries of Mass Spectral Data

Library System	No. of Spectra	No. of Compounds	CAS Registry No.	Nomenclature
NIH/EPA MSDS (NBS) see Hel78	31,597	31,597	Yes	Chemical Abstracts
Registry Mass Spectra (Wiley) see Wil74	38,646	32,403	No	Common

Dromey (Dro76a) has described a simple indexing system for classifying
mass spectra with particular applications to fast library searching of
either full or abbreviated spectra.

The significance of the computer as a component in modern spectrometry
instrumentation cannot be over-emphasised. Its importance arises as a
consequence of both the complex nature of the spectrometer and the
complexity of the spectra that are produced. Further details of the role
of the computer in this area are presented in Bur78 and the book by
Chapman (Cha78). Recent applications of computing techniques in mass
spectrometry include Fourier Transform applications (Led80), the design of
new and improved algorithms and the further development of methods such as
MS/MS, LC/MS and GC/MS. Of these, the latter will be discussed in the
next section.

SOME SOPHISTICATED INSTRUMENTS

Just as a computer can enhance the facilities provided by an instrument so
too can instruments complement each other - particularly if the activity
of the instruments can be coordinated by means of a computer system. In
principle there is no limit to the number of instruments that can be
linked together. Consequently, instrumental rigs can be designed to meet
virtually any analytical requirements. In order to demonstrate some of
the approaches that have been used, two examples will be considered - the
interlinking of a gas chromatograph with a mass spectrometer and the basic
principles involved in a highly automated multichannel continuous flow
analyser.

(1) The GC/MS Combination

The GC/MS combination is probably one of the most versatile and sensitive
tools for mixture analysis which combines the separating power of gas
chromatography with the identification capabilities of the mass spectrometer. Usually a computer is needed to synchronise the operation of the
two instruments, control the operation of each of them and process the
large quantities of complex data that is invariably produced with each
analysis. Furthermore, because a GC/MS system is capable of producing
hundreds of spectra per day, some computer based automation of the system
and of the interpretation of spectra could be of substantial benefit. A
typical GC/MS system might be organised in a way similar to that shown in
Fig. 3.7. The construction of the instrument is quite complex, involving
many different types of interface. Some of these will be discussed in
more detail in the chapter on Principles of Interfacing.

The continuously scanning mass spectrometer can be used in various ways:

        (a) to record the mass spectra of the component eluting
             from the GC at any given time,
        (b) to record total ion current as a function of time as
             the materials elute from the chromatographic column, and
        (c) to record selective ion current as a function of time.

In the simplest mode of operation, as a compound elutes from the column
its mass spectrum is recorded. If the GC fails to separate two or more

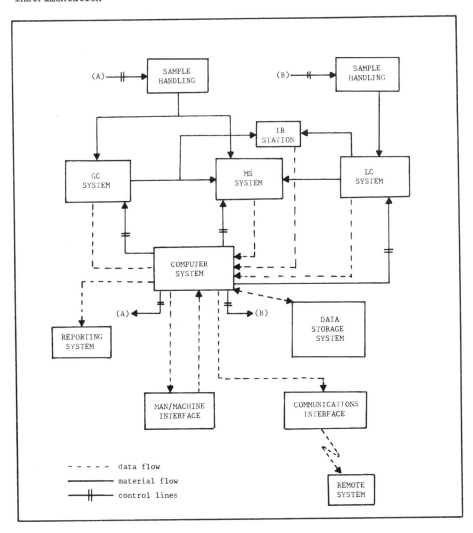

Fig. 3.7   Organisation of a combined
computer based GC/MS and LC/MS system.

different components under a given set of instrumental conditions, the MS
will produce mixed spectra. Data of this type can cause difficulties with
interpretation and confirmation due to discrepancies in the spectra.
There are usually several ways of avoiding mixed spectra during GC/MS
analysis. The use of capillary columns can significantly enhance the
ability of the chromatograph to separate components that would otherwise
elute at similar retention times. Similarly, chemical ionisation (CI)
GC/MS can facilitate the detection of co-eluting components. In addition,
careful control of packed column GC conditions can often achieve the
necessary separation. Alternatively, computer techniques based upon

tabular peak models are available to enable the extraction of mass spectra that are free of background and neighbouring component contributions. These techniques are described in the paper by Dromey (Dro76b). In addition to measuring the mass spectrum of each eluent as it emerges from the chromatography column, selective ion monitoring (Bur78) may sometimes be used to register peaks. This technique requires the dedication of the mass spectrometer for the acquisition of ion abundance data at selected masses in real time as components elute from the chromatography system.

Many descriptions of GC/MS systems have been given in the literature (for example, Bin71, Gud77, Gat78, Ner80, Hew80). These vary with respect to the different hardware available and the modes of inter-connection of the various units involved. The system described in Binks et al (Bin71) involved the use of a small computer system for on-line processing of low resolution GC/MS data and incorporated a computer based reference library system for storing, retrieving and comparing spectra. Two GC/MS systems were involved; when one was on-line to the computer the second was able to record its data on an analogue tape recorder for subsequent processing by the computer. The system described by Gates (Gat78) is similar in principle. However it uses different hardware for the actual implementation and is based upon the use of only one GC/MS system to which the computer is dedicated; sophisticated graphics facilities are provided for the presentation of results. Gates' system is designed to provide automated qualitative and quantitative analyses of 100 or more components in a complex organic mixture.

Various methods of interfacing the gas (or liquid) chromatograph to the spectrometer exist. For sample transfer, the Biemann separator (Bie71), the molecular separator (Bau78) and several others (Ner80, Hew79, Ran78) have been used. Takeuchi et al (Tak78) describe the use of a jet separator to enable the coupling of a liquid chromatograph to a mass spectrometer. Many other examples - such as the McLafferty direct split chemical ionisation method (Hew80) - also exist. Further details of high performance liquid chromatography - mass spectrometry, that is, HPLC/MS, are given in the papers by Games (Gam80a, Gam80b).

Of course, there is no reason why an infrared spectrometer should not be added to the equipment complex in order to provide supporting information that might not be available through the MS or GC system. The identification power of the MS, however, usually renders this unnecessary. From the point of view of data handling the computer now has two types of data to store in its reference library:

(1)   chromatographical retention data, and,
(2)   mass spectral data.

Obviously, as more types of data are added to the system, so the complexity of the retrieval programs increases - as does the need for the careful design of data storage strategies in order to minimise the amount of computer storage that is required for reference data. Quantitation is also an important consideration since in many cases the more sophisticated the machine becomes (in terms of its control parameters) the greater the need for ensuring their stability. Variation of control parameters (such as gas flow rate or temperature) can significantly influence calibration routines used for quantitative work. Some of the problems associated with the handling of data and the quantitative aspects of GC/MS have been discussed by Wells (Wel80). Further details of techniques for processing GC/MS data will be found in Bur78 and Cra78.

(2) The Multi-channel Continuous Flow Analyser

The principle of continuous flow analysis was described briefly in chapter
1. Further details will be found in Bau78. Continuous flow analysis
developed during the 1950s and today, like Flow Injection Analysis
(Ran81), forms the basis of many important automated instruments for wet
chemical analysis (Tec76, Tec77, Tec79). Figure 3.8 illustrates the basic
mechanism involved in most continuous flow analysers.

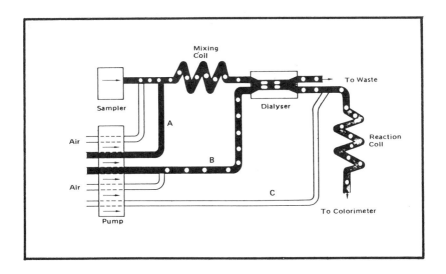

Fig. 3.8   Example of continuous flow analysis scheme
with A,B: diluents;   C: reagent stream.

Samples and reagents, segmented by air bubbles, are aspirated by means of
a proportioning pump to deliver the precise volume ratios necessary for
specific analyses. The samples, continuously following each other through
a hydraulic system, are brought together with reagents under controlled
conditions, causing specific chemical reactions which are analysed and
measured. The system continuously monitors the reaction, but the actual
measurement is made only at the steady state condition when colour
development and concentration are constant and after virtually all effects
of sample interaction have been eliminated.

The basic unit depicted above can be utilised in a multi-channel analyser
if the sampler is modified so that it accepts samples from a central
manifold system carrying the samples for analysis. The manifold is thus
able to feed many units, each running in parallel and each one
specifically designed to perform a particular type of analysis or
chemistry. The arrangement shown schematically in Fig. 3.9 illustrates
the organisation of a typical multi-channel highly parallel analyser based
upon the use of a central manifold. In this system the computer coor-
dinates the activity of the whole machine. It constantly checks all
chemistry curves, comparing them with stored, ideal curve parameters for
each specific type of analysis. At the same time it continuously monitors
reference channel data. The computer determines, verifies, calculates and

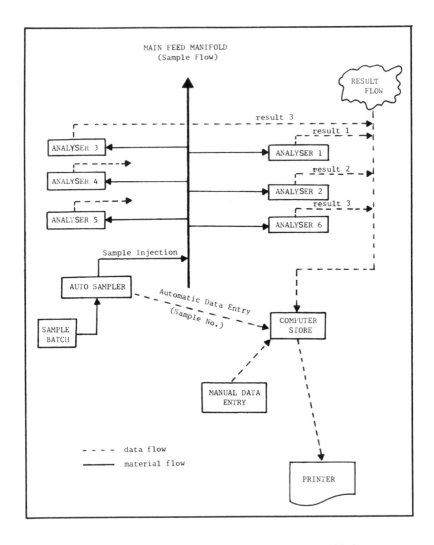

Fig. 3.9 Organisation of an automated parallel
process continuous flow analyser.

prints out the correct value for each test.   Also, provided appropriate
archival storage is available, all results produced by the system can be
stored in an archive for future reference.   Similarly, if suitable telecommunications facilities are available, the results can be transmitted to
a remote computer or can be interrogated from a remote terminal.   These
possibilities have been indicated previously in Fig. 1.11.

The way in which the flow cells are interfaced to the optical system so as
to minimise duplication of equipment is quite interesting.   In the

Technicon SMAC (Tec76) light is transmitted to and from the flow cell of the colorimeter by means of optical fibres. Through the application of 'time-sharing' techniques, each flow cell is monitored by a single photo-multiplier tube (PMT). This is achieved via the use of a scanning disk containing slots that allow the exposure of one channel at a time. The arrangement is shown schematically in Fig. 3.10.

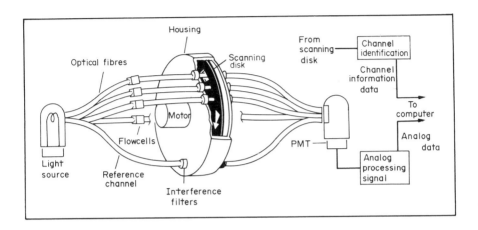

Fig. 3.10 Multiplexing of optical source and detector;
PMT: photo-multiplier tube.

The scanning process occurs at a rapid rate - typically four scans of each channel/sec and provides a minimum of 72 readings for each individual test (Tec76). The optical information obtained is transmitted to the computer for processing. A typical design of the flow cell used for optical measurements is shown in Fig. 3.11.

Flow cells are used for colorimetric measurements. Usually, one flow cell is allocated to each chemistry being monitored. Sometimes, however, two or more cells may be linked in series. This happens in the case of multiple-point enzyme analysis. Here, flow cells in series are separated by heated incubation coils. Absorbance measurements of the enzyme reactions are made at two or three time intervals and analysed by the computer.

Various types of continuous flow multi-channel analyser have been produced over the last few years. They are extensively used in the clinical and biochemical laboratories. One of the first machines of this type to gain widespread acceptance was that produced by the Technicon company in 1967 called the SMA 12/60. The acronym SMA is an abbreviation for its full name Sequential Multiple Analyser. This machine was capable of performing twelve analyses in parallel at a rate of about 60 samples, calibrations and controls per hour. The TECHNICON SMAC (Tec76) described in chapter 1 is a computerised version of this instrument and enables a substantially higher throughput of work. This machine is able to perform twenty analyses in parallel at a rate of 150 samples and standards per hour - that is, it is able to conduct 3000 tests per hour. The total amount of sample required for these tests is 450 micro-litres. Many other types of

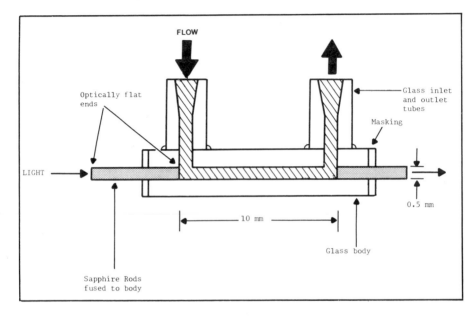

Fig. 3.11   Flow through cell used in continuous flow analysis.

automated computer controlled chemical analyser are available for a wide
variety of industrial applications.   One typical example (IQA80) has an
analysis rate of up to 900 tests per hour with results automatically
calculated and shown on a VDU and printed out on a hardcopy form.   The
machine can be programmmed for up to 32 different chemistries using up to
six reagents per test from a total of 63.

INSTRUMENTATION FOR PROCESS CONTROL

Process control in its broadest sense covers a wide variety of
applications.   The types of analyser mentioned in the last section are
important instruments for a variety of control and screening applications
in the clinical and medical areas.   They thus provide a form of process
control in the context of disease detection, prevention and treatment.   In
the manufacturing industries, as illustrated by petrochemical, food and
steel production (see chapter 1), many processes can be made highly
automatic through the use of appropriate process control instrumention
(Wig72, ISA77) used in conjunction with computer systems.   Much of the
instrumentation that is used in many process control situations can be
designed around the use of composition measurements as process variables.
Consequently, the use of on-line analysers for process control has grown
significantly.   A description of a wide range of instrumentation for
continuous analysis of chemical process systems has been presented by
Siggia (Sig59).   An introduction to automated process control and an
overview of the use of automated instruments for control applications is

contained in Wil65 (chapter 28) and Bau78 (chapter 24). Various articles (for example, San77, Hal79, Jut79) contain descriptions of the more popular type of on-line analyser such as the gas chromatograph, mass spectrometer and other types of spectroscopic and electrochemical system. Many analysers of this type have been in use for a considerable time. However, interest in extending their areas of application and versatility has increased substantially with the advent of microprocessor based systems.

CONCLUSION

Instrumentation in modern analytical chemistry is of vital importance since without it many techniques that are presently being used would not be possible. There is an extremely wide range of analytical scientific instruments available, each having a particular area of application and suited to specific types of analysis. Some of the more sophisticated of these are extremely expensive both to purchase and to run. Consequently, few laboratories could hope to support a full range. Most laboratories tend to acquire the few special purpose instruments necessary to perform their designated analytical function. These are then augmented by an appropriate selection of the more general purpose instruments. A review of the major areas of instrumental analytical chemistry will be found in the American Chemical Society's review publication "Fundamental Reviews" (Ana78).

Increasingly, within analytical chemistry computers (in various, sometimes hidden, forms) are becoming versatile electronic tools. The computer is able to improve the performance of an instrument in many ways: making easier to use, improving its precision, sensitivity or range of applicability. In this chapter some of the ways in which computer technology is being merged with analytical instrumentation have been outlined. Obviously, the treatment has not been comprehensive since space restrictions do not permit this. Hopefully, however, the reader will realise that what applies to one instrument can, in general, be applied to others - at least from the point of view of computer applications.

Although computers have figured prominently in this chapter, little has been said about their architecture or mode of operation. Any detailed discussion of the way in which computers operate has been purposely avoided. In the next chapter some of the mystique of computers will be revealed in order that their potential within the laboratory, in instruments and within process control may be more fully realised.

REFERENCES

AA875   Varian Instrument Division, 611 Hansen Way, Palo Alto, California 94303, USA, 'The Varian AA-875 Atomic Absorption Spectrometer', 1980.

Abb65   Abbott, D. and Andrews, R.S., An Introduction to Chromatography, Longmans, 1965.

Abr75   Abramson, F.P., Automated Identification of Mass Spectra by the Reverse Search, Analytical Chemistry, Volume 47, 45-49, 1975.

All63   Allen, L.C. and Johnson, L.F., Chemical Applications of
        Sensitivity Enhancement in NMR and ESR, Journal of the American
        Chemical Society, Volume 85, 2668-2670, September 1963.

Amb71   Ambrose, D., Gas Chromatography, Butterworths, ISBN: 0-408-
        70129-3, 1971.

Ana78   American Chemical Society, Fundamental Reviews, Analytical
        Chemistry, Volume 50, April 1978.

AST25   ASTM #AMD25A-s1, GC Data Compilation, Catalogue No. 1139, Altech
        Associates, 1977.

Bac76   Baczynskyj, L., Duchamp, D.J., Zieserl, J.F., Kenny, M.D. and
        Aldritch, J.B., Computer Acquisition and Processing of Metastable
        Ion Scans in a Double Focussing Mass Spectrometer, Analytical
        Chemistry, Volume 47, No. 9, 1358-1362, August 1976.

Bai75   Baiulescu, G.E. and Ilie, V.A., Stationary Phases for Gas
        Chromatography, Pergamon Press, 1975.

Bal79   Balestra, C., A Computer Controlled ENDOR Spectrometer, J. Phys.
        E: Scientific Instruments, Vol. 12, 824-827, 1979.

Bau78   Bauer, H.H., Christian, G.D. and O'Reilly, J.E., Instrumental
        Analysis, Allyn and Bacon, Inc., ISBN: 0-205-05922-8, 1978.

Bee75   Beech, G., FORTRAN IV in Chemistry - An Introduction to Computer
        Assisted Methods, John Wiley, ISBN: 0-471-06165-4, 1975.

Bey60   Beynon, J.H., Mass Spectrometry and its Application to Organic
        Chemistry, Elsevier, 1960.

Bie71   Hertz, H., Hites, R. and Biemann, K., Identification of Mass
        Spectra by Computer Searching a file of known Spectra, Analytical
        Chemistry, Volume 43, 681-691, 1971.

Bin71   Binks, R., Cleaver, R.L., Littler, J.S. and MacMillan, J.,
        Real-time Processing of Low Resolution Mass Spectra, Chemistry in
        Britain, Volume 7, No. 1, 8-12, 1971.

Bri75   Brignall, J.E. and Rhodes, G.M., Laboratory on-line Computing -
        An Introduction for Engineers and Physicists, Intertext Books,
        ISBN: 0-7002-0258-7, 1975.

Bri79   Brignall, J.E. and Young, R., Computer Aided Measurement, J.
        Phys. E: Scientific Instruments, Volume 12, 455-463, June 1979.

Buc69a Buchanan, B. and Feigenbaum, E.A., HEURISTIC DENDRAL: A Program
        for Generating Explanatory Hypotheses in Organic Chemistry,
        Chapter 12, 209-254, in Machine Intelligence, Volume 4, Eds:
        Meltzer, B. and Michie, D., Edinburgh University Press, 1969.

Buc69b Buchanan, B.G., Sutherland, G.L. and Feigenbaum, E.A., Rediscover-
        ing some Problems in Artificial Intelligence in the Context of
        Organic Chemistry, Chapter 14, 253-280, in Machine Intelligence,
        Volume 5, Eds: Meltzer, B. and Michie, D., Edinburgh University
        Press, 1969.

Buc79   Buchanan, B., Issues of Representation in Conveying the Scope and
         Limitations of Intelligent Assistant Programs, Chapter 20,407-425,
         in Machine Intelligence, Volume 9, Eds: Hayes, J.E., Michie, D.
         and Mikulich, L.I., John Wiley, ISBN: 0-470-26714-3, 1979.

Bur78   Burlingame, A.L., Shackleton, C.H.L., Howe, I., and Chizov, O.S.,
         Mass Spectrometry, 346R-384R, in 'Fundamental Reviews',
         Analytical Chemistry, Volume 50, April 1978.

Cha69   Chapman, J.R., Computers and Mass Spectrometry, Chemistry in
         Britain, Volume 5, No. 12, 563-567, December 1969.

Cha78   Chapman, J.R., Computers in Mass Spectrometry, Academic Press,
         ISBN: 0-12-168750-3, 1978.

Cly70   Clymer, A.B., Indirect Measurement of Process Variables by
         Minicomputer, IEEE Trans. Ind. Electron. Contr. Instrum., Vol.
         IECI-17, 358-362, June 1970.

Cra78   Cram, S.P. and Risby, T.H., Gas Chromatography, 213R-243R, in
         'Fundamental Reviews', Analytical Chemistry, Vol.50, April 1978.

Cou77   Cournoyer, R., Anderson, D.H. and Shearer, J.C., Fourier Trans-
         form Infrared Analysis below the One Nanogram Level, Analytical
         Chemistry, Volume 49, 2275-2277, December 1977.

Cro76   Crockett, I.L. and Mikkelson, L., The Microcomputer-Based Chroma-
         tograph and its Future, Journal of Chromatographical Science,
         169-172, Vol. 14, April 1976.

Cof78   Coffey, P., Mattson, D. and Wright, J., A Programmable GC/FT-IR
         System, American Laboratory, 126-132, May 1978.

Dav74   David, D.J., Gas Chromatographic Detectors, John Wiley, 1974.

DMS90   Varian Instrument Division, 611 Hansen Way, Palo Alto, California
         94393, USA, 'DMS-90 UV-Visible Spectrometer', Publication No.
         85-100381, January 1980.

Dro76a   Dromey, R.G., Simple Index for Classifying Spectra with Applica-
         tions to Fast Library Searching, Analytical Chemistry, Volume
         48, No. 11, 1464-1469, September 1976.

Dro76b   Dromey, R.G., Stefik, M.J., Rindfleisch, T.C. and Duffield, A.M.,
         Neighbouring Component Contributions from Gas Chromatography/Mass
         Spectrometry Data, Analytical Chemistry, Volume 48, No. 9,
         1368-1375, August 1976.

Dro80   Dromey, R.G. and Foyster, G.T., Calculation of Elemental Com-
         positions from High Resolution Mass Spectral Data, Analytical
         Chemistry, Volume 52, No. 3, 394-398, 1980.

Dul77   Dulson, W., Extension of the Low Temperature Range of a Micro-
         processor Controlled Gas Chromatrograph, Analytical Chemistry,
         Vol. 49, No. 8, 1279-1280, July 1977.

Ewi75   Ewing, G.W., Instrumental Methods of Chemical Analysis, McGraw
         Hill, ISBN: 0-07-019853-5, (4th Ed.), 1975.

Fel80   Fell, A.F., Novel Detectors for Spectroscopic Analysis, Analytical
         Proc. of the Chemical Society, Vol.17, No.7, 266-271, July 1970.

Fle63   Flett, M. St. C., Characteristic Frequencies of Chemical Groups
         in the Infrared, Elsevier Chemistry Monographs, 1963.

Gam80a   Games, D.E., Combined High Performance Liquid Chromatography -
         Mass Spectrometry, Analytical Proc. of the Chemical Society,
         Volume 17, No. 4, 110-116, April 1980.

Gam80b   Games, D.E., Applications of Combined High-Performance Liquid
         Chromatography - Mass Spectrometry, Analytical Proc. of the
         Chemical Society, Volume 17, No. 8, 322-326, August 1980.

Gan76   Ganjei, J.D., Howell, N.G., Roth, J.R. and Morrison, G.H., Multi-
         Element Atomic Spectrometry with a Computerised Vidicon Detector,
         Analytical Chemistry, Vol. 48, No. 3, 505-510, March 1976.

Gat78   Gates, S.C., Smisko, M.J., Ashendel, C.L., Young, N.D., Holland,
         J.F. and Sweeley, C.C., Automated Simultaneous Qualitative and
         Quantitative Analysis of Complex Organic Mixtures with a Gas
         Chromatography - Mass Spectrometry - Computer System, Analytical
         Chemistry, Volume 50, No. 3, 433-441, 1978.

Gib79   Gibson, T., Gas Chromatographs for Chemical Processes, 68-71,
         Focus on Engineering, Imperial Chemical Industries, June 1979.

Gra76   Gray, N.A.B., Similarity Measures for Binary Coded Mass Spectral
         Data, Analytical Chemistry, Vol.48, No.9, 1420-1421, Aug. 1976.

Gri62   Griffiths, V.S. and Lee, W.H., The Electronics of Laboratory and
         Process Instruments, Chatto and Windus Ltd., 1962.

Gro75   Grotch, S.L., Automatic Identification of Chemical Spectra. A
         Goodness of Fit Measure derived from Hypothesis Testing, Analy-
         tical Chemistry, Volume 47, No. 8, 1285-1289, 1975.

Gud77   Gudzinowicz, B.J., Gudzinowicz, M.F. and Martin, H.F., Fundamen-
         tals of Integrated GC-MS, Part III: The Integrated GC-MS Analyti-
         cal System, (Ch. 7-9), Marcel Dekker, ISBN: 0-8247-6431-5, 1977.

Hal79   Hall, J., On-line Analysers Tackle New Demands, Instruments and
         Control Systems, Volume 52, No. 8, 33-38, August 1979.

Hel78   Heller, S.R., Heller, R.S., McCormick, A., Maxwell, D.C. and
         Milne, G.W.A., Progress of the MSDC-NIH-EPA Mass Spectral Search
         System, Adv. Mass Spectrom., Volume 7, 985-988, 1979.

Hew79   Hewlett Packard, The HP5985B and HP5995A GC/MS Systems, Dec. 1979.

Hew80   Hewlett Packard, The HP LC/MS Interface for HP8985B GC/MS Systems,
         March 1980.

Hil66   Hill, H.C., Introduction to Mass Spectrometry, Heyden, 1966.

Hor76   Horlick, G. and Yuen, W.K., Fourier Domain Interpolation of
         Sampled Signals, Analytical Chemistry, Vol.48, No.11, 1643-1644,
         September 1976.

HP J80   Hewlett Packard Journal, Vol. 31, No. 2, February 1980.

INS79   INSTEM Ltd., Laboratory Products Division, Stafford Street, Stone,
          Staffs., England, Datachrome Two - System Specification, 1979.

IQA80   American Monitor Corporation Indiana, USA, The IQAS Automated
          Chemistry Analyser, product description, 1980.

IQ200   Leybold Heraeus Ltd., IQ200 Quadruple Mass Spectrometer, 1980.

ISA77   Instrument Society of America, Instrumentation in the Chemical
          and Petroleum Industries, Volume 13, Proc. of the 1977 Spring
          Industry Conference, Anaheim, California, 1977.

Jam52   James, A.T. and Martin, A.J.P., Biochem J., Vol. 50, 679, 1952.

Jen78   Jennings, W., Gas Chromatrography with Glass Capillary Columns,
          Academic Press, 1978.

Jut79   Jutila, J., Multicomponent On-stream Analysers for Process Moni-
          toring and Control, InTech, Volume 26, No. 7, 38-44, July 1979.

Koe75   Koenig, J.L., Application of Fourier Transform Infrared Spectros-
          copy to Chemical Systems, Appl. Spectrosc., Vol. 29, 293, 1975.

KON200   KONTRON Ltd., Analytic International, Bernerstrasse Sud 169,
          CH-8048, Zurich, 'Liquid Chromatography Solvent Programmer',
          Model 200, June 1980.

Lan80   Lanza, E., Golden, B.M., Zyren, J. and Slaver, H.T., An Off-line
          System for Handling Gas Chromatographic Fatty Acid Data, J.
          Chromatographical Science, Volume 18, 126-132, March 1980.

Lea73   Leathard, D.A., Applications of Digital Computers in Gas
          Chromatography, 29-86 in Advances in Analytical Chemistry and
          Instrumentation, Vol. 11 (New Developments in Gas Chromatography,
          Ed. H. Purnell), 1973.

Led64   Lederberger, J., Computation of Molecular Formulas for Mass
          Spectrometry, Holden-Day, Inc., San Francisco, 1964.

Led80   Ledford, E.B. Jr., Ghaderi, S., White, R.L., Spencer, R.B.,
          Kulkarni, P.S., Wilkins, C.L. and Gross, M.L., Exact Mass Measure-
          ment by Fourier Transform Mass Spectrometry, Analytical
          Chemistry, Volume 52, No. 3, 463-468, March 1980.

Lyn78   Lynch, P.F. and Brady, M.M., Computer Separation of Infrared
          Spectra for Analysis of Complex Organic Mixtures, Analytical
          Chemistry, Volume 50, No. 11, 1518-1522, September 1978.

Mat72   Mattson, J.S., Design and Applications of an On-line Minicomputer
          System for Dispersive Infrared Spectrometry, Analytical
          Chemistry, Vol. 49, No. 3, 470-478, March 1972.

Mat76   Mathews, R.J. and Morrison, J.D., Comparative Study of Methods of
          Computer Matching Mass Spectra - Part II, Aust. J. Chem., Volume
          29, 689-693, 1976.

116   Computers in Analytical Chemistry

McD78   McDonald, R.S., Infrared Spectrometry, 282R-299R, in 'Fundamental
         Reviews', Analytical Chemistry, Volume 50, April 1978.

McL73   Kwok, K.S., Venkataraghavan, R. and McLafferty, F.W., Computer
         Aided Interpretation of Mass Spectra, III A Self Training Inter-
         pretive and Retrieval System, Journal of the American Chemical
         Society, Volume 95, 4185-4194, 1973.

McL74   McLafferty, F.W., Hertel, R. and Villwock, R., Probability Based
         Matching of Mass Spectra, Organic Mass Spectrom., Volume 9,
         690-702, 1974.

McL75   Pesyna, G.M., McLafferty, F.W., Venkataraghavan, R. and Dayringer,
         H.E., Statistical Occurrence of Mass and Abundance Data Values in
         Mass Spectra, Analytical Chemistry, Volume 47, 1161-1164, 1975.

McL76   Pesyna, G.M., Venkataroghavan, R., Dayringer, H.E. and McLafferty,
         F.W., Probability Based Matching System using a Large Collection
         of Reference Mass Spectra, Analytical Chemistry, Vol. 48, No. 9,
         1362-1368, August 1976.

McR77   McReynolds, W.O., Gas Chromatographic Retention Data, Catalogue
         No. 1020, Altech Associates, 1977.

McR70   McReynolds, W.O., Characterisation of Some Liquid Phases, J.
         Chromatographical Science, Volume 8, No. 12, 685-691, 1970.

Mee61   Meehan, E.J., Optical Methods of Analysis, 2707-2838, in Treatise
         on Analytical Chemistry, Eds: Kolthoff, I.M. and Elving, P.J.,
         Part I, Volume 5, John Wiley, 1961.

Mel63   Meloan, C.E., Elementary Infrared Spectroscopy, Macmillan, New
         York, 1963.

Met79   Mettler Instrument Corporation, Box 71, Highstown, NJ 08520, USA,
         'Weighing, Analysing, Automating', Product Catalog, July 1979.

MIR80   Foxboro Company, c/o Wilks Infrared Centre, P.O. Box 449, S.
         Norwalk, CT 06856, USA, The MIRAN-80 Computing Quantitative
         Analyser, 1979.

Nak62   Nakanishi, K., Infrared Absorption Spectroscopy - Practical,
         Nankodo Company Ltd., Japan, 1962.

Ner80   Nermag S.A. (Formerly Ribermag S.A.), 49 quai du Halage, 92500
         -rueil (Paris) France, Ribermag S.A. R10-10 GC/MS, July 1980.

Nic80   Nicolet Instruments Ltd., Budbrooke Road, Warwick, CV3 5XH, UK,
         Product specifications for the MX-1 and 3600 Fourier Transform
         Infrared Spectrometers, July 1980.

Nov75   Novak, J., Quantitative Analysis by Gas Chromatography, Marcel
         Dekker, 1975.

Ori79   Orion Reseach Incorporated, 380 Putnam Avenue, Cambridge, Mass.
         02139, USA, Microprocessor Ionalyser/901, 1979.

Pea79   Pearce, B.G. and Shackel, B., The Ergonomics of Scientific Instru-
         ment Design, J. Phys. E: Scientific Instruments, Volume 12,
         447-464, June 1979.

Per75   Wilkins, C.L., Perone, S.P., Klopfenstein, C.E., Williams, R.C.
         and Jones, D.E., Digital Electronics and Laboratory Computer
         Experiments, Plenum Press, ISBN: 0-306-30822-3, 1975.

Per77a   Perry, J.A., Bryant, M.F. and Malmstadt, H.V., Microprocessor Con-
         trolled Scanning Dye Laser for Spectrometric Analytical Systems,
         Analytical Chemistry, Volume 49, No. 12, 1702-1709, 1977.

Per77b   Perrin, D.D., Recent Applications of Digital Computers in Analyti-
         cal Chemistry, Talanta, Vol. 24, 339-345, Pergamon Press, 1977.

Per680   Perkin Elmer Corporation, Main Avenue, Norwalk, Connecticut 06856,
         USA, The 680 Series Ratio Recording Infrared Spectrometer, 1980.

Pri80   EG and G Princeton Applied Research, P.O. Box 2565, Princeton,
         N.J. 08540, USA, Polarographic Instrumentation, 1980.

Pur62   Purnell, J.H., Gas Chromatography, John Wiley, 1962.

Pye304   Pye Unicam Ltd., York Street, Cambridge, CB 1 2PX, UK, Pye Series
         304 Gas Chromatograph, 1980.

Ran78   Randall, L.G. and Wahrhaftig, A.L., Dense Gas Chromatograph/Mass
         Spectrometer Interface, Analytical Chemistry, Volume 50, No. 12,
         1703-1705, October 1978.

Ran81   Ranger, C.B., Flow Injection Analysis: Principles, Techniques,
         Applications, Design, Analytical Chemistry, Volume 53, No. 1,
         20A-32A, January 1981.

Ree80   Reese, C.E., Chromatographical Data Acquisition - Part I, J.
         Chromatographical Science, Volume 18, 201-206, May 1980.

Ree80a   Reese, C.E., Chromatographical Data Acquisition - Part II, J.
         Chromatographical Science, to be published.

Rob70   Robinson, J.W., Undergraduate Instrumental Analysis, Marcel
         Dekker, 1970.

San77   Sandford, J., On-stream Analysers become Faster and More Reliable,
         Instruments and Control Systems, Volume 50, No. 3, March 1977.

Sas68   Sasaki, S.I., Abe, H., Ouki, T., Sakamoto, M. and Ochiai, S.,
         Automated Structure Illucidation of Several Kinds of Aliphatic and
         Alicyclic Compounds, Analytical Chemistry, Volume 40, 2220-2221,
         1968.

Sev75   Sevcik, J., Detectors in Gas Chromatography, Elsevier, 1975.

Sig59   Siggia, S., Continuous Analysis of Chemical Process Systems,
         John Wiley, 1959.

Ste80   Steiner, G.C., A Microcomputer System for Spectrophotometric Data
         Processing, Hewlett Packard Journal, Vol. 31, No. 2, 29-31,
         February, 1980.

Str73   Strobel, H.A., Chemical Instrumentation: A Systematic Approach to
         Instrumental Analysis, Addison Wesley, ISBN: 0-201-07301-3, 1973.

Syd75   Sydenham, P.H., Transducers in Measurement and Control, The
         University of New England Publishing Unit, ISBN: 0-85834-084-4,
         1975.

TA200   Mettler Instrument Corporation, Box 71, Hightstown, N.J. 08520,
         USA, TA2000c Thermoanalyser for Simultaneous TG-DSC, 1980.

Tak78   Takeuchi, T., Hirata, Y. and Okumura, Y., On-line Coupling of a
         Micro Liquid Chromatograph and Mass Spectrometer through a Jet
         Separator, Analytical Chemistry, Volume 50, No. 4, 659-660,
         April 1978.

Tec76   Technicon Instruments Corporation, 511 Benedict Avenue, Tarrytown,
         New York, USA, The Technicon SMAC High Speed Computer Controlled
         Biochemical Analyser, product description, 1976.

Tec77   Technicon Instruments Corporation, 511 Benedict Avenue, Tarrytown,
         New York, USA, The Technicon SMA II Computer Controlled
         Biochemical Analyser, product development, 1977.

Tec79   Technicon Instruments Corporation, 511 Benedict Avenue, Tarrytown,
         New York, USA, The Technicon STAR System: Sequential Test Analyser
         for Radioimmunoassay, product description, 1979.

Van79   Vanderwielen, A.J., Specificity in Quality Control, Industrial
         Research/Development, January 1979.

Vog61   Vogel, A.I., A Text-Book of Quantitative Inorganic Analysis,
         (3rd Ed.), Longmans, 1961.

Wel80   Wells, D.E., Micropollutant Analysis by Gas Chromatography - Mass
         Spectrtometry, Analytical Proceedings of the Chemical Society,
         Volume 17, No. 4, 116-120, April 1980.

Wig72   Wightman, E.J., Instrumentation in Process Control,
         Butterworths, ISBN: 0-408-70293-1, 1972.

Wil65   Willard, H.H., Merritt, L.L. and Dean, J.A., Instrumental Methods
         of Analysis, Van Nostrand (4th ed.), 1965.

Wil74   Stenhagen, E., Abrahamson, S. and McLafferty, F.W., Registry of
         Mass Spectral Data, (magnetic tape), Wiley-Interscience, New
         York, 1974.

Wu78   Wu, A.H.B. and Malmstadt, H.V., Versatile Microcomputer Controlled
         Titrator, Analytical Chemistry, Vol.50, No.14, 2090-2096, 1978.

4
Computers: Large, Medium and Small

INTRODUCTION

A computer is a device that is able to store and process information. A definition as broad as this will encompass a wide variety of machines. Some form of taxonomy is therefore required. One common classification is based upon the physical principles of operation and the type of process involved in performing a calculation. A scheme of this type enables several classes of computer to be enumerated: mechanical, biological, fluidic, chemical and electronic. Of these, the most well known are those of the electronic variety. This type of computer represents information (and operations on that information) in terms of electronic signals.

Depending upon the nature of these signals - whether they are continuous or quantised into discrete values - electronic computers may be divided into two broad categories: analogue and digital. Each of these has its particular advantages and special areas of application. However, the digital computer is probably the more well known of the two - particularly since the advent of inexpensive microcomputer systems. As the title of this chapter suggests, computers may be broadly classified into three varieties - large, medium and small - depending upon their physical size, cost and ability to store and process information.

This latter taxonomy is only very approximate since as computer technology progresses so the physical size and cost of computers decreases. At the same time, their ability to store and process information continually increases. Thus, as a consequence of these trends, a machine that might have been classified as a large computer five or ten years ago might well be classified as a small computer by current standards. This continual reduction in the size of computers has been due to the new types of technology that is used to fabricate their components - the most important of which is the switch. Within a computer the most fundamental process upon which everything else depends is high speed switching.

It is via electronic switching that computers are able to store and process data and convert it into useful information. The earliest computers used electromechanical relays or thermionic valves to implement

these switching operations. Nowadays, solid state switches such as
transistors are used. These changes in technology have brought about a
marked increase in the reliability of computers and a significant
dimunition in their size and energy consumption. The differing
requirements and capabilities of the various technologies that have been
used to fabricate computer systems are compared in Table 4.1.

TABLE 4.1 Switching Technologies - Time and Speed Comparisons

Switch Type	Mode	Speed	Energy/Power
Light switch	Mechanical	milliseconds	watts
Relay	Mechanical	milliseconds	deciwatts
Valve	Electronic	microseconds	milliwatts
Transistor	Electronic	nanoseconds	microwatts
Josephson Junction	Electronic	picoseconds	nanowatts

The most modern computers use integrated circuits (commonly known as IC's
or 'chips') as the medium to hold the switching elements that are
necessary for the storage and processing of data. A modern computer is
likely to contain in excess of 30,000 switching elements. These are of
two basic types: those used for signal processing (type A, say) and those
used for data storage (type B). Type A are more complex than type B and
require more physical storage space. Over the last ten years there has
been a substantial increase in the density of switching elements in
integrated circuits as can be seen from the following comparisons:

1969	signal processing	6 elements per chip
	data storage	600 elements per chip

1980	signal processing	900-1000 elements per chip
	data storage	64000+ elements per chip

In order to achieve these developments a number of different types of
integrated circuit technology have been used. The major ones are listed
in Table 4.2.

TABLE 4.2 The Major Types of Integrated Circuit Technology

Type	Description	Complexity of Chip
SSI	Small Scale Integration	simple gates and flip-flops
MSI	Medium Scale Integration	gates and flip-flops linked to produce more complex building blocks such as shift registers, counters, adders, timing elements and so on.
LSI	Large Scale Integration	entire systems on a chip, for example, an arithmetic logic unit or a complete micro-processor.
VLSI	Very Large Scale Integration	large number of systems and cells per chip.
ULSI	Ultra Large Scale Integration	

Obviously, there is a limit to how small computers can become. Such
factors as the allowed proximity of switching elements, the problems of
heat removal, the speed of light (which determines how big a computer can
be), reliability and the availability of appropriate manufacturing
processes all need to be taken into consideration. Despite the rapid
progress that has been made, the limits of technology are not yet anywhere
in sight. This means that the trends in cost and size dimunition
experienced in the past are likely to continue into the foreseeable future
- particularly for medium and small sized computers.

As a consequence of the decreasing cost of computer components there has
been a marked increase in the variety of digital computer currently
available. A list of some of these different types, subdivided according
to the classification scheme inherent in the title of this chapter, is
presented in Table 4.3. Each of these should be regarded as a computational tool available for the solution of the many various problems
that arise in the laboratory, the design office or process control
environment. Although there appears on the surface to be many different
types of digital computer, the underlying modes of construction,
components and principles of operation are the same for all of them - as
will be discussed in the following section.

TABLE 4.3   Some Examples of Large, Medium and Small Computers

Microcomputers Electronic Hand Calculators	SMALL
Desk-top calculators Desk-top computers Minicomputers	MEDIUM
Super-minis Mainframe Computers Super-computers	LARGE

Before proceeding it would be constructive to formulate some 'working
definitions' of large, medium and small computers. Table 4.3 introduced
several new items of terminology. Of these, three in particular are in
common usage: mainframe, minicomputer and microcomputer. These roughly
correspond to the three classes of electronic computer that were referred
to previously by the terms large, medium and small. Super-computers
represent a special class of machine used where extremely fast speeds of
computation are required. This class of machine will be described in more
detail at the end of the chapter.

The boundaries separating each of the basic classes (micro, mini, and
mainframe) will not be sharp and (as was hinted earlier) will continually
be changing as technology advances. However, in order to provide some
guidelines to enable machines to be classified, the following simple
definitions may prove useful:

Microcomputer: often referred to as a 'single board' computer
(see Figs 4.1 and 4.2). Systems usually cost under £2000. They
are usually used in such a way that they are dedicated to a
particular application. They are thus sometimes referred to as
single user systems. They have a limited address space and their
word lengths are usually not greater than 16 bits - these terms
will be explained later. On most microcomputers there is only a
limited range of software available.

Minicomputer: these are usually much more expensive than microcomputers and offer substantially more sophisticated hardware architectures. In addition a much wider range of software facilities is provided. They are often referred to as multi-user systems since they are able to support several simultaneous users. Figure 4.3 shows a typical minicomputer arrangement.

Mainframe: these are powerful multi-user systems able to support a large number of simultaneous users. They are characterised by high computation speeds, very large stores and word lengths in excess of 32 bits. They are, however, usually very expensive to purchase and maintain. They provide extremely sophisticated facilities for program development. Figure 4.4 shows the layout of a typical mainframe computer system.

In subsequent sections of this chapter a description of the basic hardware and software of a computer system will be outlined and examples given in terms of the three classes of computer described above.

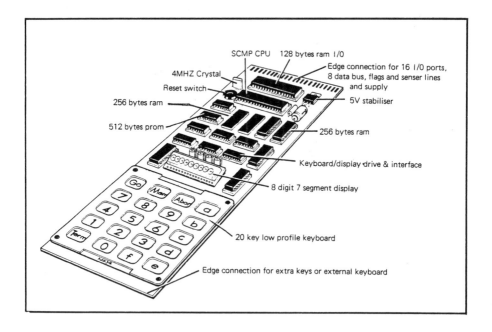

Fig. 4.1 Layout of a single board microcomputer system.

Fig 4.2   Microcomputer system designed for the hobby market.

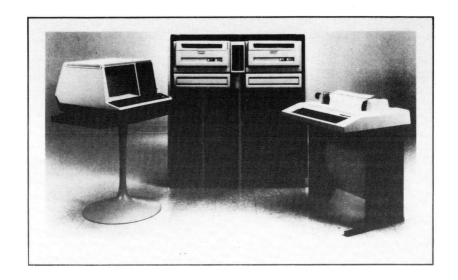

Fig. 4.3   A typical minicomputer configuration.

Fig. 4.4　Layout of a mainframe computer system.

THE FUNDAMENTAL HARDWARE ARCHITECTURE OF A COMPUTER SYSTEM

Whether a computer is classified as large, medium or small will not effect the fundamental manner in which it operates or the basic functions that it has to perform. Common to virtually all digital computer systems will be three basic hardware units interconnected by means of suitable links that enable them to communicate with each other. The three units are referred to as the MEMORY unit(s), the PROCESSOR, and the PERIPHERAL unit(s). They are interconnected by various communication channels (or buses) in a way similar to that shown in Fig. 4.5.

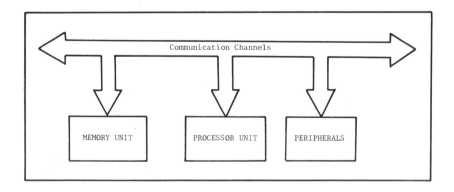

Fig. 4.5　Bus architecture for a computer system.

The basic unit shown in Fig. 4.5 may be extended in either direction by the addition of further memory units, processors or peripherals thereby providing the possibility of constructing quite complex computer systems. The type of architecture depicted in this diagram is referred to as a bus structured architecture. Many other types of basic design are possible depending upon the way in which the memory, processor and peripheral units are interconnected. Further details of the different design options will be found in any of a number of standard text books on computer architecture (Lav76, Lip78).

The memory unit is responsible for storing the data that the computer system is to process. It also stores the computer programs (or software) that are responsible for processing the stored data. As will be seen later, the memory may be of two types: **primary memory** which may be directly accessed by the processor via its internal addressing mechanisms (this may be core store or solid state memory), and **secondary memory** which is accessed indirectly via some ancillary control device. Magnetic tapes, disks and drums are typical secondary storage media.

The processor unit is responsible for performing all the control and data processing activities for which the computer is responsible. Contained within the processing unit are the electronic circuits that are responsible for the decision making and computational activities of the computer. In single processor systems the processor is often referred to as the Central Processing Unit or CPU. A more detailed description of the structure and mode of operation of the CPU will be found in books that are devoted to computer hardware (Lip78, Hea76, Hea78).

The peripheral units are those parts of the computer which enable information and data to be transmitted between the computer system and the outside world. Included here are such devices as paper-tape readers and punches, card readers, line printers, digitisers, data collection units and various types of interactive peripherals such as terminals and graphics devices. In addition, there will usually be a wide variety of telecommunications equipment to enable communication to take place between geographically remote computer systems.

In simple inexpensive computers the communication channels depicted in the previous diagram usually consist of a single multi-purpose bus that is shared by all those activities that need to communicate. The design of more sophisticated computers requires the use of multiple buses, each one being dedicated to a partcular type of function. Most computers of this type have at least:

> (a)   a DATA bus,
> (b)   a CONTROL bus, and,
> (c)   an ADDRESS bus.

These are illustrated in Fig. 4.6 for a simple microcomputer system.

The function of the DATA bus is to enable data to flow between one location within the computer and another, for example, between the memory unit and the processor. The ADDRESS bus allows the processor to specify the address of the location within the memory which is to receive data or which is to have its contents read. An important parameter associated with memory access is the cycle time of the memory unit. Essentially, this is the minimum time that the computer requires in order to retrieve an item of data from the memory. Memory speeds vary quite considerably;

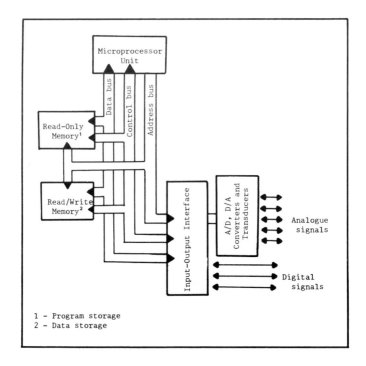

Fig. 4.6 Bus arrangement in a typical microcomputer system.

often the speed required will be dictated by the particular application for which the computer is being used. In order to control and coordinate the activity of the computer, the processor uses the CONTROL bus to send signals to other units and also sense conditions that arise in different parts of the system.

Most often the buses described above will take the form of a parallel set of electrical conductors (wires or thin strips of copper on an insulated board) that interlink the various system components. Frequently, the term bus width is used to refer to the number of parallel conductors that the bus contains. Each individual conductor carries one digital signal level or bit of information. The most common parallel bus widths are 4, 8, 12 and 16 bits (for microcomputers) with larger bus widths for minicomputer and mainframe machines. The bus width will depend upon the function it is to perform. In microcomputer systems the most common data bus width is 8 bits and the address bus width is usually 12 or 16 bits. Sometimes higher bus widths can be achieved by utilising a given bus several times in sequence; for example, in order to transmit a 32 bit signal on a 16 bit bus two groups of 16 bits could be sent in succession. Sometimes the control bus and address bus share the same physical hardware interconnections. Obviously, when buses can be physically shared the cost of the computer system is reduced. Cost reduction can also be achieved by keeping the bus widths as small as is consistent with the efficient operation of the computer system.

In Fig. 4.6 several different types of storage unit are illustrated.
R̲ead O̲nly M̲emory or (ROM) is used to hold computer software (systems
programs such as a simple operating sytem and a language translator
such as BASIC). In addition, ROM can also be used to hold special types
of application programs written by the user - particularly those developed
for dedicated process control applications or for inclusion in microcomputer based instrumentation.

The Read/Write memory shown in Fig. 4.6 is most commonly referred to as
R̲andom A̲ccess M̲emory - abbreviated to RAM. Usually, this type of
storage is used to hold the data that the processing unit is to manipulate
and transform. This data could be experimental results obtained from some
experiment. Alternatively, it might be a collection of instructions
that inform the processor how to perform the transformations/manipulations
on the stored experimental data in order to convert it into meaningful
results. Such collections of instructions are called user programs.
Usually these programs will be written in a programming language such as
BASIC or PASCAL.

The A/D (analogue to digital) and D/A (digital to analogue) converters
shown in Fig. 4.6 are used to interconvert signals between their analogue
and digital forms. Many of the analytical instruments, transducers and
other equipment to which computers are attached provide analogue signals.
These need to be converted to digital form before they can be processed by
the digital circuits of the computer. Similarly, many of the devices that
the computer has to control (magnetic valves, motors, potentiometers and
so on) require analogue signals for their operation. Consequently, the
computer's digital signals will need to be converted to analogue form
before they can be utilised for control purposes.

The Memory Structure

Both data and computer programs are stored within the memory of the
computer system. The memory of a computer consists of a series of
contiguous storage locations that are capable of holding the bit patterns
that represent the information to be stored. A memory location is
characterised by four important parameters:

    (a) its size,
    (b) its type,
    (c) its actual address, and,
    (d) its contents.

The size of an individual location is usually capable of storing a pattern
of 8 binary bits. This basic unit of storage is referred to as a byte.
The size of a computer memory is often defined in terms of the total
number of data bytes it is able to contain. Typically, the upper limits
for a microcomputer might be 64 - 128 kilobytes; for a minicomputer this
range might be extended upwards to 0.5 - 1 megabyte and for mainframe
machines the address space would extend into many megabytes. The term
address space is used to refer to the size of the memory that a computer
can directly address - that is, the address values it is able to generate.

Sometimes it is convenient to be able to address larger basic units of
storage. In addition to byte addressing, many computers permit word
addressing techniques. A word is a collection of consecutive bytes that
start at a particular type of address location called a word boundary.

Word capability often facilitates the storage and manipulation of data items in more meaningful units than would be permitted with a simple byte structure. One popular computer series (Str75) supports the following type of byte allocation strategy:

byte   a single byte

half-word   two consecutive bytes

full-word   four consecutive bytes

double-word   eight consecutive bytes

Half-words and full-words are used to store numeric integer quantities while full-words and double-words are used to hold numeric real items held in fixed-point or floating-point form. Non-numeric information (such as character strings) is usually held in contiguous collections of bytes, each character being stored in a separate byte.

Larger collections of contiguous memory locations also need to be addressed and referenced. For this purpose, two frequently used units are the block and the page. Their exact size often depends upon the particular computer system being considered. Thus, the SC/MP microcomputer (Hor79) has a page size of 4096 bytes while that of the Commodore PET system (COM79) is 256. The page size is often determined by the hardware architecture and the addressing mechanisms that the computer employs. These will be described in the appropriate technical documentation for the computer system concerned.

The variety of memory types that may be used in a computer system allows a great number of design options. An overall memory system might employ core store, bubble memory, solid state memory and so on. If it is solid state then it might be dynamic or static (Cam79), it may be Read Only, Random Access, Programmable Read Only Memory (PROM), Electrically Alterable Programmable Read Only Memory (EAPROM) and so on. A further discussion of memory types will be presented later.

The position of a particular storage location within the available memory space is specified by its address. In its simplest, this is just a decimal, octal or hexadecimal number that corresponds to the actual physical address of the location within the memory. Because of the difficulties of remembering numeric addresses, a technique called symbolic addressing is frequently used. This enables the user to associate meaningful names with different parts of the memory space. The correlation between symbolic names and physical addresses is then held in a special conversion table. A simple program called a 'look-up' routine can then be used to interconvert between the user's symbolic addresses and the numerical ones used by the machine.

In most computers the memory space is divided logically into several different sections according to the functions it is designed to perform. Certain areas are reserved for use by the computer itself, other parts are freely available to the user. Thus, different parts of the memory can be used to hold different types of data and information. A diagram that shows the way in which various parts of the memory are being used is called a memory map. Table 4.4 shows a typical example - that used in the Commodore PET microcomputer system (COM79, Don80).

TABLE 4.4   Memory map for the Commodore PET microcomputer

BLOCK #	TYPE	START ADDRESS	FUNCTION
0	RAM	$0000	Working, text, variable storage
1	RAM	$1000	Test variable storage (8K only)
2	...	$2000	Expansion RAM
3	...	$3000	Expansion RAM
4	...	$4000	Expansion RAM
5	...	$5000	Expansion RAM
6	...	$6000	Expansion RAM
7	...	$7000	Expansion RAM
8	RAM	$8000	Screen memory (1K)
9	...	$9000	Expansion ROM
10	...	$A000	Expansion ROM
11	...	$B000	Expansion ROM
12	ROM	$C000	BASIC (principally statement interpreter)
13	ROM	$D000	BASIC (principally math package)
14	ROM	$E000	Screen editor
	I/O	$E800	All internal CBM I/O
15	ROM	$F000	OS diagnostics

In those situations where a rigid structure and function is not imposed upon the memory, the contents of a particular location can usually be completely variable. At one instant in time a given location may be employed to store an item of data such as the temperature of a reaction vessel or a pH value. At some later time this same location may hold a machine instruction that is part of a computer program that is used to control the temperature of a reaction vessel or activate a device for measuring the pH of a solution.

THE FUNDAMENTAL SOFTWARE ORGANISATION OF A COMPUTER SYSTEM

Software is the term that is used to refer to the collection of computer programs that is contained within a computer system and which is generally responsible for controlling the overall activity of the computer. The most important of these items is probably the operating system which is often referred to by the synonyms executive and monitor program in microcomputer systems. The computer operating system acts as an interface between the computer user and the hardware/software contained within the machine. Usually the operating system makes available a series of commands that the user may issue in order to tell the computer what operations it is to perform, for example,

RUN   – to activate a stored program,

ALLOCATE   – to allocate an area of the computer's memory space for the storage of data or programs,

COPY   – to copy information from one area of the computer's memory to another,

FREE   - to free an area of the memory that was previously
              being used to store information; the freed storage
              is returned to a common pool of unused space.

An area of computer memory that is used to hold data or programs is
referred to as a **file**. Thus, to create a file called JACK to hold
experimental data produced in an experiment, the user might issue the
command ALLOCATE JACK to the operating system - normally, facilities will
exist to enable the user to specify how big the file is to be, its
organisation, security level and so on. The command FREE JACK would
release the memory space occupied by the file. The released storage then
becomes available for some other purpose. Similarly, the command COPY
JACK TO LIZ would make a duplicate copy of the file JACK in a new area of
store called LIZ. If a file called PROGM21 contains a computer program
for monitoring the temperature of a water bath during an experiment then
the command RUN PROGM21 would cause the program contained in the file
PROGM21 to be activated. When the user wishes to develop a new program
the command RUN BASIC is typed. This activates the BASIC interpreter that
the computer provides. Once developed a command such as SAVE NEWPROG24
might be used to save the program in the file NEWPROG24 for future use.
Usually the computer operating system will contain a wide variety of
commands of this type each having a particular function. The operating
system thus controls/interfaces with a large number of other software
items. The general categories of software subsystem that are commonly
available are illustrated in the diagram presented in Fig. 4.7.

The file management system is responsible for the creation, editing,
archiving and general security of the files created within the computer
system. Editing facilities permit the user to change and modify the
contents of stored data or programs while archiving techniques are used to
store copies of files on de-mountable storage devices such as magnetic
tapes or disks. Storage of archived files is performed in such a way as
to enable them to be easily retrieved when they are required. Many
computer systems make available special software items called data base
management packages that enable users to construct quite sophisticated
relationships between the data items that are stored in the computer file
space. Packages of this type will be discussed in more detail later.

In many important applications, especially within large organisations,
there is a growing need to be able to interlink individual computer
systems together in order to create structures that are more reliable and
more useful than single isolated systems. Such integrated collections of
computers are referred to as computer networks - they will be discussed in
more detail in chapter 12. When networks are used special types of
communication software are required in order to enable messages and files
to be transmitted from one location to another. Particularly important in
this context is software to maintain security of the information that is
transmitted over the network. To provide the necessary privacy a variety
of data encryption software is available - see chapter 7.

In order that the user of a computer can construct programs and application software for the particular problems that are to be solved, the
computer system will usually provide a wide variety of program development
facilities. Undoubtedly, the most heavily utilised of these will be the
language translators. Their purpose is to translate a program written
in some user orientated language into the basic **machine code** that is
stored within the computer's memory and which is processed by the central
processing unit.

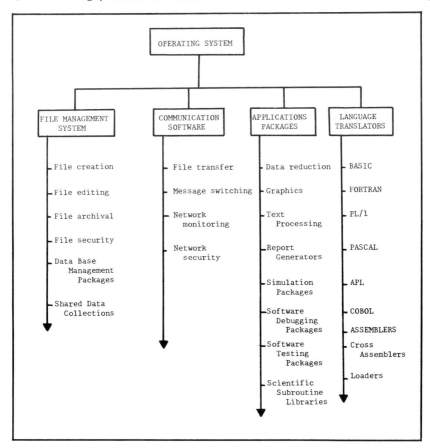

Fig. 4.7   General software organisation in a computer system.

The most common language translators are probably BASIC, FORTRAN and COBOL.   Each of these languages is referred to as a high level language.   They enable the computer user to prepare computer programs using statements that are easily understood because of their similarity to English language sentences.   High level languages are usually machine independent and permit program portability between different computer systems.   Wherever possible languages of this type are the preferred tool for software development because of the rapidity with which they enable programs to be written.   FORTRAN and BASIC are the most widely used systems for scientific applications.   The simplicity of BASIC is illustrated in the program shown in Fig. 4.8.   This is designed to calculate the average of a series of numbers that are presented to it interactively via the keyboard of a video display terminal.

Further details on BASIC and a description of how to write programs using this high level language will be found in text books that deal with computer programming (Spe75, Mon74, Mon78).   Programming in other

```
10   REM   PROGRAM TO COMPUTE AVERAGES
20   PRINT "HOW MANY DATA ITEMS?"
30   INPUT N
40   SUM=0
50   FOR I= 1 TO N
60   INPUT "NUMBER"; N
70   SUM=SUM+N
80   NEXT I
90   AV=SUM/N
100  PRINT "AVERAGE OF NUMBERS"; AV
110  STOP
```

Fig. 4.8   Computation of average by means of a BASIC program.

languages such as FORTRAN, PASCAL and PL/I is equally easy. A summary of
high level languages available for microprocessor applications is
contained in Taylor and Morgan's book (Tay80).

Prior to using a high (or low) level language for constructing computer
programs a great deal of thought needs to be given to designing the algorithms and data structures that are needed. A detailed description of
algorithms and data structures will be presented in chapter 9. However,
at this stage it is important to realise that the power of a high level
programming language - and, hence, its ease of use - will depend
critically upon the types of data and program structures that it allows
its user to create.

Most algorithmic languages allow the construction of programs which
consist of a series of statements that are executed sequentially in the
order in which they are written. The majority of statements will cause
some form of computation to be performed on the program's data structures.
Others will cause data to be moved from one location in the computer's
memory to another. Data transformation (through computation) and data
movement are fundamental operations that all high and low level languages
cater for.

Because certain types of computation will need to be repeated over and
over again some form of looping facility is a necessary pre-requisite for
the construction of efficient programs. Usually, iterative loops can be
created that are executed again and again until some specified condition
is met. The simple program shown in Fig. 4.8 contains an example of a
loop - a FOR loop. The loop body (statements 60 and 70) is executed
repeatedly until such a time that the value of I exceeds the value of N.
When this condition is met the loop terminates and the next statement in
sequence (statement number 90 in Fig. 4.8) is then executed.

The ability to perform conditional branching is another important facility
that a programming language must provide. Branching provides the means
whereby sections of a program may be skipped over if certain pre-defined
conditions are/are not met. There is a variety of mechanisms that enable
this type of facility to be implemented. The simplest approach is one in
which a condition is evaluated; if the requirements of the condition are
met then a jump is made to another part of the program - as specified in
the statement's label operand field; should the evaluation fail the branch

is not made and the statement following the condition would then be
executed. The majority of programming languages (of the high level
variety) implement conditional branching by means of an IF statement.
Some typical examples are shown below:

IF X>36 THEN GOTO ERROR;   (PL/I)

IF (X-2.1) 40,50,60   (FORTRAN)

IF X>23 AND Y<34.7 THEN 365   (BASIC)

By means of sequential flow, looping, conditional branching and the
ability to perform arithmetic computation most algorithms can easily be
expressed in one or other of the available high level programming
languages. However, when the algorithms relate to data management an
additional linguistic primitive needs to be considered - the input/output
capabilities of the language. Statements that deal with input and output
provide special types of data movement facilities - between the computer's
main memory and its peripheral devices or ancillary storage media. A more
detailed discussion of the basic input-output requirements of a
programming language is presented in chapter 10 which deals with data
bases and information systems.

The degree of difficulty of the programming task increases substantially
when an **assembler** has to be used. Assemblers use mnemonic codes to
represent the basic machine instructions processed by the CPU of the
computer. In contrast to high level languages, assemblers are highly
machine dependent and require their user to have a detailed understanding
of the hardware architecture of the computer system being used. A simple
example of a program written in assembler for the National Semiconductor
SC/MP microprocessor (Nat77) is shown in Fig. 4.9.

```
.TITLE  DISP, 'SIMPLE EXAMPLE PROGRAM'
; DISPLAYS THE WORD "dAtA" ON THE NIXIE DISPLAY
;
START: LDI   X'0D     ; LOAD ACCUMULATOR
       XPAH  1        ; TRANSFER TO POINTER 1 HIGH
       LDI   X'00     : LOAD ACCUMULATOR
       XPAL  1        ; TRANSFER TO POINTER 1 LOW
LOOP:  LDI   X'5E     ; LOAD LETTER D INTO ACCUMULATOR
       ST    3(1)     ; WRITE TO DISPLAY DIGIT 4
       LDI   X'77     ; LOAD LETTER A INTO ACCUMULATOR
       ST    2(1)     ; WRITE TO DISPLAY DIGIT 3
       LDI   X'78     ; LOAD LETTER T INTO ACCUMULATOR
       ST    1(1)     ; WRITE TO DISPLAY DIGIT 2
       LDI   X'77     ; LOAD LETTER A INTO ACCUMULATOR
       ST    0(1)     ; WRITE TO DISPLAY DIGIT 1
       JMP   LOOP     ; JUMP BACK TO REFRESH DISPLAY
.END   START
```

Fig. 4.9 Appearance of an assembler language program.

If this program was executed on appropriate hardware it would cause the
word 'dAtA' to be displayed in the rightmost four digit positions of an
8-digit-7-segment Nixie display (Jer77) attached to the microcomputer (see
Fig. 4.1). It can be seen that the program consists of a series of
statements each of which consists of a mnemonic operation code (such as
LDI, ST, JMP) followed by an operand (such as X'00, 0(1), LOOP and so

on) which in turn is followed by a comment field (commencing with a ;)
which explains what the particular instruction is intended to do. Each of
the operation codes defines a basic operation that the computer is to
perform - such as loading a value into an accumulator, storing the
contents of the accumulator into another memory location and so on. This
type of assembler instruction will be described in more detail later.

Figure 4.10 shows a listing of the contents of the computer memory
produced as a result of the translation of the assembler program (Fig 4.9)
into machine code. The program has been assembled into a series of
contiguous RAM locations starting at hexadecimal address OF20. All the
addresses in this listing and the memory contents are expressed in
hexadecimal notation (Cam79).

MEMORY ADDRESS	MEMORY CONTENTS
OF20	C40D
OF22	35
OF23	C400
OF25	31
OF26	C45E
OF28	C903
OF3A	C477
OF2C	C902
OF2E	C478
OF30	C901
OF32	C477
OF34	C900
OF36	90E8

Fig. 4.10   Format of the machine code produced by the
program shown in Fig. 4.9.

When developing software for microprocessor and minicomputer systems, it
is often desirable to use special types of language translators known as
cross-assemblers or cross-compilers. This type of translator is
usually available on mainframe computer systems so that the user has the
power and facilities of a large computer as an aid in developing software
for microcomputer applications. Typical examples of this type of software
have been described by Lubbers (Lub80).

In addition to language translators, communication software and file
management programs, the computer will invariably provide a range of
application packages that are designed to satisfy common user needs.
These packages will vary in capability. They will include software for
performing simple data reduction of experimental data, text processing,
report generation as well as providing facilities for the production of
high quality graphics. Bearing in mind some of the likely possibilities,
Fig. 4.11 shows a logical memory map that depicts the layout of the
software components within a typical computer system. Obviously, the
exact organisation of the software will vary from one computer to another.
Furthermore, the completeness of the software system that was presented

in Fig. 4.7 will depend upon both the class of computer involved (large, medium or small) and the particular system within a given class. As a general rule the availability of software and program development facilities decreases in going from a mainframe to a minicomputer or micro system.

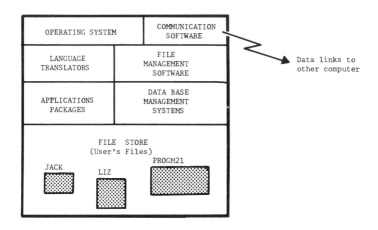

Fig. 4.11 Symbolic memory map for a computer system.

THE MICROPROCESSOR AND MICROCOMPUTER

One of the attractive features of modern computer technology is the ease with which it is possible to design and construct individual systems to meet particular requirements. Because of this there is a large number of microprocessor and microcomputer products and systems available. Fundamental to all of these systems is the basic microprocessor element. This usually takes the form of a single integrated circuit (or chip) similar in appearance to that shown in Fig. 4.13. This basic chip does not acquire the status of a microcomputer until appropriate external circuitry, memory and suitable peripheral devices are added. The logical arrangement of a simple microcomputer is shown in Fig. 4.12.

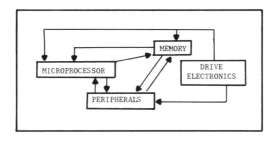

Fig. 4.12 Basic components of a microcomputer system.

The memory is required in order to store programs and data. The peripherals enable the system to be programmed and allow the entry of data into the memory. The simplest type of peripheral is a numeric keypad (hexadecimal or decimal) or alphanumeric keyboard. Alternatively, simple transducers or on-line instruments might act as special types of peripheral. The drive electronics are required in order to provide the necessary power supplies for the chips and other components. They also provide a source of timing signals by which the operations of the computer may be synchronised. The power of any microcomputer system will depend upon the capability of the basic microprocessor chip from which it is constructed. Sometimes several processor chips will be incorporated into a given microcomputer in order to achieve particular performance characteristics. The Superbrain microcomputer (Sup80) is an example of a system of this type. It incorporates two Zilog Z80 microprocessor chips - one which acts as a standard CPU while the other functions as a peripheral controller that supervises the activity of the two disk drives that the system contains.

The microprocessor chips themselves will usually be classified as 8-bit or 16-bit depending upon their design. Some of the more popular chips are listed below:

Motorola M6800	MOS Technology 6502
Zilog Z80	Texas Instruments TI 1100
Intel 8080	Texas Instruments TI 9980
Ferranti F100-L	National Semiconductor SC/MP
Fairchild F8	Digital Equipment Corporation LSI/11

A description of most of these will be found in the authoritative work by Osborne and Kane (Osb78). This is in three volumes: one gives a basic introduction to the technology, another deals with the architecture of the more popular microprocessors while the third describes the variety of support chips that are available. As an example of a microprocessor chip the National Semiconductor SC/MP chip has been chosen for further study. This particular example has been selected because of its simple architecture and the ease with which it can be incorporated into other systems. The basic principles that are presented in the following discussion will generally apply to other microprocessors even though the fine detail may differ.

The Microprocessor - An Example - The SC/MP

SC/MP is a mnemonic for Simple Cost-effective Micro-Processor. This device is an 8 bit NMOS technology processor produced by the National Semiconductor Corporation. It is useful for laboratory instrumentation applications, process controllers, machine tool control and multiprocessor systems, that is, applications that require several microprocessors to be linked together. It is a single chip microprocessor that contains an on-chip oscillator and timing generator. It uses standard memories and peripheral components. The system is supported by a complete range of development software consisting of an editor, assemblers, loaders and a debug facility. In addition there is a Low Cost Development System (called LCDS) along with pre-packaged application cards to enable users to develop their applications systems with minimum difficulty.

The SC/MP is housed in a 40 pin, dual in-line package measuring 13 x 52 mm. A diagram showing the arrangement of the pins is presented Fig. 4.13.

Two of the pins are used to supply power to the chip:

Pin 20   labelled V_{SS} $\left.\vphantom{\begin{array}{c}a\\a\end{array}}\right\}$ Power input to chip

Pin 40   labelled V_{GG}

and two are used for timing (pins 38 and 37). A suitable capacitor is usually connected across these pins for applications which are not critical. However, where timing is an important consideration a suitable crystal must be used. Alternatively, the processor can be synchronised with an external clock by connecting appropriate drive circuits to the X1 and X2 pins of the chip.

The remaining 36 pins are used for control, addressing and data input/ output functions. In a typical application the eight input/output pins (DB0 through DB7) would be connected to a common data bus and the twelve address pins (AD00 through AD11) would be connected to an address bus. Each of the pins of the data port are bi-directional and tri-state (Lan79). A tri-state pin is one which can assume any of three states: two of these are used for signalling while the third corresponds to a high impedance (open circuit) state which makes it appear as though the pin was not connected to any bus. This is an effective way of enabling bus sharing between components. The address port is also tri-state and provides a 12-bit latched address. In conjunction with bus access and other appropriate control signals, three functions are thus implemented: (a) 8-bit data input to the processor, (b) 8-bit data output from the processor, and, (c) address and status information output from the processor.

Some of the pins that are used to carry control information to or from the chip are:

NWDS   (pin 1)   Negative Write Data Strobe - when low indicates that data from the processor is valid on the 8-bit I/O bus,

NRDS   (pin 2)   Negative Read Data Strobe - when low indicates that the processor is ready to accept data from the 8-bit I/O bus, and,

NADS   (pin 39)   Negative Address Data Strobe - when low indicates that a valid address and status outputs are present on the system buses.

Other control pins on the chip are divided into two categories, as shown below:

BREQ	Bus-request
ENIN	Enable input
ENOUT	Enable output

CONT	start-stop
NRST	Negative reset
NHOLD	wait

Those in the group on the left are used to control bus access, direct
memory access - or DMA (Cam79) - and multiprocessor control. The members
of the group on the right are used to influence the way in which the
processor operates. CONT permits suspension of operations without loss of
internal status while NRST (when low) causes all in-process operations to
be aborted.   NHOLD causes the processor to enter a wait state.   This is
used in conjunction with the CONT pin to implement single cycle/
instruction execution of the microprocessor (Cam79).   This is a useful
facility for system testing.

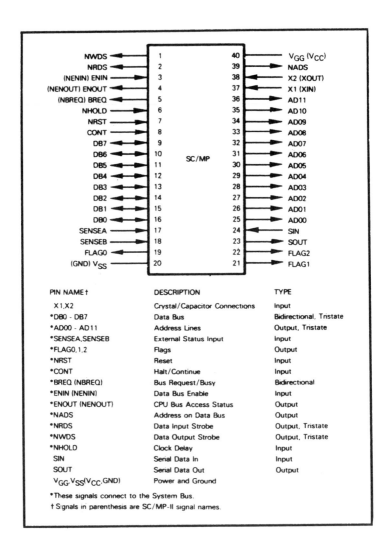

PIN NAME†	DESCRIPTION	TYPE
X1,X2	Crystal/Capacitor Connections	Input
*DB0 - DB7	Data Bus	Bidirectional, Tristate
*AD00 - AD11	Address Lines	Output, Tristate
*SENSEA,SENSEB	External Status Input	Input
*FLAG0,1,2	Flags	Output
*NRST	Reset	Input
*CONT	Halt/Continue	Input
*BREQ (NBREQ)	Bus Request/Busy	Bidirectional
*ENIN (NENIN)	Data Bus Enable	Input
*ENOUT (NENOUT)	CPU Bus Access Status	Output
*NADS	Address on Data Bus	Output
*NRDS	Data Input Strobe	Output, Tristate
*NWDS	Data Output Strobe	Output, Tristate
*NHOLD	Clock Delay	Input
SIN	Serial Data In	Input
SOUT	Serial Data Out	Output
$V_{GG}, V_{SS} (V_{CC}, GND)$	Power and Ground	

*These signals connect to the System Bus.
† Signals in parenthesis are SC/MP-II signal names.

Fig. 4.13   Pin arrangement for the SC/MP microprocessor.

The other pins on the chip permit special types of input and output operations. The SIN and SOUT pins permit serial data transfer to or from the chip under the control of a special SIO instruction (see Table 4.5). The SENSE A and SENSE B pins are TTL level (Lan79) inputs connected to bit positions 4 and 5 of the status register (see Fig. 4.15) within the CPU. These pins are read only pins. Provided interrupts are enabled by the software, SENSE A serves as an interrupt request line. The final pins on the chip: FLAG0, FLAG1 and FLAG2 are the flag outputs and correspond, respectively, to bits 0, 1 and 2 of the status register. These bits are available for user-designated functions. Each flag is TTL compatible and can drive a 1.6 milliampere load. The flags are software controlled and can be set or pulsed in a single or multiple sequence. A simplified diagram summarising the input and output capabilities of the SC/MP is presented in Fig. 4.14.

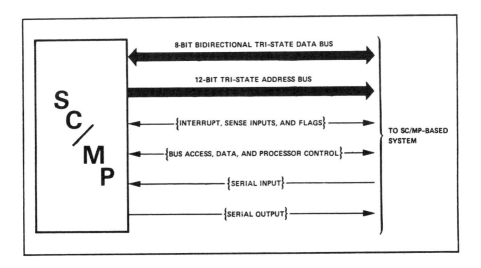

Fig. 4.14 Input and output capabilities of the SC/MP microprocessor.

The internal architecture of the SC/MP chip is illustrated in Fig. 4.15 and a summary of the operational features of the processor is presented in Table 4.5. This table includes several important parameters that need to be considered when comparing different types of computer - and different computers within a given class. The most useful comparison metrics are

> instruction repertoire,
> address space limitations,
> addressing modes,
> register types and availability, and,
> input/output capability.

The facilities offered under each of these headings will vary from one microcomputer to another and between micros, minis and mainframe systems. Tables of values for these parameters can thus be used as a useful way of comparing different computers.

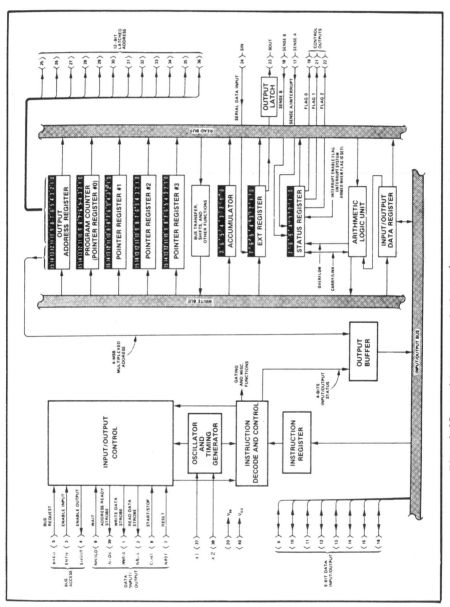

Fig. 4.15 Architecture of the SC/MP processor chip.

TABLE 4.5   Basic Features of the SC/MP Microprocessor

PARAMETER	COMMENTS
Data length	8 bits(byte)
Instruction set	46 instructions
Arithmetic	parallel, binary, fixed point, two's complement, 2 digit BCD addition
Memory	Up to 65,536 bytes of RAM/ROM
Registers	One 8-bit accumulator One 8-bit status register One 8-bit extension register Four 16 bit pointer registers (one is the Program Counter)
Addressing Modes	Program Counter Relative Indexed Auto Indexed Immediate
Input/Output and Control	16-bit address bus: 4 multiplexed bits and 12 static bits, 8-bit Bidirectional data bus

Inspection of the CPU architecture presented in Fig. 4.15 will show that the chip consists of a series of registers, buffers, data latches and processing elements such as the arithmetic logic unit and the instruction decode/control block. These are interlinked by a variety of buses and control lines that enable the CPU to act as an integrated whole controlled by programs and conditions external to the chip. A detailed description of its mode of operation and the role that each of the components plays will be found the SC/MP technical manual (Nat76) and the authoritative work by Osborne and Kane (Osb78). The following notes are intended to provide an overview of the function performed by each of the items resident on the IC so that its mode of operation may be understood.

(a)   Oscillator and Timing Generator
      - generates the basic timing signals for the various micro-
        processor control functions.

(b)   Input/Output Control Unit
      - generates input/output signals in order to control and
        synchronise bus activity.

(c)   Instruction Decode and Control
      - decodes the instruction currently being executed and provides
        all the control and gating functions required for execution of
        any particular instruction within the instruction repertoire.

(d)   Instruction Register
      - holds byte 1 of the instruction during the execution cycle.

(e)   Output Address Register
        - holds the 16-bit address. At address strobe time the 4 most
        significant bits are strobed onto the data bus (to give the page
        number of the address) and the 12 least significant bits are
        output on the address bus (to specify the offset of the required
        location within the page).

(f)   Input/Output Data Register
        - during an input cycle this register receives information from
        the 8-bit data bus; during an output cycle the contents of this
        register are transferred onto the bus.

(g)   Shift, Rotate, Transfer and Other Logical Functions
        - these circuits implement the indicated functions and also
        perform various housekeeping tasks.

(h)   Arithmetic Logic Unit
        - performs all the arithmetic and logic functions required by the
        chip.

(i)   Status Register
        - this is an 8-bit register that provides storage for arithmetic
        status (overflow, carry and so on), control and software status.

(j)   Extension Register
        - this is an 8-bit register that supports the accumulator in
        performing arithmetic, logic and data transfer operations.
        In addition, the serial I/O capability of the chip is
        implemented through the extension register.

(k)   Accumulator Register
        - this is the major processing register on the chip. It serves as
        the link between memory and all other software controlled
        registers. It is used in performing arithmetic and logic
        operations and for storing the results of these operations.
        Data transfers, shifts and rotates also use the accumulator.

(l)   Program Counter
        - this is also referred to as Pointer Register Zero. It is used
        to hold the address of the instruction being executed; it is
        incremented just before the instruction fetch cycle.

(m)   Pointer Registers 1, 2 and 3
        - these are 16-bit registers used for temporary storage and other
        miscellaneous functions as dictated by the programmer.

An understanding of the detailed nature of the way in which the above
hardware units interact is not necessary unless the user is forced into
the situation of having to write programs for the microprocessor. If this
situation arises then it is important to realise that these units
represent the basic hardware of the computer. It is this hardware which
is responsible for processing all the data which the computer is given and
for controlling all the equipment that it may be required to control. The
programmer uses the instruction set of the processor in order to control
these hardware elements and specify how they are to be used to process
items of data. As an example, suppose it is required to make the microprocessor add together three numbers that are held in particular memory

locations (at addresses 0B00, 0B01 and 0B02) and store the result in a fourth location (address 0B03). Assuming that Pointer Register 1 contains the value 0B00 and that the possibility of arithmetic overflow will not arise, a simple program to achieve the addition might appear similar to that shown in Fig. 4.16.

MEMORY ADDRESS	MEMORY CONTENTS	Comments
0F12	C100	load accumulator
0F14	F101	add in next number
0F16	F102	add in third number
0F18	C903	store result
0F1A	3F	return to monitor

Fig. 4.16 Simplified machine code program for performing addition.

The type of machine code program presented in Fig. 4.16 is difficult to write because it is extremely easy to make mistakes and it is by no means a simple task to locate any errors that are contained in such programs. Because of this, programming in an assembler or high level language is invariably preferred by most programmers. The assembler and BASIC language equivalents of the program depicted in Fig. 4.16 are shown in Fig. 4.17. There are many advantages to using BASIC even though it may result in programs that are significantly slower than equivalent programs written in Assembler (Bar81). The most attractive feature of BASIC is the speed with which operational programs can be developed. However, despite its usefulness, for many applications involving microprocessor interfacing it is far too slow and the programs must therefore be developed in the assembler language for the microprocessor being used. This will necessitate an understanding of the basic instruction repertoire for the computer and the effects that each instruction has upon the hardware units. A listing of the 46 instructions supported by the SC/MP microprocessor is presented Table 4.6. This table also provides a definition of the effect that the instruction has upon the microprocessor elements. A more detailed description of the instruction repertoire will be found in the SC/MP technical manual (Nat76) and programmer's guide (Nat77).

Programs written in machine code or developed in Assembler or BASIC using a development centre (Nat76a, Nat76b) can be loaded into a suitable memory chip and then used to control the processor in a variety of applications within the areas of process control, data logging and automation. They can be introduced into instruments in order to make them easier to use and operate. Appropriate modules can be designed to control the operational parameters of an instrument and diagnose the cause of faults when they arise. Thus, instruments can be made more intelligent by giving them memory that will enable them to store details of analytical conditions and analytical procedures. In the simplest type of application a configuration similar to that shown in Fig. 4.18 is often used. This depicts a two-chip system consisting of the SC/MP CPU and a supporting memory chip.

Obviously, the exact details of any particular configuration (microprocessor type, memory type - amount and speed, interface connections to the equipment, supporting chips and so on) will depend upon the exact

TABLE 4.6   SC/MP Microprocessor Instruction Summary

Mnemonic	Description	Object Format	Operation	Micro-Cycles
SINGLE-BYTE INSTRUCTIONS				
	Extension Register Instructions	7 6 5 4 3 2 1 0		
LDE	Load AC from Extension	0 1 0 0 0 0 0 0	$(AC) \leftarrow (E)$	6
XAE	Exchange AC and Extension	0 0 0 0 0 0 0 1	$(AC) \leftrightarrow (E)$	7
ANE	AND Extension	0 1 0 1 0 0 0 0	$(AC) \leftarrow (AC) \quad (E)$	6
ORE	OR Extension	0 1 0 1 1 0 0 0	$(AC) \leftarrow (AC) \lor (E)$	6
XRE	Exclusive-OR Extension	0 1 1 0 0 0 0 0	$(AC) \leftarrow (AC) \lor (E)$	6
DAE	Decimal Add Extension	0 1 1 0 1 0 0 0	$(AC) \leftarrow (AC)_{10} + (E)_{10} + (CY/L);(CY/L)$	11
ADE	Add Extension	0 1 1 1 0 0 0 0	$(AC) \leftarrow (AC) + (E) + (CY/L);(CY/L),(OV)$	7
CAE	Complement and Add Extension	0 1 1 1 1 0 0 0	$(AC) \leftarrow (AC) + \sim(E) + (CY/L);(CY/L),(OV)$	8
	Pointer Register Move Instructions	7 6 5 4 3 2 1 0		
XPAL	Exchange Pointer Low	0 0 1 1 0 0 ptr	$(AC) \leftrightarrow (PTR_{7:0})$	8
XPAH	Exchange Pointer High	0 0 1 1 0 1 ptr	$(AC) \leftrightarrow (PTR_{15:8})$	8
XPPC	Exchange Pointer with PC	0 0 1 1 1 1 ptr	$(PC) \leftrightarrow (PTR)$	7
	Shift, Rotate, Serial I/O Instructions	7 6 5 4 3 2 1 0		
SIO	Serial Input/Output	0 0 0 1 1 0 0 1	$(E_i) \rightarrow (E_{i-1}), SIN \rightarrow (E_7), (E_0) \rightarrow SOUT$	5
SR	Shift Right	0 0 0 1 1 1 0 0	$(AC_i) \rightarrow (AC_{i-1}), 0 \rightarrow (AC_7)$	5
SRL	Shift Right with Link	0 0 0 1 1 1 0 1	$(AC_i) \rightarrow (AC_{i-1}), (CY/L) \rightarrow (AC_7)$	5
RR	Rotate Right	0 0 0 1 1 1 1 0	$(AC_i) \rightarrow (AC_{i-1}), (AC_0) \rightarrow (AC_7)$	5
RRL	Rotate Right with Link	0 0 0 1 1 1 1 1	$(AC_i) \rightarrow (AC_{i-1}), (AC_0) \rightarrow (CY/L) \rightarrow (AC_7)$	5
	Single-Byte Miscellaneous Instructions	7 6 5 4 3 2 1 0		
HALT	Halt	0 0 0 0 0 0 0 0	Pulse H-flag	8
CCL	Clear Carry/Link	0 0 0 0 0 0 1 0	$(CY/L) \leftarrow 0$	5
SCL	Set Carry/Link	0 0 0 0 0 0 1 1	$(CY/L) \leftarrow 1$	5
DINT	Disable Interrupt	0 0 0 0 0 1 0 0	$(IE) \leftarrow 0$	6
IEN	Enable Interrupt	0 0 0 0 0 1 0 1	$(IE) \leftarrow 1$	6
CSA	Copy Status to AC	0 0 0 0 0 1 1 0	$(AC) \leftarrow (SR)$	5
CAS	Copy AC to Status	0 0 0 0 0 1 1 1	$(SR) \leftarrow (AC)$	6
NOP	No Operation	0 0 0 0 1 0 0 0	None	5

Mnemonic	Description	Object Format	Operation	Micro-Cycles
DOUBLE-BYTE INSTRUCTIONS				
	Memory Reference Instructions	7 6 5 4 3 2 1 0 7 6 5 4 3 2 1 0		
LD	Load	1 1 0 0 0 mptr disp	$(AC) \leftarrow (EA)$	18
ST	Store	1 1 0 0 1 mptr disp	$(EA) \leftarrow (AC)$	18
AND	AND	1 1 0 1 0 mptr disp	$(AC) \leftarrow (AC) \quad (EA)$	18
OR	OR	1 1 0 1 1 mptr disp	$(AC) \leftarrow (AC) \lor (EA)$	18
XOR	Exclusive-OR	1 1 1 0 0 mptr disp	$(AC) \leftarrow (AC) \lor (EA)$	18
DAD	Decimal Add	1 1 1 0 1 mptr disp	$(AC) \leftarrow (AC)_{10} + (EA)_{10} + (CY/L);(CY/L)$	23
ADD	Add	1 1 1 1 0 mptr disp	$(AC) \leftarrow (AC) + (EA) + (CY/L);(CY/L),(OV)$	19
CAD	Complement and Add	1 1 1 1 1 mptr disp	$(AC) \leftarrow (AC) + \sim(EA) + (CY/L);(CY/L),(OV)$	20
	Memory Increment/Decrement Instructions	7 6 5 4 3 2 1 0 7 6 5 4 3 2 1 0		
ILD	Increment and Load	1 0 1 0 1 0 ptr disp	$(AC), (EA) \leftarrow (EA) + 1$	22
DLD	Decrement and Load	1 0 1 1 1 0	$(AC), (EA) \leftarrow (EA) - 1$	22
	Immediate Instructions	7 6 5 4 3 2 1 0 7 6 5 4 3 2 1 0		
LDI	Load Immediate	1 1 0 0 0 1 0 0 data	$(AC) \leftarrow data$	10
ANI	AND Immediate	1 1 0 1 0 1 0 0 data	$(AC) \leftarrow (AC) \quad data$	10
ORI	OR Immediate	1 1 0 1 1 1 0 0 data	$(AC) \leftarrow (AC) \lor data$	10
XRI	Exclusive-OR Immediate	1 1 1 0 0 1 0 0 data	$(AC) \leftarrow (AC) \lor data$	10
DAI	Decimal Add Immediate	1 1 1 0 1 1 0 0 data	$(AC) \leftarrow (AC)_{10} + data_{10} + (CY/L);(CY/L)$	15
ADI	Add Immediate	1 1 1 1 0 1 0 0 data	$(AC) \leftarrow (AC) + data + (CY/L);(CY/L),(OV)$	11
CAI	Complement and Add Immediate	1 1 1 1 1 1 0 0 data	$(AC) \leftarrow (AC) + \sim data + (CY/L);(CY/L),(OV)$	12
	Transfer Instructions	7 6 5 4 3 2 1 0 7 6 5 4 3 2 1 0		
JMP	Jump	1 0 0 1 0 0 ptr disp	$(PC) \leftarrow EA$	11
JP	Jump if Positive	1 0 0 1 0 1 ptr disp	If $(AC) \geqslant 0, (PC) \leftarrow EA$	9, 11
JZ	Jump if Zero	1 0 0 1 1 0 ptr disp	If $(AC) = 0, (PC) \leftarrow EA$	9, 11
JNZ	Jump if Not Zero	1 0 0 1 1 1 ptr disp	If $(AC) \neq 0, (PC) \leftarrow EA$	9, 11
	Double-Byte Miscellaneous Instructions	7 6 5 4 3 2 1 0 7 6 5 4 3 2 1 0		
DLY	Delay	1 0 0 0 1 1 1 1 data	count AC to -1; delay = $13 + 2(AC) + 2 disp + 2^9 disp$ microcycles	13 to 131 593

ASSEMBLER Version	BASIC Version
P1=1 ADDRESS=X'0B00 LDI H(ADDRESS) XAPL P1 LDI 0(P1) ADD 1(P1) ADD 2(P1) ST 3(P1) XPPC 3	Z=A+B+C

Fig. 4.17 Comparison of BASIC and Assembler programs.

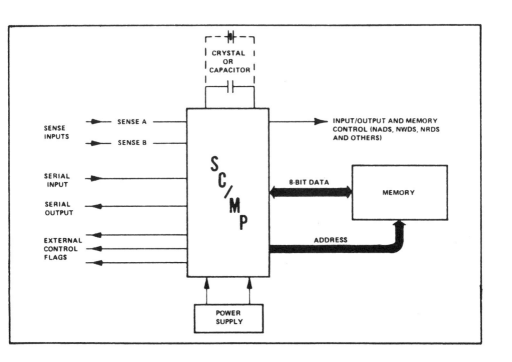

Fig. 4.18 Basic microprocessor based system.

details of the actual specific application under investigation. Despite this, however, the basic principles of what has been discussed here are generally applicable to most situations involving microprocessor based design. In this context it should be emphasised that a large number of applications that require the microprocessor approach to the solution of laboratory computing problems necessitate the use of a dedicated processor/memory system similar to that shown in Fig 4.18. This approach is illustrated in two applications which are outlined in the next section.

Using Microprocessors - Two Examples

The National Semiconductor SC/MP

Morley et al (Mor80) have designed a sequence controller for use in laboratory automation. Such a device might be used to control a sample probe, an automatic sampling valve or any other number of laboratory items whose use depends upon a repetitive cyclic process. One example is shown in the diagrams contained in Fig. 4.19 which represents the positional sequencing of a sample probe.

The probe has two vertical positions, 'up' and 'down' and two horizontal positions corresponding to 'sample' and 'wash'. A typical sequence of operations would be to raise the probe from the wash position, rotate it through 180° to the sample position, lower it into the sample in order to take a measurement and reverse the process leaving the sample probe in the wash position. After two minutes repeat the process and continue to repeat it until switched off by the operator. This is a simple process to program provided a suitable controller is available which will convert the steps of the program into electrical signals that will control the probe mechanisms. The latter will require the use of accurately controlled stepper motors (Pea77).

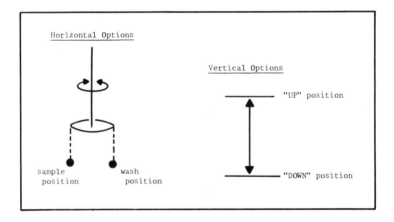

Fig. 4.19 Allowed positions for an automated sampling probe.

The controller is based upon the use of a National Semiconductor SC/MP microprocessor, 512 bytes of PROM memory (type MM5204Q) and 256 bytes of

RAM memory (type 2112N2). The PROM is used to hold the program
responsible for the operation of the controller while the RAM is used to
store the user's control sequence. This is introduced into the controller
by means of a series of eight thumbwheel switches. These are linked to a
corresponding number of light emitting diodes (LEDs) that provide a simple
read-out facility - one diode for each switch. The arrangement is shown
in the sketch presented in Fig. 4.20. The software in the SC/MP performs
two roles: (1) that of interpreting the instructions coded by the thumbwheel switches (this is program definitition mode); and, (2) execution (or
control) mode in which the previously defined program is executed. Mode 1
is activated by pressing a RESET button on the instrument's control panel
while mode 2 is invoked by a RUN switch. When in RUN mode the program
interprets the coded instructions and converts them into appropriate
input/output signals and time delays.

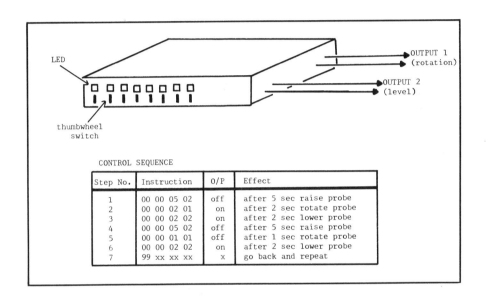

CONTROL SEQUENCE

Step No.	Instruction	O/P	Effect
1	00 00 05 02	off	after 5 sec raise probe
2	00 00 02 01	on	after 2 sec rotate probe
3	00 00 02 02	on	after 2 sec lower probe
4	00 00 05 02	off	after 5 sec raise probe
5	00 00 01 01	off	after 1 sec rotate probe
6	00 00 02 02	on	after 2 sec lower probe
7	99 xx xx xx	x	go back and repeat

Fig 4.20 Sample probe controller and a typical control sequence
entered by the user via the thumbwheel switches.

The INTEL 8085 Chip

The second example of the use of micro-chips illustrates how another
popular microprocessor IC, the INTEL 8085, may be employed in the construction of a computer based monochromator controller for use in a
spectrometer (Dal80).

One of the attractive features of this type of monochromator is the fact
that it permits wavelength programming, that is, the computer can be
programmed to direct the controller to go to any wavelength in any order.
In addition, it is possible to scan a spectrum in linear energy units by
calculating the wavelength for a given energy increment and outputting the
value as a 'move' to the controller. Scans carried out with the aid of a

microcomputer based system such as this may be made at any speed (subject
to hardware restrictions) and can be made to incorporate appropriate
delays to allow for signal averaging. A schematic of the microcomputer
section of the monochromator controller is shown in Fig. 4.21.

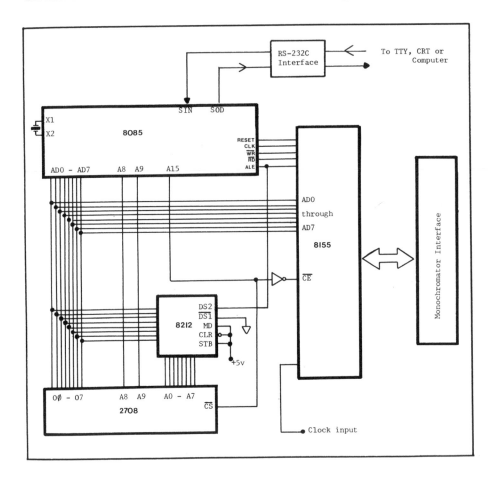

Fig. 4.21 Integrated circuits used in a microcomputer based
monochromator controller for a spectrometer.

The microprocessor chip is an INTEL 8085 (Osb78) which is connected to
three other support chips:

(a)   an INTEL 8155 static read/write memory with I/O ports
      and timer,
(b)   a 1K x 8-bit 2708 PROM chip (programmable read only
      memory), and,
(c)   an INTEL 8212 8-bit input/output port.

The control program for the CPU is stored in the 2708 PROM chip.
Retrieval of instructions from this chip is controlled by the 8212 I/O
port. The address latch enable pin on the CPU chip pulses high when
address data is being output on AD0-AD7. This address is strobed off the
bus by the 8212 I/O port which uses it to access the 2708 PROM causing the
output byte 00-07 to be passed back to the CPU. The CPU lines AD0-AD7 are
bidirectional since they are used to output the low order byte of memory
addresses and act as a bidirectional data bus. CPU pins A8 through A15
are output only lines and serve to carry the high order byte of memory
addresses. Since the 2708 is a 1K x 8-bit chip only A8 and A9 are needed
for this purpose. The 8155 read/write memory is used to store variables
associated with the monochromator control program. The programmable timer
on this chip is used to generate the waveforms necessary to control the
stepper motor (Pea77) attached to the monochromator.

The Microcomputer – An Example – The COMMODORE PET

PET is an acronym for Personal Electronic Transactor. The COMMODORE
PET is an example of a complete microcomputer system to which can be
attached a wide variety of peripherals and laboratory instruments. The
system is based around the MOS technology 6502 microprocessor (Cam79,
MOS76). PET microcomputers are available with various memory sizes,
screen and keyboard options. The most common version of the PET has a 9"
enclosed CRT capable of displaying 1000 characters arranged in terms of a
matrix having 25 lines each consisting of 40 characters. More recent
large screen versions of the microcomputer provide a line width of 80
characters. There is a brightness control at the rear of the CRT case.
Standard memory sizes are 4K, 8K, 16K and 32K bytes. The 4K and 8K
versions can be very easily expanded to 40K, of which 32K is the maximum
available for programming in BASIC. The other 8K is available for machine
code programming or floppy disk memory expansion. The appearance of a
typical PET microcomputer with its associated tape cassette unit and line
printer is shown in Fig. 4.22.

Fig. 4.22 A PET desktop computer system.

A wide range of peripherals can be attached to the PET via the various
peripheral and interface ports that are available. The computer has a
standard IEEE-488 interface (see chapter 6) and a User-Port which should

be regarded as a non-standard interface that provides 8 user programmable pins (TTL levels). These ports are discussed in more detail in chapter six in the section dealing with the PET as a laboratory computer. Two separate tape cassette ports are also provided. The sketch shown in Fig. 4.23 summarises the different types of device that can be attached to the PET microcomputer. Some of these will be described in subsequent chapters.

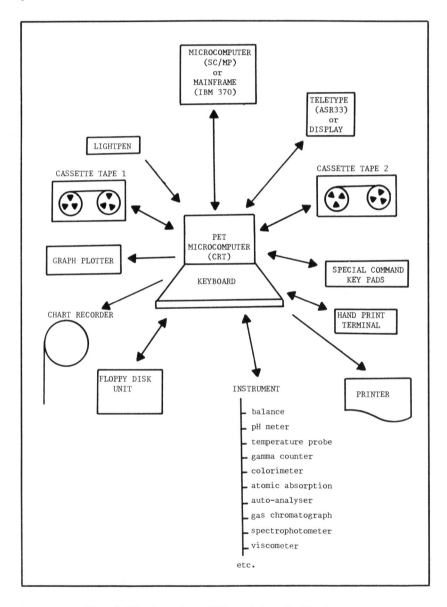

Fig. 4.23 Commodore PET peripheral attachments.

The basic minimum configuration consists of just the PET itself along with
a tape cassette unit to provide a permanent storage medium for programs
and data. Some versions of the PET have a built-in tape cassette unit,
however, these models also have a miniature keyboard in order to provide
suitable space for the tape equipment to be mounted. The cassette tape
can provide a convenient means of sending data and programs between two
different laboratories in machine readable form - where it is not feasible
to have a more sophisticated telecommunicatons link.

A wide range of software is available for the PET. The machine usually
comes supplied with a built-in BASIC interpreter (in ROM) which is the
most commonly used programming language for this micro (COM79, Don80).
Other programming systems are available such as PASCAL, LISP and
Assembler. Unlike BASIC, the PASCAL system is a compiler and so can lead
to the production of programs that run more quickly than those equivalent
ones written in BASIC. LISP is another interpretive language system which
is widely used in artificial intelligence applications. It is useful for
constructing and testing software for controlling robotic devices. For
many applications where these languages are not fast enough then programs
need to be written in assembler using an appropriate development package
(Bar81) or cross-assembler (Lub80). Should these facilities not be
available then the machine code programs need to be hand coded into the
memory using the PET's Terminal Interface Monitor (TIM).

TIM is the rudimentary operating system provided with the PET. When it is
activated - by an appropriate command typed at the keyboard - it provides
the user with the facility to inspect any area of the PET's memory, change
the contents of any RAM location, inspect the registers and execute a
stored program that is resident anywhere in memory. In addition, it
enables the user to save sections of the memory onto tape or disk and
later reload these again. Thus, knowing the format of the 6502's machine
instructions and their hexadecimal operation codes, it is an easy
(although tedious) task to construct machine code programs via the
keyboard. This process becomes much easier if a tape or disk based
assembler package is available.

Many other packages exist for the PET. These are orientated towards a
wide spectrum of application areas. Particularly useful are the system
development tools such as the disassembler (for converting machine code
back into assembler mnemonics) and the BASIC programmer's toolkit. This
latter aid comes in the form of a ROM chip that plugs into one of the ROM
expansion sockets mounted on the PET's PCB (Printed Circuit Board). It is
a valuable aid that greatly increases the speed with which BASIC programs
can be developed. This is achieved by the provision of automatic line
numbering for programs, editing and program debugging facilities. Of
course, there are many other packages available to aid the user with the
storage of results (using a data base package) and the generation of
reports via the use of word processing and report generation software.

Because of their low cost, microcomputers similar to the PET are often
referred to as 'personal computers' since they can be purchased by
individual laboratories for use in particular experiments or to perform
specific dedicated functions. A description of the use of this type of
machine in the analytical laboratory has been presented by Pierce
(Pie81). Talanta (Tal81) also contains an interesting series of articles
devoted to the application of micros in analytical chemistry.

There are many other examples of low cost microcomputer systems, for

example, the Tandy TRS-80, APPLE, Superbrain, and so on. Undoubtedly, the
PET is amongst the more popular of these. In addition to the low cost
machines, there are more expensive systems such as the Hewlett Packard
range (of which the HP-85 is an example) and those from IBM (the 5100
series). However, the increased cost of these often necessitates their
being shared between applications and this can cause difficulties. When
situations of this type arise it is often worthwhile considering the
possibility of upgrading from a microcomputer system (or series of micros)
to a more powerful minicomputer configuration. Some of the properties of
minicomputers will be discussed in the next section.

MINICOMPUTER SYSTEMS

In many ways a minicomputer may be regarded as a scaled up version of a
microcomputer system. A mini might therefore be expected to have,

1. a larger address space,
2. a greater instruction repertoire,
3. facilities to enable instructions to be optimised through
    microcoding,
4. the ability to support a larger number of peripherals,
5. peripherals having a greater capacity and speed,
6. the ability to support simultaneous processes through
    time sharing of the CPU,
7. the capacity to support a greater number of simultaneous
    users,
8. a more sophisticated range of software,
9. greater floor space and power requirements, and,
10. a significantly larger purchase price.

In general, most of these expectations are found to hold but with
exceptions in certain particular cases. For example, some of the more
powerful micros have instruction repertoires that are comparable in scope
and capability to those found in many minis. Similarly, the address space
on some of the smaller minicomputers is no greater than that available on
some of the larger micros. Despite some exceptions like these the above
list is a useful guide to the capabilities of minicomputer systems.

Minicomputers started to be developed in the late 1950s and were designed
mainly for process control applications. Since that time the range of
tasks to which they can be applied and the number of different types that
are available commercially has increased substantially. Today, some of
the more well known laboratory minicomputer systems originate from
manufacturers such as

Hewlett Packard   Digital Equipment Corporation
Perkin Elmer   Wang
General Automation   International Business Machines
PRIME   Data General
Varian   Norsk Data

who together provide a complete range of computational capability.

Of the many different small machines that have been produced one of the
most popular of those designed for laboratory and process control

applications has been the PDP series. The term PDP is an acroynm for
Programmable Data Processor, the series was introduced by their
manufacturer, Digital Equipment Corporation (DEC), in the late 1950s.
Over the years that DEC has been manufacturing this range of computers
many different versions have appeared. They differ both in their
computational capability and their underlying architecture. Some of the
different product renderings within the PDP series are listed below:

LSI-11	PDP-12
PDP-8	PDP-14
PDP-9	PDP-15
PDP-10	PDP-16
PDP-11	

These computers vary in potential from being able to perform simple on-
line control or data logging through to highly sophisticated and powerful
centralised control systems. There is thus a full range of computational
capability for the user or designer to choose from depending upon the
exact nature of the task to be performed.

The LSI-11 is a microcomputer system while the PDP-15 can form the basis
of a sophisticated multi-processor configuration. Historically, there
were PDP-1s, PDP-2s through to PDP-10s and so on. Nowadays, many of these
models are no longer manufactured. Although the PDP-8 family continues to
be one of DEC's most successful and stable product lines, it is the PDP-11
family that has provided the greatest rate of growth and development.
Within this family there are many different models each having a different
cost and performance. Some of these are indicated in Fig. 4.24.

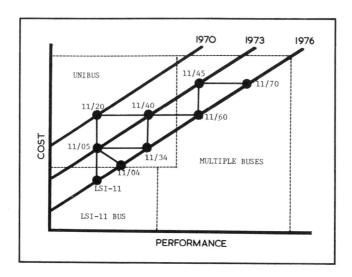

Fig. 4.24 Cost-performance comparison for some PDP-11 computers.

The PDP-11 - Hardware Summary

In general, the members of the PDP-11 family of computers differ in three
basic parameters: processor speed, memory system performance and the

power of the bus structure. By varying these three parameters, this computer series provides program compatible systems ranging from microcomputers (based upon the LSI-11) through to medium scale systems (based upon the PDP-11/70). All processors have a common physical architecture (Dig76a, Dig79) supported by a wide range of peripherals (Dig76c), operating systems and software (Dig76b) to cover many different applications.

One of the main reasons for the success of the PDP-11 is the UNIBUS concept. This is a single, high speed data bus through which all system components and peripherals communicate. Its relationship to the other parts of the system is depicted in Fig. 4.25. The UNIBUS operates in an analogous fashion to the bus systems in a microcomputer - these have been described previously (see Fig. 4.5 and Fig. 4.6). Its function is to carry control information, addresses and data between the various items that are attached to it. Physically, the bus consists of 56 parallel bidirectional lines.

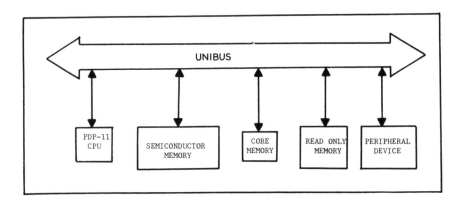

Fig. 4.25 Basic architecture of the PDP-11 minicomputer.

Another important feature of this series is the highly flexible architecture. The modular construction of all PDP-11 computers permits the user to implement a configuration that may be easily expanded as the application requirement grows. Both processors and memories are designed as plug compatible building blocks thereby providing both growth potential and ease of service replacement should the need arise. Using this type of architecture it is an easy matter to plug in new options as they become available; there is rarely any need to design new interfaces.

The attractive hardware features include a series of eight general purpose registers (the larger members of the family have sixteen), a stack capability, a vectored priority interrupt system for fast real-time response and an expandable memory capability. This hardware is controlled by means of an instruction set containing both single and double address instructions (see Fig. 4.27). These provide a means of developing highly efficient programs that manipulate the basic hardware items in optimal ways. A discussion of machine and assembly language programming techiques for the the PDP-11 series will be found in the text book by Gill (Gil78).

The memory organisation of the PDP-11 is illustrated in Fig. 4.26. It is
designed to accommodate both 16-bit words and 8-bit bytes. Words always
start at even numbered locations and are organised as follows:

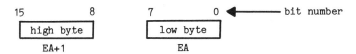

where EA is an even address value. A 16-bit word used for byte addressing
can address a maximum of 32K words. However, the top 4K locations are
reserved for peripheral and register addresses. The user therefore has
available only 28K words. With the larger machines (such as the PDP-11/55
and 11/45) it is possible to expand this limit by means of a special item
of hardware referred to as a Memory Management Unit. This device provides
an 18-bit effective memory address which permits addressing up to 124K
words of actual memory (Dig76a).

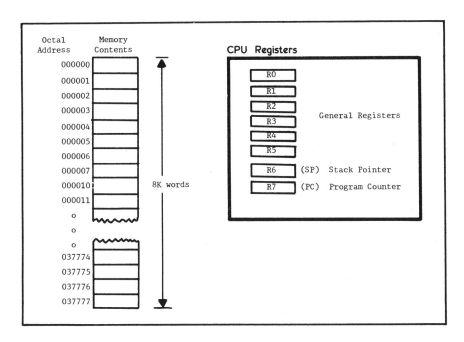

Fig. 4.26   Arrangement of an 8K word (16K byte) memory and
CPU registers for the PDP-11 minicomputer series.

A special part of the memory is used to implement the stack. This is a
temporary data storage area which allows a program to make efficient use
of frequently accessed data. A program can add or delete words or bytes
within the stack by means of simple instructions. The mechanism of
operation of the stack depends upon a 'last in - first out' concept, that
is, various items may be added to the stack in sequential order and then
retrieved or deleted from it in reverse order. Stack facilities are not

SINGLE OPERAND

General

CLR(B)	clear destination	●050DD
COM(B)	complement dst	●051DD
INC(B)	increment dst	●052DD
DEC(B)	decrement dst	●053DD
NEG(B)	negate dst	●054DD
TST(B)	test dst	●057DD

Shift & Rotate

ASR(B)	arithmetic right shift	●062DD
ASL(B)	arithmetic shift left	●063DD
ROR(B)	rotate right	●060DD
ROL(B)	rotate left	●061DD
SWAB	swap bytes	0003DD

Multiple Precision

ADC(B)	add carry	●055DD
SBC(B)	subtract carry	●056DD
SXT	sign extend	0067DD
MFPS	move byte from processor status	1067DD
MTPS	move byte to processor status	1064DD

DOUBLE OPERAND

General

MOV(B)	move source to destination	●1SSDD
CMP(B)	compare source to destination	●2SSDD
ADD	add source to destination	06SSDD
SUB	subtract source from destination	16SSDD

Logical

BIT(B)	bit test	●3SSDD
BIC(B)	bit clear	●4SSDD
BIS(B)	bit set	●5SSDD
XOR	exclusive OR	074RDD

PROGRAM CONTROL

Branch

BR	branch (unconditional)	000400
BNE	branch if not equal (to zero)	001000
BEQ	branch if equal (to zero)	001400
BPL	branch if plus	100000
BMI	branch if minus	100400
BVC	branch if overflow is clear	102000
BVS	branch if overflow is set	102400
BCC	branch if carry is clear	103000
BCS	branch if carry is set	103400

Signed and Conditional Branch

BGE	branch if greater than or equal (to zero)	002000
BLT	branch if less than (zero)	002400
BGT	branch if greater than (zero)	003000
BLE	branch if less than or equal (to zero)	003400

Unsigned Conditional Branch

BHI	branch if higher	101000
BLOS	branch if lower or same	101400
BHIS	branch if higher or same	103000
BLO	branch if lower	103400

Jump & Subroutine

JMP	jump	0001DD
JSR	jump to subroutine	004RDD
RTS	return from subroutine	00020R
MARK	mark	006400
SOB	subtract one and branch (if not zero)	077R00
SPL	set priority level	00023N

Trap & Interrupt

EMT	emulator trap	104000 - 104377
TRAP	trap	104400 - 104777
BPT	breakpoint trap	000003
IOT	input/output trap	000004
RTI	return from interrupt	000002
RTT	return from interrupt	000006

MISCELLANEOUS

HALT	halt	000000
WAIT	wait for interrupt	000001
RESET	reset external bus	000005

Condition Code Operation

CLC, CLV, CLZ, CLN, CCC	clear	000240
SEC, SEV, SEZ, SEN, SCC	set	000260

DD = destination operand of instruction
SS = source operand of instruction
R = register
N = priority level

Fig. 4.27 Instruction set for the PDP-11 series of minicomputers. The leftmost column gives the assembler language instruction mnemonic, the central column its definition and the rightmost column the operation code. The latter are given in octal with ● = 0 for a word and 1 for a byte operation.

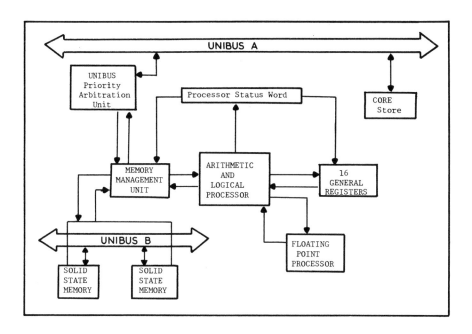

Fig. 4.28   Central processor data paths in a PDP-11/45 minicomputer\

restricted to minicomputer systems since many micros also provide this
feature. Indeed, the MCS 6502 microprocessor upon which the Commodore PET
is based provides a 256 byte stack and four machine code instructions to
facilitate the transfer of data to and from it. In the PDP-11 the user
may use any of the general registers for the purpose of creating a word/
byte stack. One of the registers, R6, is reserved for use in conjunction
with a word orientated stack.

In addition to memory capability, speed of computation is another
important criterion to be considered when choosing a minicomputer configuration. If improving the speed of computation becomes a significant
factor then it is possible to add a special hardware feature called a
Floating-Point Unit to many of the PDP-11 models (Dig76a). Its use can
result in many orders of magnitude improvement in the execution time of
arithmetic operations. This is made possible by overlapping its operation
with the CPU. In machines that do not have this option, floating point
computations have to be performed entirely by software. Addition of a
floating point unit to a system provides additional hardware (six 64-bit
floating point accumulators) that is capable of operating in single and
double precision (32 or 64 bits) floating point modes. In order to

manipulate this hardware additional assembler/machine code instructions
are provided, for example,

ADDF/ADDD	-	Add Floating/Double
LDF/LDD	-	Load Floating/Double
STF/STD	-	Store
SUBF/SUBD	-	Subtract
DIVF/DIVD	-	Divide
MULT/MULD	-	Multiply

and so on.

Combining the features offered by these additional hardware options
enables quite powerful minicomputer systems to be constructed. An example
of a system that contains these additional features is illustrated in Fig.
4.28.

The PDP-11 - Software Summary

In addition to choosing a suitable hardware configuration for a minicomputer it is also important to select an appropriate software system,
The software range from which to choose will often depend upon the
application for which it is intended and the nature of the hardware that
is to be used. Certain types of software will require particular minimal
hardware configurations to be satisfied before they can be usefully
employed. Other software items are designed for situations in which there
is a bare minimum of hardware. This latter option is important because it
is not cost effective having a sophisticated software arrangement on a
hardware system that is only going to be used for simple process control
applications. The user of the PDP-11 series has a wide range of software
options available (Dig76b). The operating systems provided by the
manufacturer include:

CAPS-11	-	cassette based program development system,
RT-11	-	foreground/background or single job operating system,
MUMPS-11	-	data management timesharing system,
RSTS-E	-	general purpose time sharing system,
RSX-11S RSX-11M RSX-11D IAM	}	real time multiprogramming systems.

Some of these are designed for single user systems while others are
orientated towards multiple user environments. In the latter case, the
number of simultaneous users that can be supported will depend upon both
the power of the processor and the amount of memory available within the
hardware configuration being used. Thus, the RSTS-E operating system can
support 24 users when running on a PDP-11/45, 32 users when running on an
11/45 and 63 when an 11/70 processor is used.

In addition to operating system options a variety of programming language environments can be provided. Typically, the most common are

> PAL-11,
> MACRO,
> FOCAL,
> Single User BASIC,
> Multi-user BASIC,
> BASIC-PLUS,
> FORTRAN IV, and,
> COBOL.

This range is continually being extended to meet users' requirements.
Many of the more modern programming languages such as APL, PASCAL, CORAL
and others are also now becoming available.

In order to provide the user with some assistance with the more common
data processing requirements the manufacturer often provides a fairly
comprehensive selection of utility programs. Included amongst these will
be software for file creation, copying and archival; text editors; memory
dump facilities and program debugging tools. Two of the most useful types
of software package that now exist are those aimed at data base management
and computer networking. Data base management packages provide the user
with the facilities necessary to store large amounts of data and then
retrieve and process this data in various ways. Networking packages
enable the user to interconnect computers that are located at different
geographical locations. These topics will be described in more detail
later.

In addition to the software provided by the manufacturer there is also a
variety of programs and packages available through user groups and other
types of software organisations such as systems houses, research centres
and so on. One such example which is becoming increasingly popular is the
UNIX system that has been developed by the Bell Telephone Laboratories
(Rit74, Tho75). UNIX is a general multi-user, interactive operating
system for use on the PDP-11/40 and 11/45 (or equivalent) computers. It
was originally developed for use on the PDP-7 and PDP-9 range of computers
(Bol81) but since these machines are no longer commercially available it
has been redesigned to run on the PDP-11 series. UNIX offers a number of
features seldom found even in much larger operating systems, including:

(1) a hierarchical file system incorporating demountable
     storage volumes,
(2) compatible file, device and inter-process I/O,
(3) the ability to initiate asynchronous processes,
(4) system command language selectable on a per user basis, and,
(5) over 100 subsystems including a dozen different programming
     languages.

Many organisations having PDP-11 computers have found that the UNIX system
offers considerably more facilities than the standard software supplied by
the manufacturer.

The PDP-11 is an extremely versatile minicomputer system having a wide
area of application both within industrial process control and in the
laboratory. Digital's PDP-11 based systems are fully integrated hardware/
software configurations that provide on-line data acquisition and analysis
for applications in the laboratory and in research centres. DEC has
produced many interesting specialised packages aimed at particular

application areas. Typical of these are the DECLAB and GAMMA-11 systems.
The DECLAB package is a hardware/software system that is designed for use
with analytical instrumentation in the laboratory. A choice of several
configurations is available; these offer the researcher a range of power
and capability in data processing, apparatus control and data analysis.
Systems of this type will be discussed in more detail in the chapter on
laboratory automation. The GAMMA-11 package is orientated towards data
handling problems in nuclear medicine. It is based upon a PDP-11/40 which
is linked to a scintillation camera. The combination of these two items
provides an invaluable tool for medical diagnosis. There are many other
interesting applications of the PDP-11 described in the literature
(Cou72).

Larger Minicomputer Systems

Referring back to the list of PDP computers presented earlier it is easy
to see that the PDP-11 is one of the smaller minicomputer ranges offered
by DEC. The larger processors can be used as building blocks to construct
more powerful minicomputer systems to meet a variety of needs. The PDP-15
family, for example, forms the basis of a range of medium scale computers
based upon a multiple processor architecture. The basic machine uses a
word length of 18 bits thereby giving it added capability with respect to
addressing and instruction repertoire. In addition, it uses special
techniques such as instruction look-ahead and memory interleaving in
order to improve performance. The basic PDP-15 architecture is shown in
Fig. 4.29.

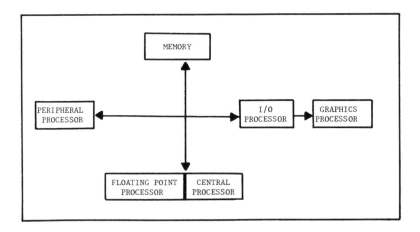

Fig. 4.29   Architecture of the PDP-15 minicomputer.

The basic motivation behind the above type of architecture is the desire
to allow asynchronous operation of as many processing elements as
possible. Thus, many of the functions of the computer (that in conventional systems required assistance from the CPU) are now each handled
by independent processing units each of which operates asynchronously. In
this way the computer is able to allow its system components (the CPU, the
I/O processor, the graphics and floating point processors) to run

independently so that performance is not dictated by the slowest element.
In the more recent models of the PDP-15 the independent autononous processor concept has been applied to the data storage system by adding a
special memory processor similar to that shown in Fig. 4.30. This
feature, combined with instruction look-ahead and memory interleaving
provides a powerful data processing system.

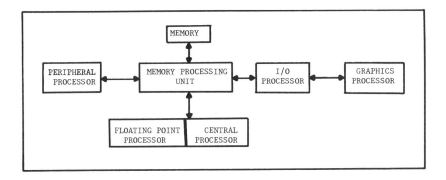

Fig 4.30 Addition of a memory processor to the basic PDP-15
minicomputer architecture in order to improve its performance.

Instruction look-ahead is a facility that attempts to anticipate future
CPU execution requirements. It does this by retrieving instructions from
memory before the CPU is actually ready to execute them. An instruction
that is fetched from memory is stored in a fast access buffer within the
look-ahead unit. When the CPU wants an instruction to execute it obtains
it from this fast buffer rather than from slower memory. While the CPU is
executing a given instruction the look-ahead unit fetches the next one
that the CPU is likely to require.

Storage interleaving is a method for improving the speed and efficiency of
a memory system by keeping more parts of it busy simultaneously. A memory
system of a computer is most often organised in terms of several
independent sections called banks each of which might contain 16K
locations. When the memory is required to read or write a word it becomes
busy for the duration of one memory cycle (about a microsecond). Only
part of the memory cycle involves the computer. Once the memory has been
given a word to store, the computer needs only wait until the memory
receives the information for storage. In reading a word the CPU must wait
about one-half the memory cycle. The rest of the cycle is controlled by
the memory itself without further CPU intervention.

In a conventional non-interleaved store a computer program tends to run in
one area of memory exclusively - see Fig. 4.31a. This means that the CPU
cannot process instructions faster than the memory cycle time. It is the
purpose of interleaving to break this bottleneck. By interleaving the
memory system, consecutive locations are spread across separate memory
banks. This permits the CPU to access one memory section while another is
completing its cycle. If more independent memory banks are involved in
the interleaving scheme, the probability of the CPU attempting to
reference a section that is still busy from a previous operation is

substantially reduced. When a memory is interleaved it is usual to
arrange for two, four, or eight-way interleaving depending upon the type
and amount of memory installed. A 4-way interleaved store is illustrated
in Fig. 4.31b.

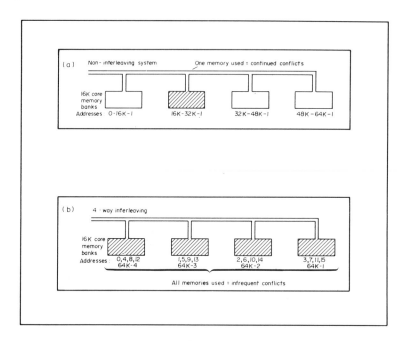

Fig 4.31 Typical memory organisation for a computer system
having (a) no interleaving and (b) a 4-way interleaved store.

One of the more attractive features of the type of computer architecture
inherent in machines similar to those depicted in Fig. 4.29 and Fig. 4.30
is the flexibility that they permit. It is a relatively easy matter to
link processors together to form multi-processor configurations. This is
ideal for computer systems requiring high reliability and improved work
throughput. Some of the possible configurations are depicted in Fig.
4.32. Diagram A shows the basic building block in terms of the three
independent processors, interleaved memory banks and bus system. In
diagram B a PDP-11 has been added to form a dual processor configuration.
The third example, diagram C, shows how the basic building block can be
repeated in a master-slave relation. Each slave has 16K of store in
common with the master system plus its own local storage. A more
sophisticated mode of connection is shown in diagram D which shows a ring
processor arrangement where each processor has memory in common with all
its neighbours.

A typical application of a dual processor configuration involving a GC/MS
analytical system is shown in Fig. 4.33. Here one processor is used to
control the operation of the gas chromatograph and mass spectrometer. In
addition, it is used to acquire data from each of these devices and then
store it in a shared memory bank that can be accessed by the second of the

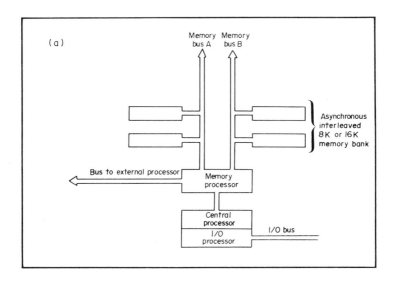

Fig. 4.32A Multiple processor computer configurations.

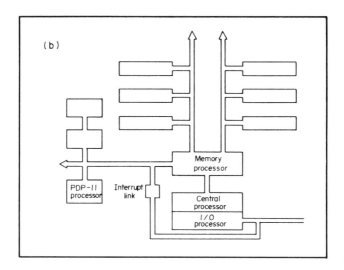

Fig. 4.32B Multiple processor configuration - mixed CPUs.

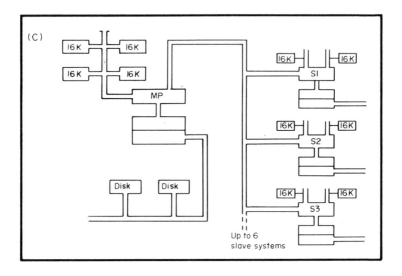

Fig. 32C Multiple processor computer configurations. MP
denotes the master processor and S1 through S3 each denote
slave processors.

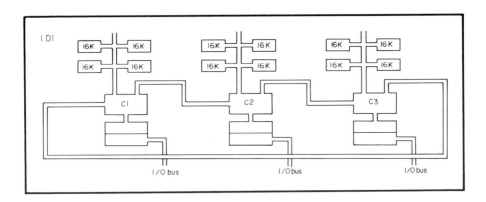

Fig. 32D Multiple processor configurations. In this diagram
C1 through C3 each represent computers in the ring system.

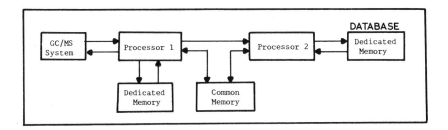

Fig. 4.33 Dual processor computer configuration used as
a basis for a GC/MS analytical system.

processors. This latter processor uses the data as a means of identifying
unknowns by applying pattern recognition techniques. These involve making
comparisons between the unassigned data values and standard reference data
held in mass spectral/retention volume libraries stored in the memory
space of the second processor. There are many applications for multiple
processor systems such as this in which the different CPUs are each
responsible for controlling different parts of a system. In addition,
CPUs can be linked together in order to improve the computation speed.
This type of application will be discussed later in the section on super-
computers.

MAINFRAME COMPUTERS

About thirty years have elapsed since the first commercial electronic
digital computer was made available to the public. The significant
technological advances that have been made since then are reflected in the
three generations of computers that have evolved:

(a) First Generation - based upon electronic valves,
(b) Second Generation - constructed from discrete solid state
     components, and,
(c) Third Generation - employing integrated circuit technology.

These different technologies were briefly outlined at the beginning of
this chapter and are further discussed by Bohl (Boh71).

The first commercially produced computer, called UNIVAC I, was sold to the
public in 1951 and was a first generation machine. Because of its
expense, both to purchase and to run, most organisations could afford to
provide only one machine of this type. This meant that some form of
centralised approach to data processing using a single main machine had to
be adopted. Today, large computers of the second and third generation
have replaced those based upon valves. However, they are still very
expensive resources and often necessitate the provision of a central data
processing service in order to justify their cost. A centralised
expensive and powerful computing configuration of this type is often

referred to as a mainframe computer. Mainframes together with the super-
computers that will be discussed in the next section constitute the class
of machines generally regarded as large computers.

Some of the basic characteristics of large machines are:

(1) long instruction word length (exceeding 32 bits),
(2) large address space and sophisticated memory management
     facilities (such as virtual storage),
(3) powerful instruction set,
(4) high speed processing capability,
(5) capable of supporting a large number of simultaneous users
     running both batch and interactive processing tasks,
(6) provides a wide range of software systems and applications
     packages,
(7) very expensive to purchase and maintain,
(8) usually highly reliable,
(9) supports a wide range of high speed storage and input/output
     peripherals, and,
(10) require teams of highly specialised staff to maintain the
     systems and provide a computing service.

There are very few mainframe manufacturers - mainly because the production
of this type of machine is so costly and can only be undertaken if a
sufficiently large market can be justified. Some of the more well known
producers of large machines are:

            IBM - International Business Machines,
            Honeywell,
            Burroughs,
            CDC - Control Data Corporation,
            DEC - Digital Equipment Corporation, and,
            UNIVAC.

Of the many different product offerings available those which originate
from IBM are amongst the most popular. One class of large machine, the
IBM System/360 (and System/370), has been used to provide mainframe
configurations in many computer installations throughout the world. The
IBM 360/370 computers were designed as general purpose machines for
commercial, scientific communication and control applications. Over the
years a range of different models were produced. These varied both in
processing speed, memory capacity and cost. Figure 4.34 shows a comparison of the relative processing power of some of the different models
in the IBM System 360/370 range.

System/360 are third generation computers and were introduced by IBM in
1964. They were orientated towards centralised batch processing
operations. Some of the many models that were produced were configured
for particular types of application, for example, fast scientific
numerical computation, high volume input/output transaction processing and
so on. These differences could be achieved through the technique of
microcoding (Boh71) whereby the implementation of the machine's basic
instruction set could be modified and optimised depending upon the type of
application to which it was to be applied. The System/370 range was
introduced in 1970 and represented an evolutionary development of
System/360 rather than an entirely novel system. The range was upwards
compatible with the 360 series in that programs written for this range
could be run on 370 machines without modification. Many new features

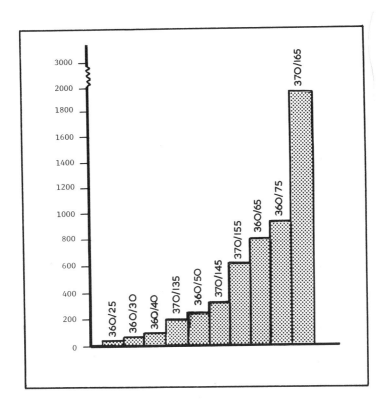

Fig. 4.34 Comparison of the processing power of IBM 360 and
370 Series computers. The unit for the vertical axis is POWU
(Post Office Work Units) per second.

appeared in the 370 architecture including virtual storage via paging
(Boh71) and various facilities that enabled CPUs to be interconnected in
order to produce highly reliable fast throughput computing systems. Two
broad types of connections are possible: tightly coupled, in which there
is a direct CPU-CPU link; and, loosely coupled, whereby the CPUs
communicate via shared memory. The organisation of a typical example of a
twin processor IBM 360/370 configuration (NUM77) is illustrated in Fig.
4.35. This is an interesting configuration in that each CPU provides an
entirely different function. One of the CPUs supports batch processing
via standard IBM software while the second provides both a batch
processing facility and an interactive terminal environment. The latter
service is not based upon standard IBM software but instead relies upon a
system developed at the University of Michigan and referred to as MTS - an
acronym for Michigan Terminal System (Boe75, Pir75).

A detailed description of the design philosophy of the System/360 range of
computers has been given by Blaaw (Bla64) and their architecture has been
outlined by Amdahl (Amd64). Details of their mode of operation, the

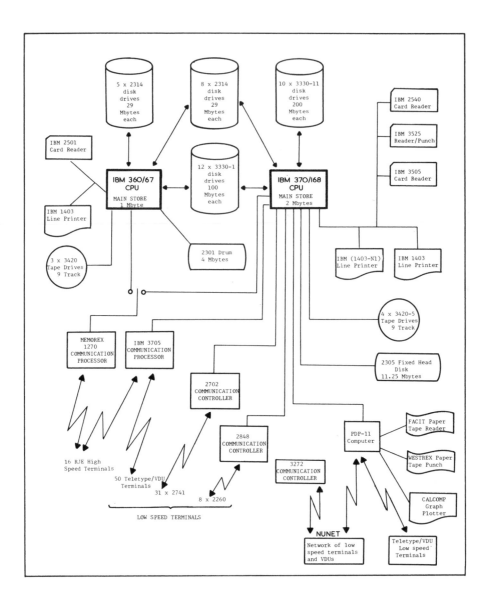

Fig. 4.35   A typical twin CPU mainframe configuration.

facilities they provide and other related technical specifications are
contained in the wide range of documentation that is available from the
manufacturer (IBM72, IBM77). Assembler language programming is dealt with
in a book by Struble (Str75). This gives a detailed description of the
machine instructions available, their format and the way in which they

control the hardware. A less detailed introduction to these topics will
be found in Bohl's book (Boh71).

The main storage of the IBM 360/370 is organised into bytes each of which
consists of eight bits. Each byte in storage has an address and the
capacity of main storage ranges from 8192 in the smaller models to
2,097,152 bytes or more on the larger ones. For arithmetic and logical
operations the bytes are grouped into words of four bytes each, half-words
of two bytes and double-words of eight bytes each. In addition to the
main storge there is a set of 16 general purpose registers, each 32 bits
in length. These registers are used in specifying addresses of operands
and in performing binary integer arithmetic and most logical operations;
their addresses are 0 through 15. Four further registers (each 64 bits in
length) are used for floating point arithmetic and have addresses 0, 2, 4,
and 6. The context of an address in an instruction determines whether it
refers to a floating point register, a general register or a location in
main storage. Implicit in the operation code for each instruction are the
addressing characteristics of each of its operands.

Arithmetic can be performed on numbers that are stored according to any of
three different representations:

> binary integer   -   these are numbers represented in binary
>   form with lengths of 16 and 32 bits,
>
> floating point   -   in which exponents and fractions are both
>   represented in binary sharing 32-bit,
>   64-bit or 128-bit storage areas, and,
>
> decimal mode   -   wherein each decimal digit is represented
>   as four bits (binary coded decimal or BCD)
>   so that two digits can be packed into one
>   byte. Decimal arithmetic can be performed
>   on numbers varying in length from 1 to 16
>   digits.

For the majority of scientific computations the numeric storage
capabilities, arithmetic processing modes and the precision with which
calculations can be made will be of significant importance. This is
particularly the case in situations where complex simulations or modelling
exercises are to be undertaken (see chapter 9). Unless the processing
power of the computer is capable of handling high speed computation it
becomes impossible to perform realistic simulations in reasonable
processing times.

In addition to the more sophisticated hardware available on mainframe
computers there is also an extremely wide range of software products.
These include a large number of different language translators (such as
BASIC, FORTRAN, COBOL, PL/I, PASCAL, CORAL, APL, ALGOL, LISP, and so on)
file editing software and applications packages for data management, data
analysis, data processing and simulation. Most scientific mainframe
systems will also provide a number of cross-compilers and cross-assemblers
that simulate products that are available on various microcomputer systems. These facilities enable software development and testing for microcomputer applications to be undertaken on a mainframe machine. This is an
extremely useful facility where it is important to minimise the development time and effort involved in implementing a microprocessor application. Once the software has been developed on the mainframe it can be

transferred to the micro system where it immediately becomes operational.

One of the most useful features of the mainframe approach to computing lies in the substantial repertoire of application packages that they are able to support. Many of these require substantial processing power and large amounts of storage and so could not easily be made available on smaller computing systems. Some examples of the comprehensive range of software packages available at a typical scientific/engineering mainframe computer installation (NUM77) are listed in Table 4.7.

TABLE 4.7 Application Packages Available on a Mainframe Computer

ATMOL	atomic and molecular orbital calculations
FACSIMILE	simulation of chemical processes
ORTEP	crystal structure plotting
XRAY	X-ray analysis
APT	automatic programming of tools
ECAP	electronic circuit analysis package
GENESYS	generalised engineering system
ICES	integrated civil engineering system
PAFEC	finite element analysis
SPSS	statistical processing
MIDAS	data analysis
GPSS	general purpose simulation
CSMP	continuous system modelling
DYNAMO	continuous system simulation
CLUSTAN	cluster analysis
GINO-F	graphics package
GHOST	graphics package
NAG	numerical algorithms library
MPSX	mathematical programming system
KWIC	keyword in context retrieval
SPIRES	data base management system
TEXTFORM	word processing package
FORMAT	word procesing package
RPG and RG	report generation systems

The number of packages of the type listed in Table 4.7 continues to increase - particularly in the areas of data management, analysis and retrieval. Sometimes these packages are orientated towards particular types of mainframe computer and, although they are often written in high level language, they are difficult to convert to run on other machines. Consequently, in order to make these packages available to a wider range of users many establishments have shown considerable interest in computer networking (Bla76) involving the linking together of different geo-graphically remote computer systems via suitable telecommunication links. These links may vary in sophistication from simple twisted wire pairs through to more advanced fibre optic systems and communication satellites. Some of these different types of link will be discussed in more detail in chapter 7. Similarly, because of their future importance, a more detailed analysis of computer networks and some application descriptions will be given in chapter 12.

Mainframe computers provide substantially greater computing power than either minicomputers or microsystems are capable of providing. However, despite the processing speeds they are able to attain, there are still many applications for which the power of the mainframe is itself insufficient. Applications that need extremely high computation rates in excess of those available through a mainframe often necessitate the design and construction of special purpose large computers referred to as super-computers. This class of large machine is discussed in the next section.

SUPER-COMPUTERS

The class of large machines referred to as super-computers has evolved as a result of the inadequacy of general purpose mainframe computers for certain types of high speed computation. In order to overcome the often severe limitations of a single CPU, attempts are sometimes made to increase the computational resources   connecting together several CPUs to form a multiple processor unit.

A multiprocessor configuration (Sat80) is one which consists of at least two processors satisfying the following conditions:

(a) the processors share global memory,
(b) they are not highly specialised and
(c) each of the processors can do significant computation individually.

A survey of multiprocessing techniques has been presented by Enslow (Ens77) and a description of the architectural strategies underlying five commercially available multiprocessor systems has been given by Satanarayanan(Sat80) who has produced an annotated bibliography on the topic. Multiprocessing systems have been developed for a variety of reasons ranging from increased reliablity through to enhanced work throughput - compared with a conventional single mainframe. It is this latter facility which is of present interest as it forms the basis for the discussion of super-computers.

Unfortunately, adding two CPUs together does not produce twice the computational speed of either one. The available throughput from dual processors is only about 1.6 to 1.8 times that of a single processor. Adding more processors (triples and quadruples) produces even lower net gains. Examination of the reasons for the relatively poor performance of this type of configuration shows that the lost power is consumed in coordination of the various concurrent activities and in preventing clashes. However, it is possible to construct multiple processor configurations that do not require this overhead by constraining the individual processors to perform the same operation simultaneously. Such systems are called array processors or super-computers. Generally, they consist of an array of identical CPUs all working synchronously under the control of a master controlling processor.

Super-computers (and array processors) consist of a large number of individual processing elements interlinked in such a way that they are able to support a large number of parallel processes. These processes are synchronised and controlled in ways that enable the super-computers to gain considerable speed increases compared with a conventional mainframe system. In order to illustrate how this is achieved, consider the addition of two 100 element vectors. A conventional computer would undertake the task by performing 100 consecutive sequential ADD operations

- one for each pair of corresponding elements. In a super-computer the computational process could be performed much more quickly since the vector addition would be achieved by performing 100 parallel (or simultaneous) addition operations thereby producing a speed increase of about a factor of 100. Such considerations as this are important in large, frequently used data processing tasks (for example, crystallographic problems), data retrieval from large collections of experimental data (via pattern matching techniques) and simulation experiments involving large numbers of entities (such as simulation of molecular effects for molecules containing 100-1000 atoms).

Many parallel processing systems have been proposed and a considerable number of these actually constructed (Ens74) for use in pattern recognition, associative processing, optical processing, maximum likelihood calculations, signal processing and the solution of coupled differential equations. A research report describing large-scale vector/array processors has been written by Paul (Pau78). It attempts to classify systems of this type according to architecture and machine organisation. Detailed descriptions of the mode of operation of several of the more famous super-computer systems are given.

Examples of array processors/super-computers include the Distributed Array Processor (DAP) produced by the UK's International Computers (Gos79), the ILLIAC IV produced by Burroughs (Slo71), PEPE (Cra72), C.mmp (Wul72), CLIP-IV (Duf76) and CRAY-I (Met78). The speeds of array processors are expressed in terms of MIPS (Millions of Instructions Per Second) or MFLOPS (Millions of Floating-point Operations Per Second); these two units are used interchangeably. A comparison of some of the speeds of various computers is shown in Table 4.8.

TABLE 4.8   Comparative Speeds of Some Super-computers

SYSTEM	SPEED (MFLOPS)		Reference
	Maximum	Average	
Attached Processor AP 120/190	12	5.9	Sug80
CRAY-I	160	23.5	Sug80
STAR-100	40	16.8	Sug80
CDC-7600	10	3.3	Sug80
ILLIAC IV	80	9.1	Sug80
DEC PDP 11/70	0.2	0.1	Sug80
IBM 370/168	3.0	0.75	Sug80
DEC LSI/11 microcomputer		0.02	Lyk75
IBM 370/158		0.4	Lyk75
IBM 360/65		0.6	Lyk75
IBM 360/195		6.0	Lyk75

In Table 4.8 both maximum and average speeds are quoted for the results taken from Sugarman (Sug80). Unfortunately, Lykos (Lyk75) does not specify whether the values cited represent average or maximum speeds.

Of the super-computers listed above, the most famous is probably the ILLIAC IV. This machine was built by the University of Illinois in conjunction with the Burroughs Corporation almost ten years ago. It consists of an ensemble of 64 'slave' computers capable of executing between 100 and 200 million instructions per second - recent advances in software design techniques have raised this figure to 300 MIPS. Each of the processing elements of ILLIAC IV is a powerful computing unit in its own right. It can perform a wide range of arithmetical operations on numbers that are 64 binary digits long; these numbers may be in any of six different formats. Each of the processing elements has its own memory capable of storing 2048 64-bit numbers.

The C.mmp system (Carnegie Mellon Multiple Mini Processor) consists of a symmetric set of up of 16 minicomputers (Dec PDP-11s) with an equivalent set of memory units all of which are interconnected through a cross-bar switch. This arrangement enables any processor to access any memory. In addition, up to 16 separate and simultaneous processor memory connections are possible.

PEPE - Parallel Element Processing Ensemble - is a highly parallel machine consisting of a general purpose computer (CDC 7600) and many processing elements. The system was designed by the US System Development Corporation and built by Burroughs. It is used by the US government for the analysis of ballistic missile defence radar data.

The design of a special purpose multiprocessor system called NEWTON has been described by Wilson (Wil75). This system is being constructed in order to help the chemist or molecular biologist investigate and understand the molecular dynamics, detailed time evolution and structure of biomolecular systems. The computer is to be used to find the set of force functions which describe inter-atomic forces as functions of atomic positions in an N-atom biomolecular system. Significant amounts of computation are involved in solving the equations that describe these chemical systems. Consequently, high speeds of computation are required if they are to be solved in a reasonable time scale. Typical speeds anticipated for the NEWTON system are in the region of 100 MIPS.

One of the major objectives of the super-computer is to produce high speeds of computation. In the majority of machines this has been achieved through some form of parallelism. An alternative approach to the speed problem involves the utilisation of a new type of technology - the Josephson Junction. Conservative estimates (Mat80) of the speed of a computer based on this type of technology indicate that speeds in the order of 100 million instructions per second might become available for general purpose computers. However, until this type of technology is harnessed commercially, many computer manufacturers will continue to build super-computers for those applications that warrant their use.

The current status and likely future trends of large computers of this type are discussed in a series of articles ('Supersystems for the 80s') contained in a special issue of the IEEE magazine COMPUTER (IEE80). A description of some of the application areas of super-power computers has been presented by Sugarman (Sug80) who suggests that one of the most important areas for scientific computers of this type is simulation and modelling (see chapter 9). It is envisaged that during the 1980s computers having speeds in excess of 1000 MIPS will be needed for many of the applications that are now envisaged. Further details on a variety of super-computers and multi-processing systems will be found in a report edited by White (Whi76).

CONCLUSION

There are many different types of digital electronic computer. They range in power from small microcomputers through to highly sophisticated super-computer systems. Each different type of machine is best suited to a particular type of application. In applying this technology to the solution of computational problems it is important to match requirement against processing capability. Having done this it is then a simple matter to choose a machine (or combination of machines) that is most capable of meeting the computational requirement. Although computers differ in their sizes and speeds the fundamental principles upon which their operation is based do not usually depend upon these parameters.

When assessing computational requirements it is important to remember that software considerations are as important as hardware factors. While the hardware for a particular application must be appropriate so must the software. Consequently, it is best to think in terms of the optimal solution to a laboratory computing problem in which the most appropriate combination of hardware and software is selected.

The advent of the microcomputer and its suitability for free-standing dedicated applications now enables considerable intelligence to be built into laboratory instruments and machines of various sorts via the intro-duction of a CPU, memory and stored programs. Indeed, the introduction of microprocessors into analytical instruments - ranging from an automatic pipette through to a GC/MS system - enable these tools to become increasingly smart (Mau78) making them easier to use, more reliable, and often, much safer. The falling costs of minicomputers and mainframes (and associated peripheral hardware) has enabled two substantial developments to take place: firstly, more and more laboratories are acquiring their own minicomputers to help administer and organise the running of the laboratory - this is particularly the case in large scale laboratories that employ a substantial number of scientists and technical support staff; secondly, an increasing amount of computer storage is becoming available within the laboratory to enable the automatic acquisition of large volumes of experimental data (see chapter 5) and its processing and manipulation using readily available packages (see chapter 9). Cheap readily available storage also enables the provision of a wide range of facilities that were hitherto not possible. Typical of these are the storage of experimental methods and laboratory safety/emergency procedures in on-line computer systems that can be interrogated via suitable terminal devices.

Two important trends appear to be of future importance. Both involve some arrangement of multiple CPUs. The first requires their localisation, the second their distribution. In this chapter the first of these has been discussed. The improved performance of multiple processors has been outlined. In view of its potential data processing capabilities, special purpose hardware (in the form of super-computers and array processors) is likely to assume an increasingly important role in the future. These devices process data at speeds which greatly exceed those possible with conventional mainframes. Consequently, they will enable many important tasks to be undertaken that were previously not feasible. The second mode of CPU interconnection forms the basis of distributed computing. This involves connecting together computers that are located at widely different geographical locations. The falling cost of communication facilities coupled with the greater degree of standardisation in the area of computer interfaces will mean that this mode of CPU connection will

become much easier. Using these techniques sophisticated computer networks containing micros, minis, mainframes and super-computers can be
constructed (Lyk75, Bla76). This approach to computer interconnection
will be discussed in more detail in chapter 12.

REFERENCES

Amd64   Amdahl, G.M., Blaauw, G.A. and Brooks, F.P., Architecture of the
         IBM System/360, IBM Journal of Research and Development,
         Volume 8, 87-101, April 1964.

Bar81   Barker, P.G., A Comparison of Sort Times Using a
         Microcomputer, Interactive Systems Research Group, Department
         of Computer Science, Teesside Polytechnic, County Cleveland,
         UK., February 1981.

Bla64   Blaauw, G.A., Brooks, F.P., Stevens, W.Y., Amdahl, G.M. and
         Padegs, A., The Structure of System/360, IBM Systems Journal,
         Volume 3, No. 2, 119-196, 1964.

Bla76   Blanc, R.P. and Cotton, I.W., Computer Networking, IEEE
         Publication No. 36, IEEE Press, 1976.

Boe75   Boettner, D.W. and Alexander, M.T., The Michigan Terminal
         System, Proceedings of the IEEE, Volume 63, No. 6, 912-918,
         June 1975.

Boh71   Bohl, M., Information Processing, (2nd Ed.) Science Research
         Associates, Inc., ISBN: 0-574-21040-7, 1971.

Bol81   Boldyreff, C., With UNIX You can Compute Without Programming,
         Practical Computing, Volume 4, Issue 2, 77-78, February 1981.

Cam79   Camp, R.C., Smay, T.A. and Triska, C.J., Microprocessor Systems
         Engineering, Matrix Publishers, Inc., Portland, Oregon, ISBN:
         0-916460-26-6, 1979.

COM79   COMMODORE Business Machines, PET User's Manual, Publication
         No. 320856-3, June 1979.

Cou72   Coury, F.F., A Practical Guide to Minicomputer Applications,
         IEEE Press, ISBN: 0-87942-0005-7, 1972.

Cra72   Crane, B.A., Gilmartin, M.J., Huttenhoff, J.H., Rux, P.T. and
         Shively, R.R., PEPE Computer Architecture, IEEE COMPCON '72,
         57-60, September 1972.

Dal80   Dalle-Molle, R. and Defresse, J.D., A Microprocessor based Mono-
         chromator Controller, Journal of Automatic Chemistry, Volume
         2, No. 2, 76-80, ISSN: 0142-0453, April 1980.

Dig76a   Digital Equipment Corporation, PDP/11 - 04/34/45/55 Processor
         Handbook, 1976.

Dig76b   Digital Equipment Corporation, PDP-11 Software Handbook, 1976.

Dig76c   Digital Equipment Corporation, PDP-11 Peripherals Handbook,
          1976.

Dig79   Digital Equipment Corporation, PDP-11 Processor Handbook,1979.

Don80   Donahue, C.S. and Enger, J.K., PET/CBM Personal Computer
          Guide, Osborne/McGraw-Hill, ISBN: 0-931988-30-6, 1980.

Duf76   Duff, M.J.B., CLIP - An Array Processor for Image Processing,
          191-203, in Multiprocessor Systems, (Ed., White, C.H.),
          Infotech International, ISBN: 8553-9290-8, 1976.

Ens74   Enslow, P.H. (Ed.), Multiprocessors and Parallel Processing,
          John Wiley and Sons, New York, 1974.

Ens77   Enslow, P.H., Multiprocessor Organisation - A Survey, ACM
          Computing Surveys, Volume 9, No. 1, 103-129, 1977.

Gil78   Gill, A., Machine and Assembly Language Programming of the
          PDP-11, ISBN: 13-541870-1, 1978.

Gos79   Gostick, R.W., Software and Algorithms for the Distributed Array
          Processor, ICL Technical Journal, Volume 1, No. 2, 116-135,
          1979.

Hea76   Healey, M., Minicomputers and Microprocessors, Hodder and
          Stoughton, ISBN: 0-340-20113-4, 1976.

Hea78   Hartley, M.G. and Healey, M., A First Course in Computer
          Technology, McGraw-Hill, ISBN: 0-07-084080-6, 1978.

Hor79   Hordeski, M.F., Microprocessor Cookbook, Tab Books, Blue Ridge
          Summitt, Pa 17214, ISBN: 0-8306-1053-7, 1979.

IBM72   IBM Corporation, IBM System/360 and System/370 Bibliography,
          Form No: GA22-6822-18, February 1972.

IBM77   IBM Corporation, IBM System/370 Bibliography, Form:
          GC20-0001, September 1977.

IEE80   Institute of Electrical and Electronic Engineers, Inc., Super-
          systems for the 80s, IEEE COMPUTER, Volume 13, No. 11,
          November 1980.

Jer77   Jermyn Ltd., Sevenoaks, Kent, UK., Electronic Components
          Catalogue, page 319, 1977.

Lan79   Lancaster, D., TTL Cookbook, Howard W. Sams, Inc., ISBN:
          0-672-21035-5, 1979.

Lav76   Lavington, S.H., Processor Architecture, NCC Publications,
          ISBN: 0-85012-154-X, 1976.

Lip78   Lippiatt, A.G., The Architecture of Small Computer Systems,
          Prentice-Hall International, ISBN: 0-13-044750-1, 1978.

Lub80   Lubbers, C.E., Microprocessor Cross Assemblers User's Guide,
          University of Michigan Computing Centre, February 1980.

Lyk75   Lykos, P., Computer Networking in Chemistry, American Chemical
         Society Symposium Series, No. 19, ISBN: 8412-0301-6, 1975.

Mat80   Matisoo, J., The Superconducting Computer, Scientific
         American, Volume 242, No. 5, 38-53, May 1980.

Mau78   Maugh, T.H., Microprocessors: more Instruments are becoming
         Smart, SCIENCE, Volume 199, No.4335, 1323-1324, 24th March
         1978.

Met78   Metz, W.D., Midwest Computer Architect Struggles with Speed of
         Light, SCIENCE, Volume 199, No. 4327, 404-409, 27th January
         1978.

Mon74   Monro, D.M., Interactive Computing with BASIC - A First
         Course, Edward Arnold, ISBN: 0-7131-2488-1, 1974.

Mon78   Monro, D.M., Basic BASIC - An Introduction to Programming,
         Edward Arnold, ISBN: 0-7131-2732-5, 1978.

Mor80   Morley, F., The Use of a Simple 8-bit Micro as a Flexible
         Sequence Controller for Developing Laboratory Automation,
         Journal of Automatic Chemistry, Volume 2, No. 1, 19-21, ISSN:
         0142-0453, January 1980.

MOS76   MOS Technology Inc., MCS6500 Microcomputer Family Hardware
         Manual, Publication No. 6500-10A, January 1976.

Nat76   National Semiconductor Corporation, SC/MP Technical
         Description, Publication No. 4200079B, September 1976.

Nat76a   National Semiconductor Corporation, SC/MP Low Cost Development
         Centre, Publication No. 42001005A, November 1976.

Nat76b   National Semiconductor Corporation, SC/MP Microprocessor
         Applications Handbook, Publication No. 420305239-001A, Oct.1976

Nat77   National Semiconductor Corporation, SC/MP Microprocessor
         Assembly Language Programming Manual, Publication No. 4200094C,
         January 1977.

NUM77   NUMAC Documentation Group, Introduction to NUMAC, University
         of Newcastle Upon Tyne, UK, September 1977.

Osb78a   Osborne, A., An Introduction to Microcomputers: Volume 1 -
         Basic Concepts, Osborne and Associates, Inc., ISBN:
         0-931988-02-0, 1976.

Osb78b   Osborne, A. and Kane, J., An Introduction to Microcomputers:
         Volume 2 - Some Real Microprocessors, Osborne and Associates,
         Inc., ISBN: 0-931988-15-0, 1978.

Osb78c   Osborne, A. and Kane, J., An Introduction to Microcomputers:
         Volume 3 - Some Real Support Devices, Osborne & Associates,
         Inc., ISBN: 0-931988-18-7, 1978.

178   Computers in Analytical Chemistry

Pau78   Paul, G., *Large Scale Vector/Array Processors*, IBM Research
           Report, RC7306, (31432), Systems Technology Dept., T.J. Watson
           Research Centre, Yorktown Heights, NY10598, September 1978.

Pea77   Peatman, J.B., *Microcomputer-Based Design*, McGraw-Hill
           Kogakusha Ltd., ISBN: 0-07-049138-0, 1977.

Pie81   Pierce, T.B., Newton, D.A. and Huddleston, J., Microcomputer
           Design for Analytical Research, *Chemistry in Britain*, Volume
           17, No. 3, 122-129, March 1981.

Pir75   Pirkola, G.C., A File System for a General Purpose Time Sharing
           System, *Proceedings of the IEEE*, Volume 63, No. 6, 918-924,
           June 1975.

Rit74   Ritchie, D.M. and Thompson, K., The UNIX Time Sharing System,
           Comm. ACM., Volume 17, No. 7 365-375, July 1974.

Sat80a   Satyanarayanan, M., Commercial Multiprocessing Systems, *IEEE
           COMPUTER*, Volume 13, No. 5, 75-96, May 1980.

Sat80b   Satyanarayanan, M., Multiprocessing - An Annotated Biblio-
           graphy, *IEEE COMPUTER*, Volume 13, No. 5, 101-116, May 1980.

Slo71   Slotnick, D.L., The Fastest Computer, *Scientific American*,
           Volume 224, No. 2, 76-87, February 1971.

Spe75   Spencer, D.D., *A Guide to BASIC Programming*, Addison-Wesley
           Publishing Company, ISBN: 0-201-07106-1, 1975.

Str75   Struble, G.W., *Assembler Language Programming: The IBM System/
           360 and 370*, Addison-Wesley, ISBN: 0-201-07322-6, 1975.

Sug80   Sugarman, R., 'Superpower' Computers, *IEEE SPECTRUM*, Volume
           17, No. 4, 28-34, April 1980.

Sup80   Intertec Data Systems Products, Corporate Headquarters, 2300
           Broad River Road, Columbia, South Carolina, 29210, USA.,
           SUPERBRAIN Microcomputer Technical Specification, 1980.

Tal81   Various authors, Microprocessors in Analytical Chemistry,
           Talanta, Volume 28, No. 7B, 487-546, July 1981.

Tay80   Taylor, D. and Morgan, L., *High Level Languages for Micro-
           processor Projects*, National Computing Centre Publications,
           United Kingdom, ISBN: 0-85012-233-3, 1980.

Tho75   Thompson, K. and Ritchie, D.M., *UNIX Programmer's Manual*, (6th
           Ed.), Bell Telephone Laboratories, May 1975.

Whi76   White, C.H., (Ed.), *Multiprocessor Systems*, Infotech Inter-
           national, ISBN: 8553-9290-8, 1976.

Wil75   Wilson, K.R., Multiprocessor Molecular Mechanics, 17-52, in
           Computer Networking in Chemistry, (Ed., Lykos, P.), American
           Chemical Society Symposium Series, No. 19, ISBN: 8412-0301-6,
           1975.

Wul72   Wulf, W.A. and Bell, C.G., C.mmp - A Multi-Mini-Processor,
           AFIPS Conference Proceedings, (FJCC), Volume 41, Part II,
           765-777, 1972.

5

Data Collection

INTRODUCTION

Most analytical investigations require the analyst to make one or more measurements of several different experimental variables. On some occasions these measurements may be just routine observations of the values of standard process conditions used to control or monitor some industrial or manufacturing process. At other times the analyst may be more interested in studying the changes in these variables in response to variations in selected control parameters. The periods for which observations are made may vary from minutes or hours to weeks, months or years. Such observations usually produce large quantities of experimental data. The process of collecting this data, be it manual or automatic, is called data collection.

In the very early days of analytical chemistry the human eye, the brain, a pencil and a notebook were all that were available to record the results of analytical experiments. This type of manual data collection system is still in use today in a large number of laboratories for those experiments to which the technique is suited. However, there are some significant drawbacks to the use of manual procedures for the collection of data:

    (i)   automatic or mechanical techniques may be more cost effective,

    (ii)   there may be significant problems of reproducibility of results from operator to operator,

  (iii)   the human is limited in the rate at which he/she is able to handle data recording operations,

   (iv) the human is prone to making transcription errors and may also be influenced by distracting stimulii, and,

    (v)   a human operator can become bored and tired when involved in handling large numbers of observations over an extended time period.

As a means of overcoming some of the above problems arising from human limitations in data recording, there have been significant developments in all types of scientific measuring equipment. Over the years there has

been a substantial improvement in the facilities that instruments provide
in order to facilitate data capture. Nowadays, many can automatically
record data, process it electronically and then display the final results
on an appropriate display device such as a strip chart recorder. Because
of the increasingly frequent need to be able to send this data to other
remote locations, store it for extended periods of time and process it in
various ways, many modern instruments store all the acquired data
electronically. This is achieved through the use of appropriate memory
systems controlled by embedded on-board microcomputers. Final results are
then displayed on CRT devices rather than on conventional 'hard-copy'
units. These types of developments in instruments were briefly described
in Chapter 3.

Data collection is referred to by a variety of other closely related
names. Two of the more commonly used synonyms are 'data logging' and
'data acquisition'. Each of these terms embodies the idea of collecting
data associated with some observable process. For example, this may
involve measuring the values of process variables in a manufacturing plant
or recording environmental parameters in a pollution monitoring
experiment. In both cases the basic underlying process is the same. This
is depicted schematically in Fig. 5.1.

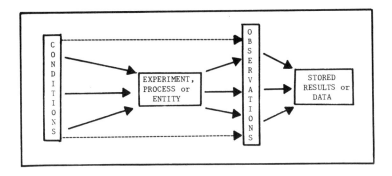

Fig. 5.1 Basic model for the data collection process.

The purpose of data logging is thus to record changes in selected
dependent and control variables as a function of time. In order that the
process variables may later be analysed for trends and dependency upon
process conditions, the latter must also be recorded along with the
process data. Of course, in 'real-time' control applications the results
of the observations must be acted upon spontaneously. The information
derived from the data must be used in such a way that the conditions
affecting the process can be changed so as to 'steer' the process in the
direction that it has to proceed in order to achieve particular predefined
outcomes.

The design of data collection systems can be an involved and specialised
process. Frequently the steps required will be determined by the nature
of the application that is the subject of the investigation. Laboratory
systems will probably have different characteristics from those of systems
intended for non-laboratory process monitoring situations. The latter

will also show considerable variety depending upon the environmental
conditions that are likely to prevail and the characteristics of the data
to be collected. Some approaches to different data collection systems
will be discussed later in the chapter.

Of the many different possible approaches to data acquisition, two broad
categories exist: manual and automatic. Manual systems will not be
discussed in any detail even though the data acquired might ultimately be
processed by a computer or introduced into a computer system for storage
purposes. This type of system is a special case of the simple model
presented in Fig. 5.2.

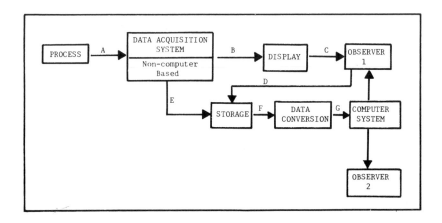

Fig. 5.2 Data flow involved in recording scientific measurements.

The data flow depicted by the sequence A,B,C,D, ... denotes manual data
collection. The values produced by the instruments are recorded manually
by the observer. This may be achieved by storing the results in written
form in a laboratory notebook or by entering them into a desk-top computer
system using a simple keyboard device. Addition of an outlet port to the
instrument enables a simple form of automatic data recording (path E) to
be implemented. In this situation a paper tape punch, a cassette tape or
flexible disk might be used - these will be described in more detail later
in this chapter. Data recorded in this fashion can later be processed by
a computer system.

The greater part of this chapter is concerned with the more highly
automated aspects of data collection that involve (or necessitate) the use
of some form of computer. This will usually form an integral part of the
acquisition equipment - either for control purposes or to aid the data
collection itself. A typical arrangement for such a computer based system
is illustrated in Fig. 5.3.

In this type of system the instrumentation that forms the basis of the
data acquisition unit will probably contain its own internal memory for
use during data capture. This may also be available to aid the processing
of results prior to their display. The external memory shown in the
diagram will provide a long term storage facility for the results. This

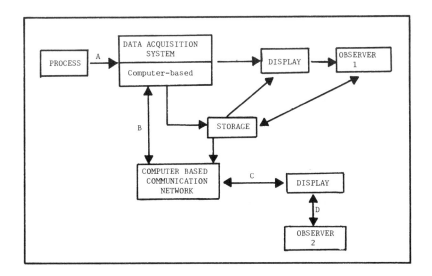

Fig. 5.3   Computer based data acquisition system.

memory may be hierarchically organised if the application warrants this
type of approach. At any post-analysis stage the observers are able to
manipulate the stored data in many different ways - perhaps replotting it
with different scale factors or after applying some statistical
corrections, etc. Pathway A,B,C,D enables a remote user to receive the
experimental results, and, if need be, control the data acquisition
equipment from a remote location. Equipment of this type has great
potential in a significant number of automated data collection
applications.

SOURCES AND TYPES OF DATA

(A) Sources of Data

Data can originate from a wide variety of sources and can be produced by
manual or automated methods. On-line data collection is nowadays a
routine operation in a large number of laboratories. Various instruments
have been used for this purpose: pH meters and ion sensitive electrodes,
mass spectrometers, gas chromatographs, UV/VIS spectrophotometers, NMR and
X-ray spectrometers, radiation counters, polarographs, thermal equipment,
electronic balances and a host of others. It was emphasised in Chapter 3
that one of the most fundamental parts of an instrument is its detection
system. This provides the basis for measurement and usually determines
many important instrumental parameters such as sensitivity, selectivity,
accuracy and precision.

From the point of view of data collection, conventional instruments
contain three essential components: the detector itself, the detector
electronics and the display system. Within such an instrument it is

possible to add circuitry to sample the time varying signals originating
from any of these units. As shown in Fig. 5.4, sample point A can be used
to take out signals directly from the detector. Sample point B provides a
means of obtaining signals that have been conditioned in various ways in
order to make them more suitable for manipulation by the display device.
Finally, sample point C enables copies of the displayed signals to be
recorded and stored away for future or immediate use.

Increasingly, a greater number of modern instrument designers appreciate
the need to be able to connect instruments to computers for on-line data
aquisition. Consequently, many instruments are nowadays provided with an
external data interface that can be directly connected to a computer
system. This dispenses with the need to take out signals from either of
the points A or B shown on the right of Fig. 5.4. However, there are
situations where it might be desirable to maintain an arrangement for
sampling from point C. As a result of the availability of external data
ports, the aquisition of data from an analytical instrument becomes a much
easier task than it was some years ago. However, in situations where
these ports do not exist there will be a need to attach probes at an
appropriate point inside the instrument. The exact point of attachment
will depend upon the nature of the detector and its associated electronic
circuits.

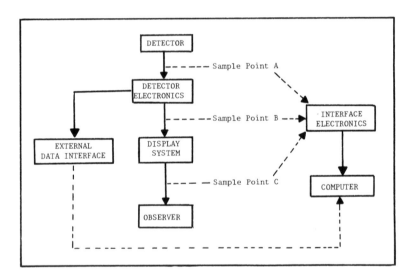

Fig. 5.4 Possible sources of signals/data in a conventional instrument.

The detector electronics will usually consist of some form of pre-
amplifier that either magnifies a voltage change or converts an output
current into a voltage suitable for driving a digitiser (She73) or a
display device such as a recorder or visual display unit. In very many
cases these circuits are based upon the use of operational amplifiers
(Bra69, Sch72, Nei79, Orr80, Ree80). Sometimes, however, auto-ranging
amplifiers are used. This type of amplifier automatically adjusts its
amplification factor in order to keep the output voltage within certain

limits while simultaneously generating a coded digital output which indicates the amplification factor being used. The output from the amplifier is usually passed to an analogue filter circuit designed to reduce the noise level in the signal. This process is called signal conditioning. The unit containing the detector electronics may also perform some form of signal modification (for example, rectification, integration, differentiation or attenuation) depending upon the exact nature of the detector, the requirements of the display system and the observer. Once an appropriate analogue signal has been obtained from the detector it is converted to digital form by a suitable analogue to digital conversion technique. Some instruments provide digital data output ports and others, that involve electronic counting circuits, will normally produce digital data directly. However, where conversion is necessary many techniques are available (Hoe68, Hna76). These may involve the use of a complicated circuit or a single IC chip depending upon the precision, speed and conversion frequency required. Analogue to digital conversion will be discussed in more detail later in this chapter. The book by Carrick (Car79) is a useful guide to methods or overcoming some of the problems associated with connection computers to various sorts of instruments.

In addition to obtaining data directly from on-line analytical instruments, several other sources of data are likely to exist. Some of these will undoubtedly involve the use of semi-automatic data aquisition techniques. Situations of this type arise from two major sources. Firstly, when it is not feasible to modify an instrument for direct computer interfacing. Secondly, in cases where the analytical techniques involved are not suitable for the direct involvement of a computer system and some manual assistance is necessary in order to effect the analysis. Typical examples of techniques that do not permit facile automatic data collection would include thin layer chromatography, paper chromatography and zone electrophoresis. These methods of analysis usually produce results that need some form of manual manipulation in order to produce data suitable for entry into a computer system. Intermediate processing techniques might involve weighing areas of chromatography or chart recorder paper, the use of a planimeter, digitiser, photodensitometer, particle counting equipment and other types of sensitive measuring device. All of these latter machines provide examples of secondary sources of data that can be input to the computer subsequent to any intermediate manual manipulation of the primary results of analysis.

(B) Types of Data

The types of data that are most often encountered in analytical chemistry fall into two broad classes:

    (A) DIGITAL, that is, discrete quantised values such as the pH of a solution at a particular instant in time, the rate constant for an inversion reaction or the radioactivity count from a radiocarbon or tritium labelled compound at a specific point during its decay, and,

    (B) ANALOGUE, or continuously varying values such as the absorption of a sample as a function of wavelength in an infrared/UV spectrum or the changes in current associated with a GC flame ionisation detector.

Some examples of the different types of data that are likely to be encountered in the analytical laboratory were briefly outlined in chapter 1. Gas chromatography and spectroscopy are probably two of the most widely used analytical techniques currently available. The signals produced and recorded be each of these are characteristically analogue in nature. A typical example of a chromatogram was presented in Fig. 1.8. Sometimes digital data may be obtained from a gas chromatograph if the signals produced by the detector are processed by a computer integrator in order to obtain the areas under the peaks. Another illustration of analogue data was presented in Fig. 1.9 where the visible absorption spectrum of the ferrous iron and o-phenanthroline complex was shown. Although both graphs were each representations of analogue quantities they differed from each other in two ways: the values plotted and the shape of the curves. In the first example detector response was plotted as a function of time while in the second case the response was plotted against wavelength. The trace of the gas chromatogram showed much greater periodicty (that is, passing through successive states of high and low detector response) than did the UV spectrum.

Digital data was encountered in several examples - the determination of iodine by radioimmunoassay and the results produced by liquid scintillation counting. Other examples include values produced by instruments such as refractometers, specific gravity meters (densitometers) and equipment for measuring boiling points and freezing points. Successive readings of a burrette during a titration represent yet another form of digital data. Where laboratory instrumentation is concerned, analogue signals seem to be far more commonplace than any other.

Some further illustrations of analogue data are presented in the diagrams contained in Fig. 5.5. A and B each represent electron spin resonance spectra (New73) and illustrate a very useful technique - that of plotting the first derivative curve of the absorption signal. First and second derivative curves are often used in order to improve the signal-to-noise ratio in a spectrum. Example C shows the type of results contained in a typical AC polarogram (Smi72). Electroanalytical methods can produce a wide variety of different types of data (Lev64, Bau78, Str73) depending upon the exact details of the measurements being made.

Example D shows the results of liquid chromatography analysis (HPLC) for sorbate and benzoate in soft drinks at fixed (left hand chromatogram) and variable (right hand chromatogram) wavelengths. The latter analysis was performed using a UV/VIS variable wavelength detector that was programmed to effect a wavelength change between eluents in order to optimise the sensitivity for each material. The peak on the right shows the benzoate absorption at 223nm and that on the left shows the sorbate absorption at 262nm. Notice the 'spike' occurring between the two peaks in the right hand chromatogram. This reflects the point at which the wavelength change takes place. Considerable care must be taken to omit this signal as noise when using an automatic data acquisition system.

These examples illustrate some of the wide variety of data that has to be catered for in automated data collection equipment. Fortunately, there are two basic guidelines that are often helpful when working in this area: data from similar sources will usually be similar in type (for example, mass spectra are all very similar in their basic structure as are chromatograms and infrared spectra) and, usually, only two analytical variables are involved in the data acquisition process - detector response as the dependent variable and time or some related control variable as the other.

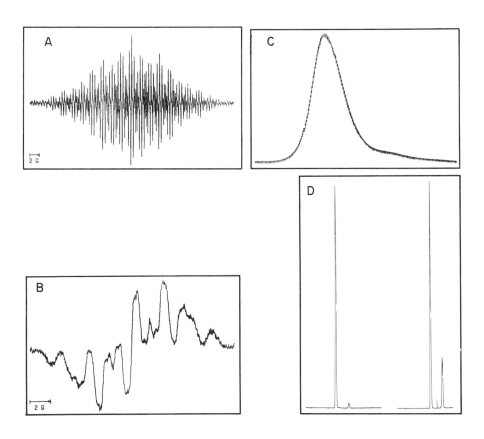

Fig. 5.5. Examples of analogue data.

BASIC PRINCIPLES INVOLVED

An introduction to the concepts involved in computer based data acquisition systems will be found in text books on instrumental chemistry (Per75)
or microcomputer based instrumentation (Bib78). Introductory articles are
also contained in journals dealing with automated analysis (Ree80) or
control systems (Asm80). Review treatments within particular areas of
analytical chemistry are covered in articles by Cram and Risby (Cra78,
Cra80), McDonald (McD78, McD80) and Burlingame et al (Bur78, Bur80). The
simplest type of data acquisition device is one which captures a signal
and stores it for later use. Such a device for recording transient
signals has been described by Betty and Horlick (Bet77). Their system is
based upon the use of an analogue-to-digital converter (ADC), digital
memory and a digital-to-analogue convertor (DAC). The equipment is shown
schematically in Fig. 5.6.

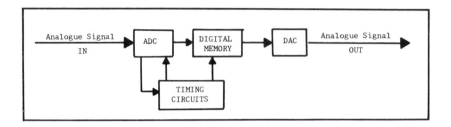

Fig. 5.6   Use of ADC's (analogue-to-digital) and DAC's
(digital-to-analogue converter) in data capture operations.

An arrangement such as this may be used to collect data at high speed and
then read it out later on a slow speed device such as a strip chart
recorder. Alternatively, it may be used to obtain data at slow speed for
later read-out on a high speed device (such as an oscilloscope) or
transfer to a computer. For long term storage the analogue signals that
are captured can be stored in analogue form on magnetic tape. However,
for a large number of applications it would probably be preferable to
retain the acquired data in digital form since this aids its facile
manipulation by a computer system or easy transportation in a digital
computer network. The basic relationships between the hardware components
involved in a typical data acquisition system assoicated with an analytical instrument are illustrated in Fig. 5.7.

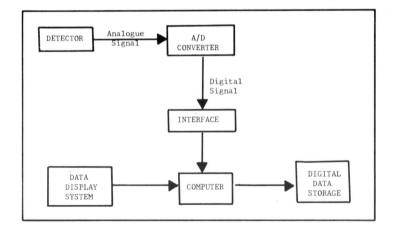

Fig. 5.7   Hardware components necessary for data
capture from an instrument.

The characteristics of the detector will depend upon the type of instrument that is used to perform the measurements being taken. Some examples have been mentioned in previous chapters. The present discussion assumes that the detector will produce an analogue signal. The first task to be undertaken is its conversion to digital form. This will be discussed in the following section.

Analogue to Digital Conversion

The function of the analogue-to-digital converter is to perform conversions of the analogue signals at precisely defined time intervals so that the computer receives a series of digital values that represent the incoming analogue signal waveform. This scheme is illustrated in Fig. 5.8.

The points on the curve (indicated by crosses) are those at which the analogue waveform is sampled. The column of figures on the right are the corresponding digital outputs produced by the ADC. These values assume that the horizontal part of the analogue waveform represents digital zero. The resolution of the conversion device will determine how closely the digital values on the right hand side represent the waveform on the left. This will determine the number of binary output digits (bits) that the converter is able to produce. With four bits, all input signals have to be mapped onto sixteen possible output states while with 8 bits the number of output states available is 256. Greater resolution is thus obtained with ADCs that have a large number of output states. Many commonly available ones have 12 output bits that correspond to a total of 4096 states.

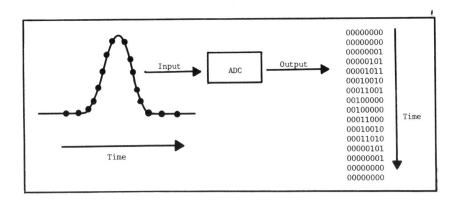

Fig. 5.8   The analogue-to-digital conversion (ADC) process.

A variety of analogue to digital conversion techniques exist - the sample and hold successive approximation converter, the dual slope converter, the single slope ADC, the voltage to frequency converter and so on. Some of the advantages and disadvantages of each of these have been compared by Reese (Ree80). Circuits for a number of different analogue to digital converters are given in Perone's book (Per75). For certain applications inexpensive single monolithic integrated circuit chips are available. The pin-outs for a typical 8 bit successive approximation analogue to digital converter chip are shown in Fig. 5.9 (RSC80). The conversion time for such a chip is about 10 µsecond.

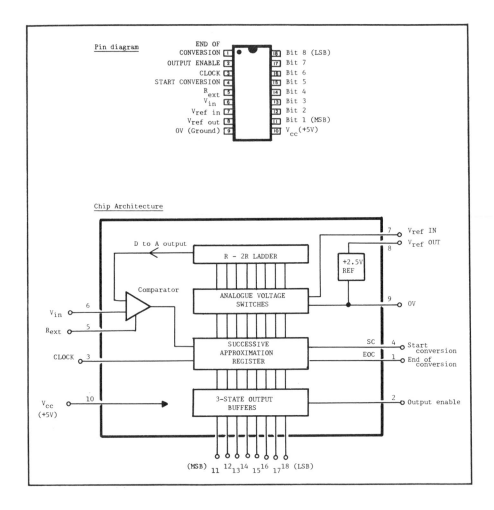

Fig. 5.9 Pinouts and architecture of an ADC integrated circuit.

The successive approximation A/D converter is probably one of the most commonly used even though it is susceptible to noise. An explanation of its mode of operation has been given by both Perone (Per75) and Reese (Ree80). Like most other practical A/D converters it is based upon the use of a comparator to match the input voltage with sums of appropriately weighted fractions of an internal reference voltage. In the method of successive approximations each bit is tested in turn, commencing with the most significant. If the result of adding the contribution to the bit exceeds the value of the incoming signal then the bit is set to zero, otherwise, it is set to one. By the time the least significant bit has been tested and set in this way the best approximation will have been accomplished.

The performance of an analogue-to-digital converter is usually specified by means of its speed and accuracy product. For all A/D circuits there is a characteristic settling time that increases with the number of bits of precision. Because of the finite conversion time inherent in an ADC there will be some uncertainty associated with the digitised value.

Problems arising from this effect can be minimised by preceding the converter with a sample and hold amplifier - hence, the name sample and hold ADC. Another source of problems arises from the fact that there is only a finite number of states associated with the output side of the converter. Consequently, the final value produced is always an approximation to the true analogue signal since increments smaller than that represented by the least significant bit cannot be accommodated. This effect, combined with possible problems associated with signal sampling speed can induce signal noise/distortion into the system if considerable care is not taken. An excellent illustration of these effects has been given by Brignall and Rhodes (Bri75).

Sampling Rate Considerations

There are several approaches to the way in which data may be acquired for subsequent analysis, a selection of these is listed below:

(a)   a single measurement may be made at a given instant
      in time - this may be performed in duplicate in order
      to ensure reliablity,
(b)   many measurements are made as a result of periodic
      sampling as a function of time,
(c)   cause and effect sampling in which a slight change is
      made to a system and its effect observed,
(d)   some fixed number of observations are to be taken
      during a given time span.

Situations such as those described in (b) and (c) may often necessitate data being collected at quite high acquisition rates varying from below 2 KHz up to 1 MHz or higher. In cases such as this it is important that the optimum sampling frequency is used if reliable recordings of analogue waveforms are to be obtained in digital form without loss of information and with minimum wastage of computer resources. Constraints are thus imposed on the data acquisition process due to sampling rate considerations and computing requirements. If the sampling frequency is too low then important information in the signal can be lost - this is shown in an extreme case in diagram (A) of Fig. 5.10. Alternatively, the signal can become very distorted - as shown in diagram (B). An illustration of these effects is given in Perone's book (Per75, experiment 12). Similarly, if the sampling frequency is too great, this can cause problems for the computing system that has to store and organise the large amounts of data that may be produced. When storing this data a memory buffer will usually be used. As this becomes full its contents will need to be transferred to some secondary storage device. This could be a relatively slow process that might impede the overall performance of the system. Its effects can be overcome by increasing the number of buffers available and using them in some form of alternating sequence. In the example shown in diagram (C), digitised data is stored in memory until the first buffer becomes full. When this happens, incoming data is then stored in the second buffer while the contents of the first one are transferred to secondary storage - typically, magnetic tape or paper tape.

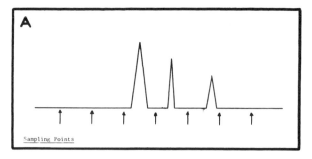

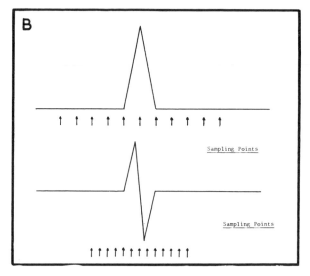

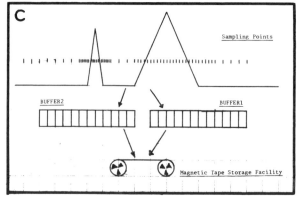

Fig. 5.10 Sampling rate considerations during data capture.
Case A illustrates how data may be lost; case B shows how
distortion might arise; and case C depicts a strategy that
involves the use of different sampling rates depending upon
the characteristics of the signal.

In order to ensure that no information is lost from the incoming signal it is possible to use the sampling theorem (Bib78) to determine the minimum sampling rate that may be used. This minimum data acquisition rate is that which enables all the frequency and amplitude information in the unknown signal to be captured. Sometimes this is referred to as the Nyquist sampling frequency. If the analogue waveform is to be accurately reconstructed from the acquired data then it is customary to sample at frequencies many times that of the Nyquist value.

Some valuable practical considerations relevant to the digitisation of analogue signals have been described by Kelly and Horlick (Kel73). In this work studies were made of sampling interval, sampling duration, quantisation level, digitisation time, aperture time and random variations in sampling rate. The work also tabulates the maximum sampling interval and minimum number of samples needed to digitise triangular, exponential, Lorentzian and Gaussian peaks for given values of maximum absolute error.

The importance of sampling rate cannot be over-stressed since incorrect results can easily be obtained unless considerable care is taken. A common source of error can arise from the phenonemon known as aliasing. This manifests itself when the sampling rate is less than twice the signal frequency. Suppose Fs is the sampling rate for a given data acquisition experiment; if the signal being sampled contains signals above Fs/2, then they will be folded back to fall in the 0 to Fs/2 spectrum, thus causing spurious results to appear. Perone (Per75) demonstrates this effect for the NMR spectrum of mesitylene and shows the magnitude of the inaccuracies that can be produced. Of course, in addition to producing spurious frequency shifts, aliasing can also cause amplitude changes.

Some indication of the computing problems also needs to be mentioned. When sampling at high data acquisiton rates or at low rates over extended periods of time, large volumes of data are likely to be produced. Consequently, wherever possible, there is a need to economise on storage space by compacting the data in various ways using both hardware and software techniques. In many cases, monitoring a detector output will produce virtually invariant baseline signals for a high percentage of the observation period - this is particularly the case with chromatographic peaks that are sharp and well resolved. Such invariant baseline data values need not be stored. Similarly, the number of samples that need to be taken will often depend upon the shape of the signal being recorded - sharp peaks requiring more sampling than broad slowly changing peaks. There is a case, therefore, for implementing techniques to enable variable data acquisition rates to be implemented. Leyden et al (Ley77) have described a voltage controlled oscillator device that enables the use of signal-proportional acquisition rates. It involves a simple hardware system that evaluates data significance and adjusts the rate of data sampling accordingly. This is a useful device since it permits savings to be made both in computer memory space and in processing time. Prior to the availability of hardware units such as this, compaction of data had to be performed by equivalent software routines.

The Computer, Interface and Data Store

Within a data acquisition system the interface usually provides the means by which the on-line computer is connected to the instrument(s). The exact nature of the interface will depend upon both the type of computer and the type of instrument(s) involved. In the next chapter the nature of interfacing will be described in more detail.

The most rapid means of storing acquired data is in the memory space of
the computer itself. However, this has a limited size and, depending upon
the rate of acquisition and the amount of data to be collected, it can
easily become exhausted. Consequently, at suitable points during the data
collection process it is necessary to transfer the data held in computer
memory to appropriate ancillary storage devices. The secondary storage
devices that can be attached to the computer in order to enable the long
term storage of the acquired data will be of two basic types: magnetic and
non-magnetic. Examples of the former include cassette tape, flexible disk
and magnetic cards. Paper tape, punched cards and optical disk are
typical examples of the non-magnetic variety. Data storage devices will
be discussed in more detail in a later section of this chapter.

The type of computer that is used in a data collection system will, like
the interface, depend upon the overall functions that are to be performed
during data collection and subsequent to it. From the discussion
presented in the last chapter it will be apparent that three basic types
of computer could feasibly be employed: a mainframe, a minicomputer or a
microprocessor. As a general rule there are three points worth bearing in
mind:

(a)   mainframes are not usually used directly for data
       acquisition purposes,
(b)   minicomputers are most often used when large volumes
       of data are involved and many instruments are to be
       serviced simultaneously, and,
(c)   microcomputers are usefully employed in applications
       that require uninterrupted use of a processing unit.

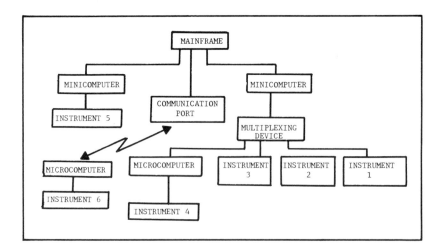

Fig. 5.11 Possible uses of the mainframe, minicomputer and
microcomputer in instrument control and data acquisition.

Obviously, there are no clear cut rules to follow - the choice of computer will depend directly upon the application concerned. The advantage of the microcomputer is that it is inexpensive and can therefore be dedicated to a particular instrument or experimental rig. If the power of the micro is not sufficient for the application concerned, then a minicomputer may need to be used. Some sophisticated instruments such as the mass spectrometer and the GC/MS may require a dedicated minicomputer system if a substantial amount of data acquisition, control and computer processing is to be performed. Usually, however, when a minicomputer is used in the laboratory it will often be highly multiplexed in order that it can service several instruments. Some of these possibilities are illustrated in Fig. 5.11.

Some of the approaches that have been used for data collection will be summarised in the next section.

APPROACHES TO DATA ACQUISITION

When designing a data collection system many important factors need to be taken into consideration, for example,

(a)   the duration of the experiment or monitoring period,
(b)   the rate at which samples are to be acquired,
(c)   the types of instrumentation involved,
(d)   the method of sampling,
(e)   the way in which the acquired data is to be processed,
(f)   the required system reliability, availability and responsiveness,
(g)   the types of observation being made,
(h)   whether the system is to be used for control purposes, and,
(i)   the financial resources available.

Many more factors could be listed but these are sufficient to indicate that real-time data collection is by no means a simple process. Over the years a number of different approaches to the problem have evolved. The technical details of these are adequately described in the literature (Car79, Ros75, Per75, Cou72, Bin71, Jon69, She73). Some of the frequently used techniques are listed below:

(1)   use of a Digital Multi-Meter (DMM),
(2)   use of an oscilloscope,
(3)   use of data logging equipment,
(4)   use of card-cage systems,
(5)   'home-made' computer based systems,
(6)   use of a laboratory data system, and,
(7)   special process control systems.

In this section each of the methods listed above will be briefly outlined. Wherever possible, appropriate references to the technical literature will be given in order that the reader can obtain the fine detail of the methods cited.

Use of a Digital Multi-Meter (DMM)

One of the most useful advantages of the DMM is its portability and ease
with which it can be interfaced to a desk-top computer. In addition, it
also offers relatively inexpensive approaches to data collection. The
function of the digital multi-meter is to perform analogue to digital
conversion. Most meters of this type produce parallel binary coded
decimal (BCD) output along with various control and timing information. A
simple example of an automated data handling system based upon the use of
a digital multi-meter has been described by Larsen (Lar73). The system is
shown schematically in Fig. 5.12.

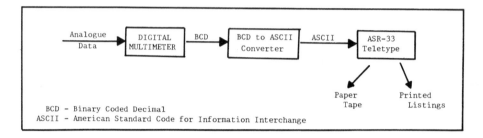

Fig. 5.12 Use of a digital multi-meter for data acquisition.

The BCD signal produced by the multi-meter is converted to a standard code
called ASCII (American Standard Code for Information Interchange) that is
sent to an ASR-33 teletype. This enables the original data to be listed
on printer paper or punched onto paper tape for storage purposes. Larsen
has used this equipment for acquiring data from controlled potential
coulometry experiments, thermal data collection and for the acquisition of
spectroscopic data.

The most popular type of digital instrument for data acquisition work is
probably the digital voltmeter. Busch et al (Bus78) have used an instrument of this type in the construction of a microprocessor controlled differential titrator. The arrangement of equipment is depicted in Fig. 5.13.

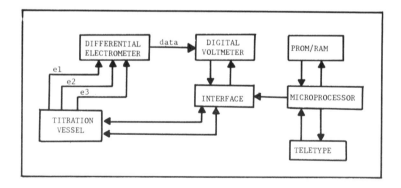

Fig. 5.13 Use of a digital voltmeter for microprocessor
based data capture.

Data Collection   197

In this sketch e1, e2, and e3 represent the sensing electrodes that are
the sources of the data to be acquired. The DVM is used to measure the
analogue voltage produced by the electrometer. The output from the
digital voltmeter is passed (via an interface) to the microprocessor which
stores the data and processes it in various ways. In addition, the microprocessor unit is responsible for controlling the addition of titrant to
the vessel during the analysis. Arrangements such as that depicted above
are becoming increasingly popular in a wide range of laboratories.

Use of an Oscilloscope

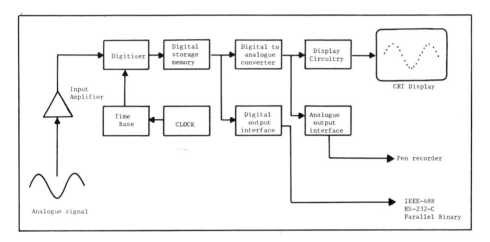

Fig. 5.14 Basic principle underlying the digital
storage oscilloscope.

The mode of operation of the oscilloscope and its use as a measuring
instrument has been described by Van Erk (Erk78). The most familiar type
of oscilloscope is probably that of the analogue variety. These are sometimes referred to as storage tube oscilloscopes because of the way in
which they store the wave form - either on a charged grid behind the
screen or on the screen itself using a special type of phosphor. In the
past, oscilloscopes were used mainly for the display of rapidly changing
wave forms or transients that could not be followed using conventional
recording devices. In order to obtain permanent records from the analogue
oscilloscope it was customary to attach some form of photographic equipment to it. This enabled displayed graphic images to be captured on film
for later development. Nowadays, however, since the advent of the digital
storage oscilloscope the process is much easier. Consequently, the use of
this instrument as a data collection device is growing substantially. The
digital storage oscilloscope digitises the analogue input wave form into a
digital signal which is then stored in semiconductor memory and later converted back to analogue form for display on a conventional cathode ray
tube. The arrangement is shown schematically in Fig. 5.14. Some of the
newer types of instrument have built-in secondary storage devices
(flexible disks) that permit extensive amounts of data to be stored -
typically up to 32 wave forms each with 1024 points. The useful features
of this type of oscilloscope lie in the ease with which data can be processed and, if need be, transferred by means of appropriate communication

links, such as the IEEE-488 and RS-232-C (see later), to other digital
units such as a microprocessor or minicomputer system. Shackil (Sha80)
has described and compared some of the commercially available digital
storage oscilloscopes.

As an illustration of its use, the work of Joshi and Sacks (Jos79)
provides a noteworthy example. They describe how a digital storage oscilloscope may be utilised to take rapid measurements in a droplet generator
system for atomic absorption spectroscopy. The oscilloscope is used to
simultaneously record the output from two monochromators and then plot out
the results on an analogue X-Y plotter.

Use of Data Loggers

A data logging device is a piece of equipment (possibly based upon the use
of a microprocessor) that is used to record data from one or more sources
as a function of time. The acquired data is not required for real-time
control purposes and it may be either analogue or digital in form - or
combinations depending upon the types of transducers involved. Once it
has been recorded the data can later be retrieved and analysed by either
manual or automatic means. Because it can be used in a variety of
locations (in the laboratory, on the factory floor, in an open air experiment, etc) data logging equipment has to be of rugged construction.
Furthermore, it has to be designed for reliable, unattended operation over
long periods of time in situations that may cause it to become exposed to
severe environmental conditions. Although some data loggers have been
designed around the use of paper tape (She73), the majority are based upon
the use of multi-channel magnetic tape recorders - mainly because of the
large data storage capacity of this medium. Both cassette tape and reel-
to-reel systems are commonly employed, the latter having the greater
storage capacity. In both cases, recording standards are available
(ANSI-ECMA) which means that recorded data is easily transported between
systems. During the recording process (which may involve analogue or
digital techniques) each channel on the tape is allocated to a particular transducer. However, a variety of other recording strategies may be
used. Many of the more modern data loggers are programmable either via an
on-board keypad system or by means of a desk-top computer which may be
disconnected after the device has been programmed. Programmable data
loggers enable sophisticated data capture strategies to be set up (Fri80,
Hol80a, Mac80).

An introduction to the theory and practice of modern instrumentation tape
recording has been presented in a recent publication by EMI (EMI78). A
wide variety of data logging examples are given along with a considerable
amount of practical advice on how to choose an appropriate system. Data
logging equipment is available from a number of commercial sources. EMI
(EMI80), Solartron (Sol80), Hewlett Packard (Hew80) and Microdata (Mic80)
are typical vendors of this type of equipment.

Use of Card Cage Systems

In many applications the centre of a data aquisition system is the
computer. Thus one way of building a system is to connect a collection of
measuring instruments and programmable output devices to such a machine.
This approach is desirable for applications that require extra high precision. An alternative way of achieving data collection is through the

use of card cage devices. A card cage is essentially just a chassis into
which can be installed a variety of functional I/O modules usually in the
form of printed circuit cards.

The two approaches to data collection outlined above are summarised in
Fig. 5.15.

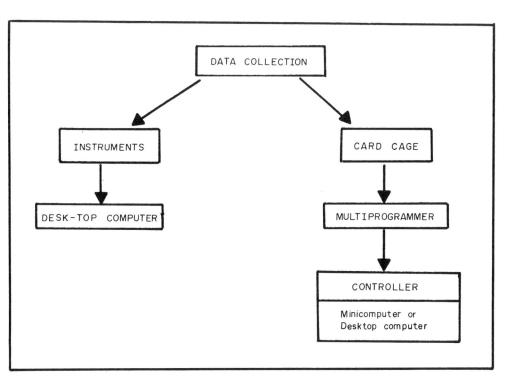

Fig. 5.15 Hardware approaches to data collection:
programmable instruments versus card cage systems - or both.

Usually, a wide variety of plug-in cards are availablle to cover most data
acquisition and control applications, for example,

             - voltage, current and resistance measurements,
             - frequency measurements,
             - digital inputs/outputs,
             - special interrupt inputs,
             - low level and high level scanners,
             - digital timer clocks,
             - relay outputs,
             - variable current/voltage outputs,
             - stepper motor outputs,
             - pulse outputs,
etc.

In addition, most systems provide facilities to enable user bread-
boarding, that is, the attachment of temporary experimental cards
containing patched circuits that are to be evaluated. These are useful
when non-standard interfaces are being designed and incorporated into a
system.

A multiprogrammer similar to that shown in Fig. 5.15 is usually used to
supervise the data acquisition and control activities of the individual
cards contained within a card cage. It can be programmed so the controller unit so that it can run and control data acquistion autonomously
while the controller attends to other functions. The following example
illustrates the type of instructions that the user of a typical system
(Hew80) might incorporate into a program in order to initiate a data
collection operation:

```
WRT   723, "OP,1, -5T, WA, 10T, IP,2T"
RED   72301,A
```

Here an output instruction (OP) commands a D/A converter card in slot 1 of
the card cage to provide an analogue stimulus of -5 volts to the system it
is controlling. This is followed by a wait instruction (WA) causing the
multiprogrammer to wait 10 seconds before taking the associated reading
from an input card in slot 2 of the card cage. Finally, the input data is
read back to the controller.

The controller in such a system might be a desk-top computer or a minicomputer. A desk-top system offers all the necessary parts of a computer
in a single package. It includes displays, keyboard, mass storage,
printer and a complete operating system with a programming language. Most
have enhanced program editing facilities making them very easy to use.
They are portable and relatively inexpensive. In contrast, minicomputers
are more powerful and flexible. They offer a modular solution with
several languages, multiple terminals, multi-tasking facilities and the
ability to handle large data bases. They are very fast and their modular
approach allow designers to tailor systems to individual needs. Desk-top
computers suit those applications where a great deal of programming is to
be done or where user interaction is required. Minicomputers are suited
to situations where high speed and large data bases are needed and where
there is less on-line programming and interaction.

An interesting example of the card cage approach to data acquistion has
been described by Farley et al (Far80). In this work various printed
circuit cards were interconnected by means of a S-100 bus (see next
chapter) to provide a microcomputer control and data acquistion system for
a micro-wave optical spectrometer. The system contained a 100 trace, 22
slot mother-board into which daughter-boards containing various logic
circuits could be inserted. Typical daughter-boards included: parallel
I/O cards, a CPU board, memory boards and cards for stepper motor control.

Home-made Computer Based Data Collection Systems

This category of data acquisition equipment is used to include any form of
system that has been built from fundamental (computer based) building
blocks. This approach is a useful way of obtaining an end product that
cannot be purchased commercially. Alternatively, the method is employed
as a means of reducing system cost - purchasing a ready made system may be

prohibitively expensive. During the 1970's a number of data acquisition systems based upon the use of minicomputers appeared in the literature. Towards the end of this decade a substantial number of systems based upon the use of microcomputers also started to appear. Presently, there is a growing number of both types, each with their particular advantages and applications areas.

Some of the early examples of data acquistion systems employing minicomputers have been described by Burlingame et al (Bur70 - high resolution mass spectrometery), Weyler (Wey70 - X-ray fluorescence spectrometry), Hewitt (Hew70 - NMR spectrometry), Cotton (Cot70 - gas chromatography) and Chuang et al (Chu70 - infrared spectometry). A simple system for real time data acquisition from a gas chromatograph using a minicomputer (a Hewlett Packard HP-2115A) has been described by Perone (Per71). The arrangement of equipment is illustrated in Fig. 5.16.

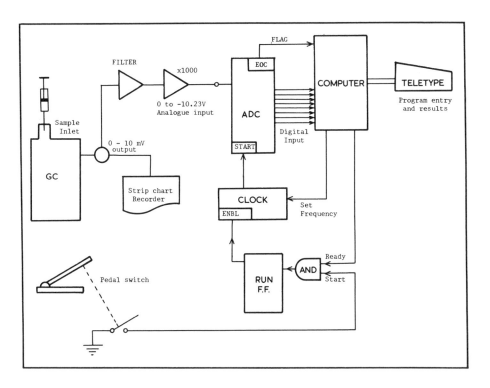

Fig. 5.16 Arrangement of equipment for computer based real-time data acquisition from a gas chromatograph.

The output from the GC is digitised and stored in the computer for later use. Because of storage limitations, baseline data between peaks is rejected by the software. Chromatographic peaks are then recorded using 20 - 30 storage locations for a ten peak chromatogram.

A more sophisticated system for collecting data from up to 10 live rats during drug screening tests has been described by the Swedish drug company Hassle AB (Has80). This system is based on a similar arrangement to that used by Laffan et al (Laf72); each rat is fitted with a catheter that is attached to its aorta. This enables blood pressure to be monitored throughout the duration of an experiment. Results are collected using a DEC PDP-8 computer which is used to continuously monitor the experimental conditions and results for each of the rats for periods up to 10 hours. In this example the use of the computer greatly enhances both the rate and reliablility of data collection while reducing its cost. Another interesting example of minicomputer based data collection involving a telemetry system has been described by Schofield (Sch75). In this example the equipment is used to monitor water quality and involves the use of instruments to measure dissolved oxygen, turbidity, pH value, temperature and the concentration of ions such as nitrate and gaseous pollutants like ammonia. The central station equipment consists of a Texas Instruments 960A minicomputer and associated peripherals. Each of the out-stations contains a hardware interface with a water quality and telephone network modulator. At pre-determined times a program in the central computer causes the telephone numbers of the out-stations to be dialled in turn. Each water quality monitor then transmits its data over the public switched telephone network to the central site for processing. A similar system is in use in Finland for conservation studies (Koh80).

Milano and Kwang-Yil (Mil77) have described a novel acquistion system based upon the use of a fibre optic probe. The arrangement of their equipment is shown in Fig. 5.17.

The minicomputer (a Data General Nova II) is used to perform data acquisition, reduction and analysis of signals produced by a fibre optic probe that is attached to a solid state diode array spectrometer. The system performs, simultaneously, analysis of hemoglobins in whole blood in vivo. Measurements can be made within 30 seconds with a precision of about 1%.

Digitisation of instrument traces is an often used data acquistion technique in a large number of laboratories. Conventional approaches have tended to require the use of expensive special purpose digitisation equipment. An alternative technique has been devised by Delaney and Uden (Del78). They have used a minicomputer system (DEC PDP-11) to digitise data using an X-Y servo-recorder. When in use the set of data to be digitised (spectrum, chromatogram or other analogue trace) is positioned on the plotter bed. The recorder pen is then manually moved to selected points on the trace and digitisation accomplised by pressing a push-button trigger. The authors have used their system to digitise a library of 500 vapour phase infrared spectra in terms of peak position and intensity. Of course, the system could easily be used to collect the data as results are being plotted on the recorder.

Many other interesting minicomputer based systems have been described in the literature. A selection of these is listed below in order to illustrate the variety of applications involved:

(a) the control of vidicon camera and data acquisition from it using a DEC PDP-8/m minicomputer (Goe79),

(b) the control and data acquisition from mass spectrometers using a Data General Nova 3/12 minicomputer (Wei79),

(c) implementation of Fourier transform ion cyclotron resonance mass spectrometry using a Data General Nova 3/12 (Hun79),

(d) control and acquisition of data from a multi-element
     emission/ fluorescence spectrometer using a DEC PDP-11
     system (Ul179), and,

(e) the characterisation of chemical systems using continuous
     flow spectrophotometers, computers and graphics equipment
     (Fra79, Fra79a).

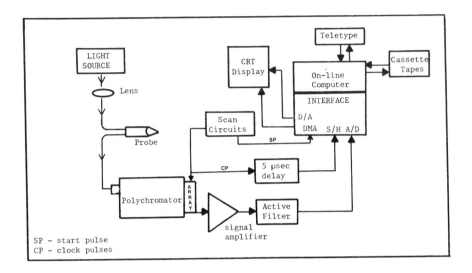

Fig. 5.17   Data acquisition by means of a fibre optic probe.

The last of these examples is interesting since it involves the use of two
PDP-11 computers - one for data acquisition/control and the other for
graphic display of results. The first of these is a PDP-11/45 and has 28
Kwords of memory, 5.0 Mbytes of on-line disk storage and 128 Kwords of
on-line flexible disk storage. It has a 128 channel ADC giving 14 bits
precision (plus sign) and a throughput rate of 40 KHz. The second
computer is a DEC PDP-11/40. The PDP-11/45 is connected by a high speed
parallel communications link to the PDP-11/40 and there is a general
purpose in-house interface for the valves, pumps, mixers, monochromators,
pH meter, digital thermometer and digital electronic balance that is used
for calibration purposes. The PDP-11/40 is used to run the graphics
system which has a 17" CRT, a light pen and an incremental plotter.

Minicomputers provide powerful solutions to the problem of data acquistion
and control at reasonable cost. However, for many applications the cost
of a machine of this type usually cannot be justified. Consequently, the
use of microprocessor based systems is increasing substantially - particularly where a dedicated processor is required for use with small pieces of
laboratory equipment. Examples of microprocessor based data acquisition
units have been described by Fornili (For80), Cosgrove et al (Cos80) and
Holtzman (Hol80). The system outlined by Holtzman is typical of many that
are currently being employed. The arrangement is shown in Fig. 5.18.

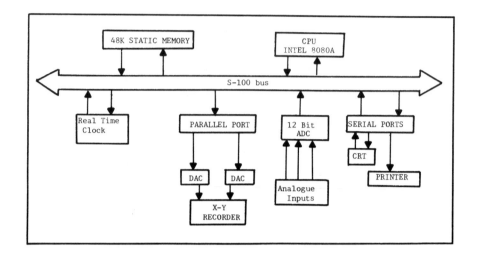

Fig. 5.18   Holtzman's microprocessor based data acquisition unit.

A similar system for conducting potentiometric titrations has been described be Gampp et al (Gam80). In this system the microprocessor controls the addition of reagent, monitors the pH until it becomes constant and stores the constant value. The data obtained is recorded on magnetic tape be means of a cassette recorder interfaced to the system via a standard interface (RS-232-C). The data can later be input from the cassette recorder to a desk-top computer for further mathematical processing.

Various other examples are described in the literature: Veazey and Nieman have outlined the use of a microcomputer data acquisition system to measure clinically important organic reductants using a stopped flow chemiluminescence apparatus (Vea79); Lilley (Lil80) has outlined the way in which a microprocessor based signal processing system can be used for collecting sonic data in acoustic emission experiments; Lloyd et al (Llo80) have shown how a microcomputer (DEC LSI-11) can be used as a part of a Fourier transform photoacoustic visible spectrometer while Borders and Birks (Bor80) have described a high speed pulse amplifier/discriminator (PAD) for photon counting experiments - the output from the PAD is counted for a period of time which is controlled by a microcomputer (INTEL 8080) which then collects the data from the counters (and other sensors), processes it and prints the results.

The examples cited above are just a few of the rapidly increasing number of computerised data collection units based upon the use of micro-computers. Unfortunately, the use of large scale integration (LSI) technology within analytical measuring equipment is not without its problems. These are adequately summarised in the following quotation taken from Wonsiewicz (Won78):

'the vision of the automated laboratory has promise:
computers control equipment, collect data, and analyse
and display results. The experimenter freed from
tedium, devotes more energy to creative pursuits,
presumably research and development. Unfortunately,
the vision has proved to be a mirage for more than one
experimenter who, after a year of learning the
mysteries of hardware and software, finds the control
of experiments as far away as ever'.

Betteridge (Bet80) has outlined some of the practical considerations
associated with the use of microprocessors. In many cases, interfacing
presents the biggest obstacle. This topic will be addressed in the next
chapter where some of the standard interfaces and interfacing techniques
will be described. However, in addition to the technological problems
there are the associated human factors accompanying the use of automated
data acquisition and control of experiments. These have been debated by
Malmstadt (Mal80) in his article entitled "Push button analysis". In this
paper Malmstadt discusses the changing role of the analyst as a consequence of the rapidly evolving computerised data collection techniques.
This matter will be pursued in greater depth in chapter 8 on laboratory
automation.

The Use of Commercial Laboratory Data Systems

As a result of the increasing use of computer based instrumental
techniques, powerful laboratory data systems are becoming more widely
available from commercial sources. These provide facilities for data
capture, data processing and data analysis. In addition, they usually
come supplied with libraries of computerised reference data (data bases)
to aid the interpretation of experimentally observed results. Various
types of data station are available - some are general purpose, others are
orientated towards particular areas of application such as infrared, NMR
or mass spectroscopy.

The E-935 data acquisition system from Varian (Var80) is typical of many
commercial systems. It provides large memory storage (128-256 Kbytes),
rapid scanning facilities (up to 64 KHz), a standard interface port (HP-IB
- IEEE-488), an interactive visual display unit and two programming
languages (BASIC and ASSEMBLER) which enables the analyst to develop
in-house programs for data manipulation. Many examples of the use of
commercial systems are described in the literature. A few of these are
summarised below.

(a) DECLAB is a system supplied by Digital Equipment Corpora-
     tion. Harnly et al (Har79) used this system (running on a
     PDP 11/34) to collect data from a multi-element atomic
     absorption spectrometer.

(b) DIGILAB FTS-NMR/3 and JEOL EC/100 Data Systems have been
     used for NMR determination of sucrose in sugar beet juices
     (Low79), while a

(c) Finnigan Incos 2000 GC/MS Data System has been used to
     record the mass spectra produced by direct pyrolysis of
     cellulose and its derivatives in a mass spectrometer
     (Fra79b).

(d) Data Translation's LAB-DATEX Data System is a highly
     modular system in that users plug in only those logic
     boards that are appropriate to their needs (Cav80). It is
     based on a DEC LSI/11 microcomputer and is able to acquire
     data at speeds up to 135 KHz directly to memory.

Systems such as those described in (a) through (d) above are becoming
increasingly popular even though they may not always have the throughput
of home-made systems.

Special Process Control Systems

Data acquisition and sampling are important considerations in automated
process control systems. An elementary discussion of on-line instrumentation for control purposes was presented in chapter 3. Further details
will be found in Bau78 and Bib78. A typical arrangement of sensing and
control elements for use in process monitoring applications is illustrated
in the sketch contained in Fig. 5.19.

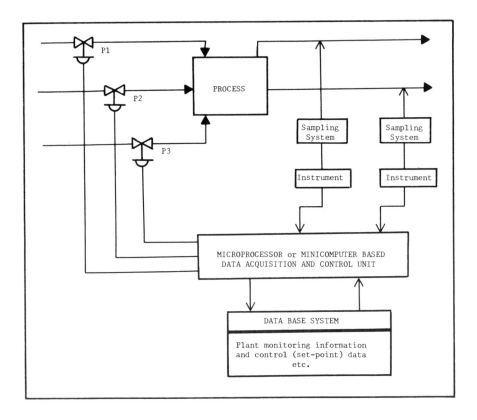

Fig. 5.19 On-line data acquisition for process control applications.

In this illustration the instruments monitor the input(s) and output(s) to the processing units responsible for carrying out some production process. The instruments involved may be of a conventional design or may be highly computerised through the incorporation of microprocessors. Based upon the information stored in the data base associated with the data system, the control unit is able to manipulate process controllers P1, P2 and P3 in such a way that control of the process is maintained. The data system is able to keep a continuous record of observational data collected during the plant monitoring operations. This type of data is often useful for retrospective analysis of plant performance.

In analytical process control, some of the advantages of using micro-processor based systems include:

(a) the ability for continuous dynamic self checking,
(b) the ease with which systems can be duplicated or triplicated, thereby increasing reliability,
(c) a substantial increase in display information via VDUs and colour graphics,
(d) the easier recording of permanent records of system performance,
(e) a considerable reduction in size compared with conventional control elements, and
(f) the construction of cheaper systems.

Some of the recent developments in data capture for process control are described in articles by Walls (Wal80), Welfare (Wel80), Hutchinson (Hut80) and Chard (Cha80).

STORAGE DEVICES

The storage facilities for computer systems are often divided into two broad types:

(a) primary storage such as core store and semiconductor memory, and
(b) secondary storage devices like magnetic tape, disks, paper tape and so on.

Primary storage is regarded as being that part of the memory system that is permanently attached to the computer and which the central processing unit can directly address and access. The basic architecture of the computer system and the addressing mechanism that it employs will place an upper limit on the amount of primary storage that may be used. This is an important consideration when considering the use of primary memory for data acquisition purposes. For smaller computers, typical memory sizes vary between 8 Kbytes and 256 Kbytes. Larger computers have primary stores ranging from about 0.5 Mbyte up to 10 Mbytes or more. Because it is based on the use of semiconductors and involves no moving parts, this type of storage offers the fastest access times and is usually the first choice for high speed data acquisition experiments. Typical access times for commonly available RAM (Random Access Memory) chips vary between 20 - 450 nsec.

The term secondary storage is used to describe the sections of the memory
system which the computer cannot directly address and which are not
permanently connected to the system. Devices in this category include
magnetic drums (permanently attached but not directly addressable),
magnetic disks (which may be either fixed or exchangeable, rigid or
flexible), magnetic tapes and magnetic cards. Other, non-magnetic, storage
media include optical disks, computer output on microfilm (COM), punched
paper tape, punched cards, graph plotters, printed listings and graphic
displays that might be attached to photographic equipment. The relationship of some of the above storage devices to the types of computers they
support is shown in Fig. 5.20.

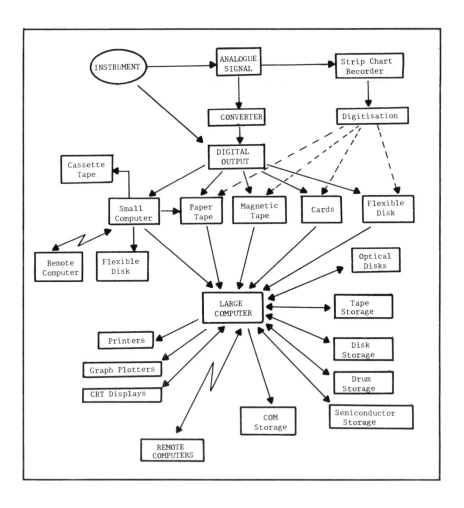

Fig. 5.20 Relationship of storage devices to the
computers they support.

Data Collection 209

Collectively the various storage units shown in this diagram are referred
to as peripheral devices. Descriptions of the different types of peripherals and their basic mode of operation are give in the book by Wilkinson
and Horrocks (Wil80).

Semiconductor Storage

Commonly available semiconductor storage is of three main types:

(a) RAM - Random Access Memory (static and dynamic),
(b) ROM - Read Only Memory, and
(c) EPROM - Eraseable Programmable Read Only Memory.

For data storage during data acquisition only the RAM is applicable since
the others do not permit data to be written into them - other than when
they are initially programmed. IC chips are available that offer a
variety of memory sizes, word lengths and access speeds and are fabricated
in a variety of different integrated circuit technologies. Some examples
of the different types of memory chips available are listed in Table 5.1.
Using appropriate combinations of these memory chips as basic building
blocks it is possible to construct a memory system to suit any particular
type of application. For example, the fast data acquisition system
described by Cosgrove et al (Cos80) utilises a 16 bit parallel 4096 word
read/write memory capable of providing a maximum data input rate of 2MHz
per 16 bit word. The microprocessor based unit outlined by Fornili
(For80) uses a 1024 byte x 8 bit RAM memory as a temporary store during
the acquisition of data from a spectrometer for subsequent punching on
paper tape. Similarly, the memory system built by Yarnitzky et al (Yar80)
is organised as a 16 x 1024 bit memory. This can hold up to 32 polaro-
grams for averaging purposes in square wave voltammetry instrumentation.
Many other examples of the use of digital memory data acquisition have
been described in the literature.

TABLE 5.1 Typical Organisational Details for some RAM Memory Chips

	Size	Organised As	Access Speed
(a)	1024	128 x 8	450 ns
(b)	1024	1024 x 1	350 ns
(c)	1024	256 x 4	450 ns
(d)	4096	1024 x 4	300 ns
(e)	4096	4096 x 1	450 ns
(f)	16384	16384 x 1	200 ns

One of the advantages of building a home-made memory system is that it
enables the storage limitations inherent in microcomputer/minicomputer
architecture to be overcome - if this presents a difficulty for particular
types of data acquisition. Further detailed discussions on semiconductor
memories of various types will be found in the book by Rony and Larsen
(Ron79).

CAC - H

Magnetic Storage Media

Of the different types of secondary storage devices, the magnetic media
are more popular than the non-magnetic variety since they offer easy to
use, high speed data storage and retrieval facilities. When considering
the use of this class of storage three important factors need to be taken
into account: speed of access, type of access (random or seqential) and
storage capacity. In terms of speed of access the following general
relationship holds:

semiconductor memory$>$ magnetic drum $>$ disk$>$ tape

A magnetic drum consists of a cylinder whose curved suface is coated with
a magnetic material. It is rotated at high speed around its longitudinal
axis. A series of magnetic heads held close to the surface of the drum
are able to read/write information from/to this surface as a series of
parallel magnetic tracks of information. Each head in the drum unit is
rigidly fixed and monitors the information on only one track. The
arrangement of a typical drum is illustrated in Fig. 5.21 - the tracks are
shown as dotted lines.

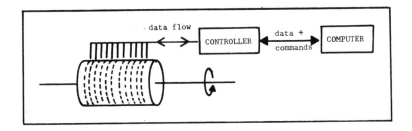

Fig. 5.21   Data flow in a magnetic drum system.

Because information transmission between the computer and the drum does
not involve any physical movement of the read/write heads the transfer
speed can substantially exceed that obtainable with most disk systems.
The maximum time that it takes to retrieve an item of information is
simply the rotational time of the drum. This time is the same for all
items wherever they are located on the surface. Drums are usually only
used on large mainframe computers. Further details on magnetic drums (and
other storage devices) will be found in Boh76. Table 5.2 lists some of
the important storage characteristics of the IBM 2301 drum (IBM73, IBM70).

TABLE 5.2   Storage Details for a Typical Magnetic Drum System

Transfer Rate:	1.2 Mbytes/sec
Number of Tracks:	800 (200 addressable)
Rotational Delay:	8.6 ms (average)
Capacity:	4.09 Mbytes
Rotational Speed:	3500 rpm

Data Collection   211

A more popular form of direct access storage is available via magnetic
disks. These are available in two forms: exchangeable and fixed. The
latter enables transfer speeds approaching those of drums to be achieved.
As their name implies, fixed disk systems do not enable the storage units
they contain to be removed once they have been set up. On the other hand,
exchangeable disk systems contain demountable cartridges that may be
removed and replaced with alternative ones. This mechanism provides an
expandable storage system that is only limited by the number of disk
cartridges available.

A typical disk cartridge consists of a series of parallel surfaces (disks)
that are coated with magnetic material and mounted on a central spindle
around which the whole assembly rotates. Each of the surfaces (except
that at the top and bottom) has a read/write head which forms part of a
comb-like mechanism that is able to move in and out between the perimeter
of the disk (point B) and the spindle (point A). Motion of the head
assembly is always parallel to the recording surfaces. The arrangement is
illustrated in Fig. 5.22.

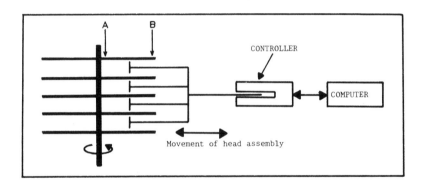

Fig. 5.22 Arrangement of the head assembly in a magnetic
disk. A and B denote the innermost and outermost data tracks,
respectively. These define the extent of head movement during
read/write operations.

As the head assembly moves inwards and outwards between the points A and B
it is only able to stop at a fixed number of points. These determine the
track positions at which information may be written and from which it may
later be retrieved. A simple arrangement of tracks is shown in Fig. 5.23.

The number of tracks will depend upon the type of disk: for large machines
between 300 and 600 are common, while for smaller computers between 30 and
100 are used.

In contrast to drums and fixed disk systems, when exchangeable disks are
used the time to access a piece of information will depend upon the track
on which it is located. Obtaining an item of data will most often involve
mechanical movement of the read/write heads. Thus, if the heads are
located over track B and the required information is on track A, the heads
would need to be re-positioned. Because this involves mechanical motion
it has the overall effect of slowing the access time down considerably
compared with drums or fixed head disks.

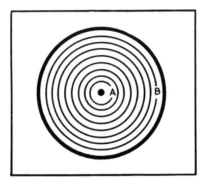

Fig. 5.23 Arrangement of tracks on the surface of
a magnetic disk. A is the innermost track and B the
outermost one.

The vertical set of tracks that can be accessed without movement of the
read/write head assembly is called a cylinder. The number of tracks in a
cylinder and the storage capacity of the disk will depend directly upon
the number of recording surfaces that are available. Large disks have
between 20 and 40 recording surfaces, otheres have 5 or 10, while flexible
disks for use on microcomputers use only 1. A comparison of some of the
storage characteristics for particular types of large exchangeable (IBM76)
and fixed (IBM77) head disks is shown in Table 5.3.

TABLE 5.3 Comparison of Exchangeable and Fixed Disk Systems

	IBM 3330	IBM 3350	
Number of Surfaces	19	30	
Tracks per Surface	411	555	
Total Capacity	100	317.5	Mbytes
Data Rate	806	1198	Kbytes/sec
Latency	8.4	8.4	ms
Seek Time	30	25	ms

In addition to the number of tracks available, the diameter of the recording surface is also important. For micros the diameter of the disk
surface (5 or 8 inches) is much less than those used on larger systems.
Microcomputers usually use flexible (or floppy) disks having a storage
capacity of between 150 K and 500 Kbytes depending upon their size, how
they are used and the density at which information is recorded (Col80).

Exchangeable disk systems have been the most popular because of their low
price. However, because of the greater precision with which they can be
made and because they do not contain any moving heads, fixed disks can
offer a substantial increase in performance over exchangeable systems.
Fixed disks have been available on large machines for many years (IBM73)
but until recently they have not been available for small machines. The

figures listed in Table 5.4 (IMI80) are typical of the type of performance
that can be expected of the fixed disk systems that are available for use
with micros and minicomputers.

TABLE 5.4 Characteristics of Microcomputer Fixed Disk Systems

Recording Capacity:	11	Mbytes
Number of Disks:	2	
Number of Surfaces:	3	
Diameter of Disks:	200	mm
Data Tracks/Surface:	350	
Track Density:	300	tpi
Recording Density:	5868	bpi
Disk Speed	3600	rpm
Transfer Rate:	648	Kbytes/sec
Minimum Access Time:	50	ms
Maximum Access Time:	100	ms
Latency Time (average):	8.3	ms

The advantages of magnetic drums and disks are (a) fast recording of and
access to data, and, (b) they may be randomly and directly accessed. That
is, the retrieval of a particular item does not require the manipulation
of any item other than the one being sort. Magnetic tape offers neither
of these advantages. Speed of access is much less than that for disks and
the mode of access is strictly sequential. Thus, in order to access an
item of data held on magnetic tape all the preceding items must be
accessed. While this is ideal for data logging it is not so useful for
subsequent random inspection of the data - as could be achieved through
the use of direct access storage devices. The distinction between random
and sequential access to data will be further discussed in chapter 10.

Two types of magnetic tape system are available for data acquisition
purposes: (a) reel to reel, and, (b) cassette tapes. Tape capacity will
depend upon a variety of factors such as tape width, number of tracks,
length and recording density. For digital recording on large machines
both 7 and 9 track tape are used (0.5 inch wide tape). On many data
loggers a larger number of tracks (14, 28, 32 and 42 - 1 inch tape) are
employed while on 0.25 inch tape 1, 2 or 4 tracks are available -
depending upon the design of the read/write head mechanism (EMI78). A
typical 0.5 inch tape might have a length of 2400 feet. If this is used
to record digital data at a density of 1600 bits per inch then it would be
capable of storing a total of 46 Mbytes of information. In contrast, a
cassette tape used on a microcomputer system (for example, the Commodore
PET) might have a length of only 132 metres (a C90 tape - 2 x 45 minute
sides) or 433 feet and would be capable of storing about 100 Kbytes on a
single side - total of 200 Kbytes in all.

The transfer rate for large digital tape systems will depend upon the
speed at which the tape moves over the head - 37.5, 75, 112.5 or 200
inches per second (Boh76) - and the recording density (800, 1600 or 6250
bits per inch). These figures give a range of 30-1250 Kbytes/sec. At the
other extreme, the simple cassettes used on microcomputers usually record
data at much slower speeds - the PET, for example, records at about 36
characters per second. Obviously, speeds and capacities will vary from
system to system. The values quoted above are used only to indicate the
types of ranges that are involved.

Non-Magnetic Storage Media

Prior to the common availability of magnetic tape, the most popular type
of non-magnetic storage medium used for scientific data acquisition has
been paper tape. Its popularity is attributable to its cheapness and
compactness compared with punched cards - another widely used data storage
medium. Many different types of data logger have been designed around the
use of paper tape devices.

Data values are recorded on paper tape by punching patterns of round holes
in parallel tracks or channels along the length of the tape. Characters
are represented by unique combinations of holes across the width of the
tape. Equipment is available to handle five, six, seven or eight
channels. Five or eight track tape is probably the most common. The
example shown in Fig. 5.24 represents a typical section of a piece of
eight track paper tape.

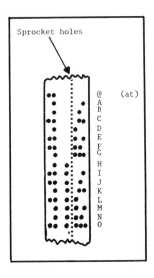

Fig. 5.24 A segment of 8-track paper tape.

Paper tape punches may be designed to operate using a variety of codes
that may involve the use of local, national or international standards
(Ros71). Local coding schemes are those that are devised for particular
special purpose data logging applications or which are unique to a
specific type of computer. It is best to avoid the use of these if data
transportability is to be ensured. National and international standards
(such as ASCII, EBCDIC, ISO, etc) are available to ease the task of data
movement via paper tape. Most commercially available paper tape equipment
usually produces punched tapes to a standard coding scheme. However, as
data logging equipment based upon magnetic tape and disk becomes more
popular, the demand for paper tape equipment is tending to decrease.

CONCLUSION

Data collection in its broadest sense is concerned with obtaining raw data from scientific or process control instruments and then transmitting this data to a central handling station for processing. The data that is obtained may be archived in order to form a historical record of the system from which it was obtained, it may also be used to perform real time control of a process, or it may be used to aid diagnosis and fault finding in an experimental or production environment.

Various approaches to the data collection task have been outlined using several different techniques. One of the most fundamental operations involved in the data acquisition process is that of converting analogue signals to their digital equivalent. The latter are often easier to transmit, store and process in computing systems. Since acquired data has to be stored, the availability of appropriate storage devices is of fundamental importance. In view of this, techniques for storing data have been briefly outlined in order that the facilities available, and their limitations, are made known.

The construction of automatic data logging equipment involving the use of a computer will involve the interconnection or interfacing of instruments and computer hardware. This topic will be discussed in the next chapter.

REFERENCES

Ana80   American Chemical Society, "Fundamental Reviews", Analytical Chemistry, Volume 52, No. 5, April 1980.

Asm80   Asmus, P., Getting Started in Data Acquisition and Control, Instruments and Control Systems, Volume 53, No. 6, 43-45, June 1980.

Bau78   Bauer, H.H., Christian, G.D. and O'Reilly, J.E., Instrumental Analysis, Allyn and Bacon, Inc., ISBN: 0-205-05922-8, 1978.

Bet77   Betty, K.R. and Horlick, G., 1024-Point Transient Recorder, Analytical Chemistry, Volume 49, No. 2, 342-344, February 1977.

Bet80   Betteridge, D., Practical Considerations in the Analytical Use of Microprocessors, Analytical Proceedings of the Royal Society of Chemistry, Volume 17, No. 5, 181-183, May 1980.

Bib78   Bibbero, R.J., Microprocessors in Instruments and Control, John Wiley, ISBN: 0-471-01595-4, 1977.

Bin71   Binks, R., Cleaver, R.L., Littler, J.S. and MacMillan, J., Real-Time Processing of Low Resolution Mass Spectra, Chemistry in Britain, Volume 7, No. 1, 8-12, January 1971.

Boh76   Bohl, M., Information Processing (Second Edition), Science Research Associates, ISBN: 0-574-21040-7, 1976.

Bor80   Borders, R.A. and Birks, J.W., High Speed Pulse Amplifier/Discri-
        minator and Counter for Photon Counting, Analytical Chemistry,
        Volume 52, No. 8, 1273-1278, July 1980.

Bra69   Brand, M.J.D. and Fleet, B., Operational Amplifiers in Chemical
        Instrumentation, Chemistry in Britain,Volume 5, No. 12, 557-562,
        December 1969.

Bri75   Brignall, J.E. and Rhodes, G.M., Laboratory On-Line Computing -
        An Introduction for Engineers and Physicists, Intertext Books,
        ISBN: 0-7002-0258-7, 1975.

Bur70   Burlingame, A.L., Smith D.H., Merren, T.O. and Olsen, R.W., Real
        Time High Resolution Mass Spectrometry, Chapter 3, 17-38, in
        Computers in Analytical Chemistry, (Volume 4 - Progress in
        Analytical Chemistry). Selected papers from the 1968 Eastern
        Analytical Symposium, Plenum Press, SBN: 306-39304-2, 1970.

Bur78   Burlingame, A.L., Shackleton, C.H.L., Howe, I. and Chizhov, O.S.,
        Mass Spectometry, 346R-384R, in "Fundamental Reviews", Analytical
        Chemistry, Volume 50, April 1978.

Bur80   Burlingame, A.L., Baillie, T.A., Derrick, P.J. and Chizhov, O.S.,
        Mass Spectrometry, 214R-258R, in "Fundamental Reviews" Analytical
        Chemistry, Volume 52, No. 5, April 1980.

Bus78   Busch, N., Freyer, P. and Szameit, H., Microprocessor Controlled
        Differential Titrator, Analytical Chemistry, Volume 50, No. 14,
        2166-2167, December 1978.

Car79   Carrick, A., Computers and Instrumentation, Heyden & Sons Ltd.,
        ISBN: 0-85501-452-0, 1979.

Cav80   Cavill, J., Essentially Modular Acquisition System for the
        Specialist, Control and Instrumentation, Volume 12, No. 7, 39-41,
        July 1980.

Cha80   Chard, A., Blasted Computer's Gone Down! What Are We Going to Do
        Now?, Control and Instrumentation, Volume 12, No. 9, 55-57,
        September 1980.

Chu70   Chuang, T., Misko, G., Dalla Lana, I.G. and Fisher, D.G., On-Line
        Operation of a PE621 Infrared Spectrophotometer - IBM/1800 Computer
        System, Chapter 7, 75-92, in Computers in Analytical Chemistry,
        (Volume 4 - Progress in Analytical Chemistry), Selected papers
        from the 1968 Eastern Analytical Symposium, Plenum Press, SBN:
        306-39304-02, 1970.

Col80   Columbia Data Products, Inc., 9050 Red Branch Road, Columbia,
        Md21045, USA. Reference Manual for Model 400 Mini-floppy Disk
        System, 1980.

Cos80   Cosgrove, T., Littler, J.S. and Stewart, K., A Microprocessor
        Controlled Fast Data Collection and Display Store, J. Phys, E:
        Sci. Instrum., Volume 13, 821-822, 1980.

Cot70   Cotton, J.M., Application of the Infotronics CRS-110/50 Computer
        Integrator Systems for On-Line GC Analysis, Chapter 6, 63-74, in
        Computers in Analytical Chemistry, (Volume 4 - Progress in
        Analytical Chemistry). Selected papers from the 1968 Eastern
        Analytical Symposium, Plenum Press, SBN: 306-39304-2, 1970.

Cou72   Coury, F.F., A Practical Guide to Minicomputer Applications.
        IEEE Press, ISBN: 0-87942-005-7, 1972.

Cra78   Cram, S.P. and Risby, T.H., Gas Chromatography, 213R-243R, in
        "Fundamental Reviews", Analytical Chemistry, Volume 50, April
        1978.

Cra80   Cram, S.P., Risby, T.H., Field, L.R. and Wie-Lu, Y., Gas
        Chromatography 324R-360R, in "Fundamental Reviews", Analytical
        Chemistry, Volume 52, No. 5, April 1980.

Del78   Delaney, M.F. and Uden, P.C., Digitiser for Generating Computer
        Readable Data, Analytical Chemistry, Volume 50, No. 14, 2156-
        2157, 1978.

EMI78   Modern Instrumentation Tape Recording: An Engineering Handbook,
        The Engineering Department, EMI Technology, Inc., 1978.

EMI80   SE Labs (EMI) Ltd., Feltham, Middlesex, UK, 1980.

Erk78   Van Erk, R., OSCILLOSCOPES - Fundamental Operation and Measuring
        Examples, McGraw-Hill, ISBN: 0-07067050-1, 1978.

Far80   Farley, J.W., Johnson, A.H. and Wing, W.H., Microcomputer
        Controlled Microwave-Optical Spectrometer, J. Phys. E: Sci.
        Instrum., Vol. 13, 848-856, 1980.

For80   Fornili, S.L., Simple Microprocessor-Based System for Data
        Acquisition and Control, J. Phys. E: Sci. Instrum., Volume 13,
        34-36, 1980.

Fra79   Frazer, J.W., Rigdon, L.P., Brand, H.R. and Pomernacki, C.L.,
        Characterising Chemical Systems with On-line Computers and
        Graphics, Analytical Chemistry, Volume 51, No. 11, 1739-1747,
        September 1979.

Fra79a Frazer, J.W., Rigdon, L.P., Brand, H.R., Pomernacki, C.L. and
        Brubaker, T.A., Characterising Chemical Systems with On-line
        Computers and Graphics: Alkaline Phosphotase Catalysed Reaction,
        Analytical Chemistry, Volume 51, No. 11, 1747-1754, September
        1979.

Fra79b Franklin, W.E., Direct Pyrolysis of Cellulose and Cellulose
        Derivatives in a Mass Spectrometer with a Data System, Analytical
        Chemistry, Volume 51, No. 7, 992-996, June 1979.

Fri80   Frith, A., Microprocessor Based Data Acquisition and Control
        System, Control and Instrumentation, Volume 12, No. 7, 31-34,
        July 1980.

Gam80   Gampp, H., Maeder, M., Zuberbuhler, A.D. and Kaden, T.A.,
        Microprocessor Controlled System for Automatic Acquisition of
        Potentiometric Data and their Non-linear Least Squares Fit in
        Equilibrium Studies, Talanta, Volume 27, 513-518, 1980.

Goe79   Goeringer, D.E. and Pardue, H.L., Time Resolved Phosphoresence
        Spectrometry with a Silicon Intensified Target Vidicon and
        Regression Analysis Methods, Analytical Chemistry, Volume 51, No.
        7, 1054-1060, June 1979.

Har79   Harnly, J.M., O'Haver, T.C., Golden, B. and Wolf, W.R., Background
        Corrected Simultaneous Multi-element Atomic Absorption Spectro-
        meter, Analytical Chemistry, Volume 51, No. 12, 2007-2014,
        October 1979.

Has80   Hassle, A.B., Doctor is that Drug Safe?, Digital Equipment
        Corporation, DECNEWS, Volume IV, No. 5, 1-3, March 1980.

Hew70   Hewitt, R.C., Computer Interface and Digital Sweep for an NMR
        Spectrometer, Chapter 5, 49-62, in Computers in Analytical
        Chemistry, (Volume 4 - Progress in Analytical Chemistry), Selected
        papers from the 1968 Eastern Analytical Symposium, Plenum Press,
        SBN: 306-39304-2, 1970.

Hew80   Hewlett-Packard Ltd., King Street, Winnersh, Wokingham, Berkshire,
        RG11 1BR.

Hew80a Hewlett-Packard, 6942A Multiprogrammer Technical Data,
        Publication No. 5952-4034, July 1980.

Hna76   Hnatek, E.R., A User's Handbook of D/A and A/D Converters, John
        Wiley, ISBN: 0-471-40109-9, 1976.

Hoe68   Hoeschele, D.F., Analog-to-Digital/Digital-to-Analog Conversion
        Techniques, John Wiley, New York, 1968.

Hol80   Holtzman, J.L., Microprocessor-based Computer for the Acquisition
        and Manipulation of Data in Rapid Kinetic Studies, Analytical
        Chemistry, Volume 52, No. 6, 989-991, May 1980.

Hol80a Holstead, D., A Friendly Fellow Indeed is the Modern Data Logger,
        Control and Instrumentation, Volume 12, No. 7, 34-37, July 1980.

Hun79   Hunter, R.L. and McIver, R.T., Mechanism of Low-Pressure Chemical
        Ionisation Mass Spectrometry, Analytical Chemistry, Volume 51,
        No. 6, 699-704, May 1979.

Hut80   Hutchinson, R., Why Wet Chemical Analysis is Going On-line,
        Control and Instrumentation, Volume 12, No. 3, 39, March 1980.

IBM70   IBM Corporation, System/360 Component Descriptions - 2820 Storage
        Control and 2301 Drum Storage, Form: GA22-2895-3, 1970.

IBM73   IBM Corporation, Instroduction to IBM Direct Access Storage
        Devices and Organisation Methods, (Student Text), Form:
        GC20-1649-06, February 1973.

IBM76   IBM Corporation, IBM 3830 - 3330 Disk Storage Reference Manual,
         Form: GA26-1592-5, 1976.

IBM77   IBM Corporation, IBM 3350 Direct Access Storage, Form:
         GA26-1638-2, 1977.

IMI80   IMI - International Memories Incorporated, 10381 Bandley Drive,
         Cupertino, CA950145, USA., Technical Specifications for the IMI7710
         Disk Drive, 1980.

Jon69   Jones, K. and Fozard, A., On-line Data Processing of Laboratory
         Data Instruments, Chemistry in Britain, Volume 5, No. 12,
         552-556, December 1969.

Jos79   Joshi, B.M., and Sacks, R.D., Circular Slot Burner - Droplet
         Generator System for High Temperature Reaction and Vapour Transport
         Studies, Analytical Chemistry, Volume 51, No. 11, 1781-1785,
         September 1979.

Kel73   Kelly, P.C. and Horlick, G., Practical Considerations for
         Digitising Analog Signals, Analytical Chemistry, Volume 45,
         518-527, 1973

Koh80   Kohonen, T., Lee-Frampton, J. and Voom, G., Monitoring a River in
         Furtherance of Conservation, Control and Instrumentation, Volume
         12, No. 1, 41-43, January 1980.

Laf72   Laffan, R.J., Peterson, A., Hitch, S.W. and Jeunelot, C., A
         Technique for Prolonged Continuous Recording of Blood Pressure of
         Unrestrained Rats, Cardiovascular Research, Volume 6, 319-324,
         1972.

Lar73   Larsen, D.G., Automated Data Logging using a Digital Logger,
         Analytical Chemistry, Volume 45, No. 1, 217-220, January 1973.

Lev64   Leveson, L.L., Introduction to Electroanalysis, Butterworths,
         London, 1964.

Ley77   Leyden, D.E., Rothman, L.D. and Lennox, J.C., Voltage Controlled
         Oscillator for Signal Proportional Data Acquisition Rates,
         Analytical Chemistry, Volume 49, No. 4, 681-682, April 1977.

Lil80   Lilley, T., Acoustic Emissions from Polymers, Part I: Chemical and
         Analytical Aspects, Analytical Proceedings of the Royal Society of
         Chemistry, Volume 17, No. 10, 432-433, October 1980.

Llo80   Lloyd, L.B., Burnham, R.K., Chandler, W.L., Eyring, E.M. and
         Farrow, M.M., Fourier Transform Photoacoustic Visible Spectroscopy
         of Solids and Liquids, Analytical Chemistry, Volume 52, No. 11,
         1595-1598, September 1980.

Low79   Lowman, D.W. and Maciel, G.E., Determination of Sucrose in Sugar
         Beet Juices by Nuclear Magnetic Resonance Spectrometry, Analytical
         Chemistry, Volume 51, No. 1, 85-90, January 1979.

Mac80   MacDonald, N., How the Microprocessor has Revolutionised the Data
         Loggers, Control and Instrumentation, Volume 12, No. 7, 43-45,
         July 1980.

Mal80   Malmstadt, H.V., Push Button Analysis, Journal of Automatic Chemistry, Volume 2, No. 3, 115-117, July 1980.

McD78   McDonald, R.S., Infrared Spectrometry, 282R-299R, in "Fundamental Reviews", Analytical Chemistry, Volume 50, April 1978.

McD80   McDonald, R.S., Infrared Spectrometry, 361R-383R, in "Fundamental Reviews", Analytical Chemistry, Volume 52, No. 5, April 1980.

Mic80   Microdata Ltd., Monitor House, Station Road, Radlett, Herts, WD7 8JX.

Mil77   Milano, M.J. and Kwang-Yil, K., Diode Array Spectrometer for the Simultaneous Determination of Hemoglobins in Whole Blood, Analytical Chemistry, Volume 49, No. 4, 555-559, April 1977.

Nei79   Neil, J.T., OP AMPS - Basic Theory and Practical Applications, Electronics Circuit Design, No. 1, 59-69, Modmags Ltd., Autumn 1979.

New73   Newman, L., The Use of the Large Computer in Chemical Instrumentation - Application to Magnetic Resonance Spectroscopy, Chapter 1, 3-41 in Volume 3 of Computers in Chemistry and Instrumentation (Spectroscopy and Kinetics), edited by Mattson, J.S., Mark, H.B. and MacDonald, H.C., Marcel Dekker, Inc., ISBN: 0-8247-6058-1, 1973.

Orr70   Orr, C.H. and Norris, J.A. (eds), Computers in Analytical Chemistry (Volume 4 - Progress in Analytical Chemistry), Selected papers from the 1968 Eastern Analytical Symposium, Plenum Press, SBN: 306-39304-2, 1970.

Orr80   Orr, T., OP AMPS - Basic Theory and Useful Circuits, Electronics Circuit Design, No. 2, 4-18, Modmags Ltd., Winter 1980.

Per71   Perone, S.P. and Eagleston, J.F., On-line Digital Computer Applications in Chemistry, J. Chem. Ed., Volume 48, No. 7, 438-442, July 1971.

Per75   Wilkins, C.L., Perone, S.P., Klopfenstein, C.E., Williams, R.C. and Jones, D.E., Digital Electronics and Laboratory Computer Experiments, Plenum Press, ISBN: 0-306-30822-3, 1975.

Ree80   Reese, C.E., Chromatographic Data Acquisition and Processing. Part 1. Data Acquisition, Journal of Chromatographic Science, Volume 18, 201-206, May 1980.

Ron79   Rony, P.R. and Larsen, D.G., Logic and Memory Experiments using TTL IC's, Book 2, Chapter 9, "Semiconductor Memories", 238-305, Howard, W. Sams & Co. Inc., ISBN: 0-672-21543-8, 1979.

Ros71   Ross, H.M., Character Codes, 209-234, in Computer User's Year Book, 1971.

Ros75   Rosner, R.A., Penny, B.K. and Clout, P.N. (eds), On-line Computing in the Laboratory, Proceedings of a Conference entitled 'On-line Computers for Laboratory Use', held at Imperial College, London, 11th-12th September 1975, ISBN: 0-903012-33-2, 1975.

RSC80   R.S. Components Ltd., 13-17 Epworth St., London EC2P 2HA, UK, Data
        Sheet No. R/4052, A/D Convertor IC, July 1980.

Sch72   Schroeder, R.R., Operational Amplifier Instruments for Electro-
        chemistry, Chapter 10, 263-350, in Volume 2 of Computers in
        Chemistry and Instrumentation (Electrochemistry: Calculations,
        Simulation and Instrumentation), edited by Mattson, J.S., Mark,
        H.B. and MacDonald, H.C., Marcel Dekker, Inc., ISBN: 0-8247-1443-4,
        1972.

Sch75   Schofield, J.W., Data Collection from Water Quality Monitors,
        362-374, in On-line Computing in the Laboratory, Proceedings of a
        Conference entitled 'On-line Computers for Laboratory Use', held at
        Imperial College, London, 11th-12th September 1975, ISBN:
        0-903012-33-2, 1975.

Sha80   Shackil, A.F., Digital Storage Oscilloscopes, IEEE Spectrum,
        Volume 17, No. 7, 22-25, July 1980.

She73   Shepherd, T.M. and Vincent, C.A., Digital Data in Chemistry,
        Chemistry in Britain, Volume 9, No. 1, 66-70, January 1973.

Smi72   Smith, D.E., Applications of On-line Digital Computers in AC
        Polarography and Related Techniques, Chapter 12, 369-422, in Volume
        2 of Computers in Chemistry and Instrumentation (Electrochemistry:
        Calculations, Simulation and Instrumentation, edited by Mattson,
        J.S., Mark, H.B. and MacDonald, H.C., Marcel Dekker, Inc., ISBN:
        0-8247-1433-4, 1972.

Sol80   The Solartron Electronic Group Ltd., Farnborough, Hampshire, GU14
        7PW, UK.

Str73   Strobel, H.A., Chemical Instrumentation: A Systematic Approach to
        Instrumental Analysis, Addison-Wesley, ISBN: 0-201,07301-3, 1973.

Ull79   Ullman, A.H., Pollard, B.D., Boutilier, G.D., Bateh, R.P., Hanley,
        P. and Winefordner, J.D., Computer Controlled Multi-element Atomic
        Emission/Fluorescence Spectrometer System, Analytical Chemistry,
        Volume 51, No. 14, 2382-2387, December 1979.

Var80   Varian Associates Ltd., 28 Manor Road, Walton-on-Thames, Surrey,
        KT12 2QF, UK., E-935 Data Acquisition System, 1980.

Vea79   Veazey, R.L. and Nieman, T.A., Chemiluminescent Determination of
        Clinically Important Organic Reductants, Analytical Chemistry,
        Volume 51, No. 13, 2092-2096, November 1979.

Wal80   Walls, R., A New Way of Using the Microprocessor in Process
        Control, Control and Instrumentation, Volume 12, No. 3, 61-63,
        March 1980.

Wei79   Weinkam, R.J. and D'Angona, J.L., Accurate Mass Measurement at Low
        Resolving Power with Mass Spectrometer - Computer Systems,
        Analytical Chemistry, Volume 51, No. 7, 1074-1077, June 1979.

Wel80   Welfare, M. and Gaffon, R.E., Analyser System Packages and Costs
        Boost Reliability, Control and Instrumentation, Volume 12, No. 3,
        33-37, March 1980.

Wey70   Weyler, P.A., A Computer Controlled X-ray Spectrograph, Chapter 4,
         39-48, in Computers in Analytical Chemistry, (Volume 4 - Progress
         in Analytical Chemistry), Selected papers from the 1968 Eastern
         Analytical Symposium, Plenum Press, ISBN: 0-340-23652-3, 1980.

Wil80   Wilkinson, B. and Horrocks, D., Computer Peripherals, Hodder and
         Stoughton, ISBN: 0-340-23652-3, 1980.

Won78   Wonsiewicz, B.C., Storm, A.R. and Sieber, J.D., UNIX Time Sharing
         System: Microcomputer Control of Apparatus, Machinery and
         Experiments, The BELL System Technical Journal, Volume 57, No.
         76, 2209, 2232, July-August, 1978.

Yar80   Yarnitzky, C., Osteryoung, R.A. and Osteryoung, J., Instrument
         Design for a One-drop Square Wave Analyser, Analytical Chemistry,
         Volume 52, No. 8, 1174-1178, July 1980.

6

Interfaces and Principles of Interfacing

INTRODUCTION

For the discussion that is to take place in this chapter the common features of analytical instruments and computers need to be emphasised. Undoubtedly, their most important similarity lies in the fact that they are both examples of machines. That is, they perform some labour saving task on behalf of their human controller. Machines, in general, have various kinds of inputs and usually produce a number of different types of output. Some of these inputs (and outputs) may take the form of control signals. These are used to synchronise the machine's activity or monitor its performance. The relationship between the inputs to and outputs from a 'generalised' machine is depicted schematically in Fig. 6.1.

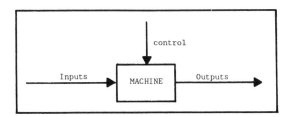

Fig. 6.1 Inputs and outputs for a generalised machine.

Over the last few years there has been an increasing demand for facilities that enable a machine of one kind to be connected to a second, possibly different, type of machine. There are several reasons underlying this important requirement. They arise from the need (a) to enable one machine to control another (see Fig. 6.2A); (b) to produce more sophisticated types of machine (see Fig. 6.2B) using simpler machines as primitive building blocks (for example, the GC/MS and GC/IR combinations discussed in Chapter 3); and (c) to implement some automated chain of processes (see

Fig. 6.2C) in which outputs from one stage are passed automatically to a subsequent stage.

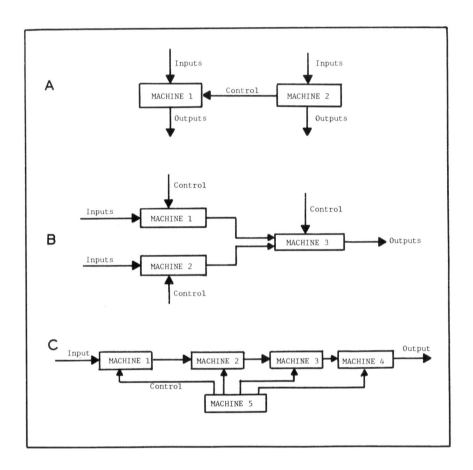

Fig. 6.2   Some possible approaches to machine inter-connection.

Because machines will usually have different structures, inputs, outputs and control requirements, it is not always an easy matter to connect one directly to the other. In a large number of cases the inter-connection will necessitate the availability of a special type of coupling device called an interface. The function of this is quite simple. It converts the outputs produced by one machine into a form that is compatible with the input requirements of another. This role is depicted in Fig. 6.3. As can be seen from this sketch, some of the outputs produced by the first machine are compatible with the corresponding inputs on the target machine. Naturally, these will not need to be transformed or processed by the interface. Only those outputs which are in some way incompatible with the input requirements of the second machine will need to be manipulated. The type of manipulation that is involved will depend upon the detailed nature of the inputs and outputs - for example, whether they are electrical signals, flowing materials, mechanical linkages or whatever.

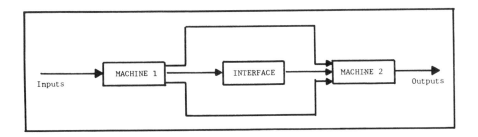

Fig. 6.3   Machine inter-connection by means of an interface.

The logical function performed by the interface can easily be seen by inspecting the diagram contained in Fig. 6.4. The upper part of this figure represents the two machines, instruments or devices that are to be linked together. Each of the boxes corresponds to an instrument. The lines within the boxes represent material or information flow requirements of the individual devices. The jagged edges are used to emphasise their dis-similarity. As can be seen from the lower diagram the interface enables the smooth inter-connection of the two (or sometimes more) instruments.

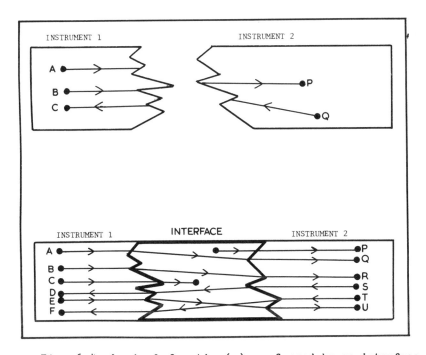

Fig. 6.4   Logical function(s) performed by an interface.

In the lower part of the diagram (Fig. 6.4) the communication lines
presented by instrument 1 to the interface are labelled A through F and
those in the second are denoted by the letters P through U. The first,
and probably major, function of the interface is to physically inter-link
a subset of lines from the set (A,B,C,D,E,F) to an appropriate subset of
the other set (P,Q,R,S,T,U). Its second function, as was mentioned above,
is that of achieving material/signal transformation. In the case of
electrical signals this may involve level shifting from one voltage to
another; in a digital communication link the interface may have to act as
a buffer in order to match the differing transmission speeds of the two
instruments. Similarly, appropriate pressure and flow regulation might be
necessary in order to achieve compatibility between the devices being
joined together. The third function of the interface is to provide
facilities for line termination and origination using dummy sinks and
sources. These are necessary in those situations where instruments are
incompatible because outputs from one cannot be matched by corresponding
inputs on its partner and vice versa. This situation is illustrated in
Fig. 6.4 by lines C and P. The former needs terminating within the
interface while the latter needs an appropriate source to be generated by
the interface. In addition, on some occasions it may be necessary to
bring out additional lines from one instrument or lead them into another.
Again, the interface provides the means for achieving all of these
requirements.

Most of this chapter is devoted to electrical interfaces although some
mention is made of other kinds. In the next section some of the currently
available standard electrical interfaces are examined.

STANDARD INTERFACES

Based upon the discussion that was presented in the previous section it is
easy to see that interfaces often appear (logically) as a series of inter-
connecting lines between two or more devices. Consequently, they are
often referred to as interface buses - by analogy with computer systems.
Over the years many different types of standard interface bus have been
proposed and developed. The majority of these cater for the requirements
of those wishing to connect peripherals or instruments to computer
systems.

Internal and External Interface Buses

There are various ways in which interfaces may be classified. Ellefsen
(Ell80) has suggested a sub-division into two basic types: (a) internal,
and, (b) external. Some examples of these are presented schematically in
Fig. 6.5.

Internal interface buses tend to be fixed or rigid in nature and are
usually contained within some instrument or machine system. Various
hardware modules within the instrument or computer plug into the interface
bus using some form of multi-way edge connectors wired according to a
common convention. An example of an internal interface bus was described
in chapter 4 - the UNIBUS which is a central part of the DEC PDP-11
minicomputer architecture. Since the development of microcomputer systems
there have been many other types of bus developed. A few of the more well

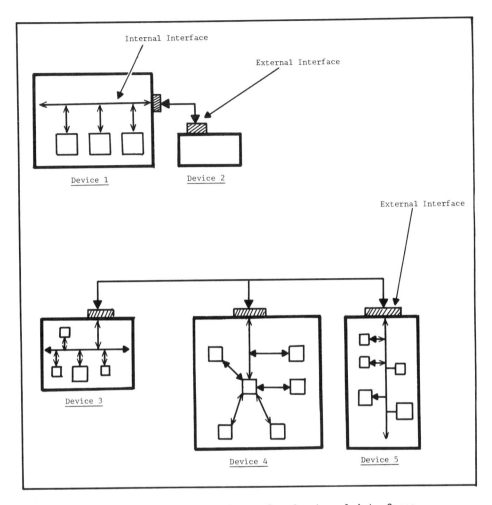

Fig. 6.5   Some examples of internal and external interfaces.

known examples of these include MULTIBUS (from the Intel Corporation), Z-BUS (from Zilog), MICROBUS (from National Semiconductor), and the S-100 bus (from Altair). Some of these will be described in more detail later.

The other category of interface bus, the external type, is often temporary in nature and usually takes the form of a flexible multi-core inter-connecting cable with appropriately wired plugs/adaptors at either end. The plugs slot into matching sockets mounted on the devices that are to be inter-connected. When several devices are to be linked together on a common bus the plugs are designed to be of a stackable variety. Various types of external standard buses exist. Three of the most well known are probably the IEEE-583 (CAMAC), the IEEE-488 (instrumentation bus) and the EIA RS-232-C interfaces. A brief description of each of these will be given later.

Parallel and Serial Interface Buses

An alternative approach to characterising interfaces is by the mode in which they transfer data signals. Two basic types can again be distinguished: (a) parallel, and, (b) serial interfaces. Those of the parallel kind are used for high speed communication between internal modules of a system (for example, microprocessor or microcomputer buses) and for system to system communication in the case of external interfaces. The IEEE-488 and 583 buses are examples of parallel interfaces as are the UNIBUS and S-100 bus. Parallel buses transfer all bits of data along separate lines simultaneously. The number of data lines will depend upon the design of the interface - eight is commonly used. Figure 6.6 shows a simple example of a parallel digital interface having four data lines.

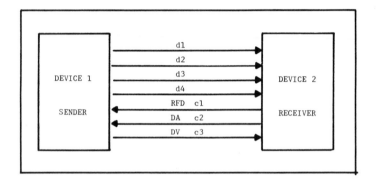

Fig. 6.6 Simple example of a parallel digital interface.

In this simple illustration, d1, d2, d3 and d4 represent the data lines across which signals are to be transmitted. So that the transmission of data between the two devices can be controlled and synchronised some additional interface management lines will be required. These are labelled c1, c2 and c3 in Fig. 6.6. The significance of these lines is as follows:

            RFD - Ready for Data   (c1),
            DA   - Data Acknowledge   (c2), and,
            DV   - Data Valid   (c3).

The three control lines enable the devices to communicate with each other according to a well-defined prototcol or 'handshaking' procedure. This protocol is usually expressed in terms of the logic state or voltage level on the individual lines as a function of time. The graphs that are used reflect the time dependent states of the interface lines and are commonly referred to as timing diagrams. Figure 6.7 illustrates a simple example of such a diagram.

When the receiving device is ready to receive data it indicates this condition to the sender by raising the signal level of the RFD line from low to high (at time t1). At the same time the DA line 'goes low' to indicate that valid data has not yet been received by device 2. As soon as device 1 detects the high state on the RFD line it assembles its data

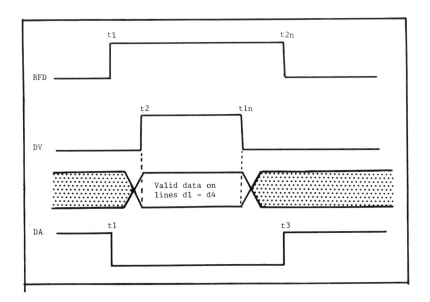

Fig. 6.7   Timing diagram illustrating data exchange between
two instruments.

ready for transmission on the d1 - d4 lines. When the data is ready to be
sent, device 1 raises the level of the signal on the data valid line (DV)
from low to high (at time t2). This now indicates that there is valid
data on the data lines which may be read by device 2. Until the DV signal
goes high the states of the data lines will be indeterminate - this will
also be the case after the DV signal goes from high to low. Thus, the
sender only guarantees to maintain valid data on the data lines during the
time that the DV line is high. Once the data has been put onto the
interface and DV goes into a high state there are two possible outcomes.
The sender can either maintain the signals on DV and d1 through d4 until
the receiver acknowledges receipt of data by driving the DA line high (at
the time t3) after which the whole process is repeated in order to
transmit the next item of data. Alternatively, the sender can maintain
the states of DV and d1 - d4 for a fixed period of time and then, if DA
has not gone high to indicate receipt of data, it can try repeating the
whole process again. In more sophisticated systems there would be
additional control lines to aid both the sender and receiver to test for
malfunction of the lines and corruption of data during transmission.
Similarly, there would be suitable protocol arrangements to enable
bidirectional tranmission of data on the bus.

When a common parallel bus is used to transmit data between many
instruments an additonal set of lines will be required (called address
lines) to enable individual instruments to be identified and isolated for
attention. An illustration of a typical bus of this type is shown in Fig.
6.8. In this example the four address lines would permit a total of
sixteen devices to communicate via the shared data lines. Each device
would have its own unique address to enable it to be identified in a data
exchange transaction.

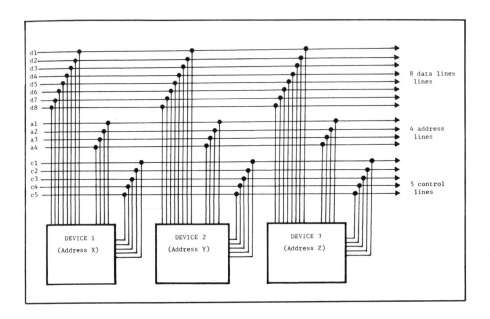

Fig. 6.8   Parallel interface bus with addressable devices.

In contrast to parallel interface buses, serial transmission systems operate using one or two connection lines in order to carry all of the necessary signals between system modules. Thus, to achieve the transmission of eight bits of data using a serial interface an arrangement similar to that depicted in Fig. 6.9 might be used.

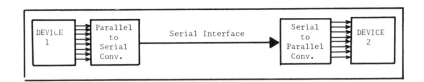

Fig. 6.9   Use of a serial interface for transmitting data.

In this example, device 1 produces an eight bit data byte that is to be transmitted to device 2 along a serial data interface. As a means of achieving this a special parallel-to-serial converter is used to serialise the data prior to its transmission. At the receiving end of the interface a serial-to-parallel converter translates the data back into parallel form for delivery to device 2. Integrated circuits are available to perform these conversions. Figure 6.10 shows the mechanism by which they operate.

In many standard serial tranmission interface schemes the data bits (b1 through b8) are framed by the addition of extra bits at the front and rear of the string to be transmitted. Usually a start bit is placed at the

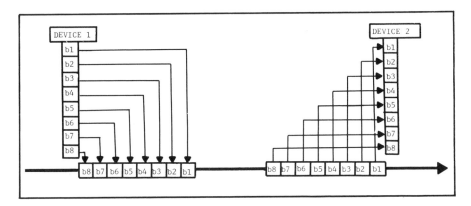

Fig. 6.10   Data flow involved in parallel-to-serial conversion.

front and one or two stop bits at the end.   These serve to alert the
receiver to the presence of data on the interface (start bit) and provide
a short time delay (stop bits) that enables the device to prepare itself
for receipt of the next unit of data.   The eight data bits are used to
encode the data to be sent between the two devices.   Two commonly used
standard coding schemes are EBCDIC (Extended Binary Coded Decimal Interchange Code) and ASCII (American Standard Code for Information Interchange).   The first of these is an eight bit code representing a 256
character alphabet while the second is a seven bit coding scheme that
encodes an alphabet of 128 symbols.   The mapping between bit patterns and
the characters they represent will be found in most books on microcomputer
interfacing (Zak79, Cam79).   In addition to the stop and start bits that
are used to frame the data a parity bit is also added in order to aid
error detection (Zak79).   Data transmission can be accomplished using
either even or odd parity.

The IEEE-488 Interface Bus

Because of the need to have some form of universal standard that would
enable the interconnection of laboratory instruments (to each other and to
computers) the International Electrotechnical Commission (IEC) formulated
recommendations regarding the requirements that such a standard should
meet (Lou74, Kno75).

As a result of these recommendations the Institute of Electrical and
Electronic Engineers published the IEEE-488 standard in 1975.   This was
slightly modified and updated in 1978.   Since its introduction this
standard has been adopted by a large number of laboratory instrument and
microcomputer manufacturers.   One leading instrument producer, Hewlett
Packard, has been closely involved with the development of this standard.
Their implementation of the IEEE-488 interface is known as HP-IB which is
an acronym for Hewlett Packard Interface Bus.   Some early descriptions of
this interface will be found in the papers by Loughry (Lou72), Nelson
(Nel72) and Ricci (Ric74).   In addition to the officially published
documentation (IEE75), descriptions of the IEEE-488 interface will be

found in appropriate microcomputer manuals (COM79), instrument handbooks (Hew79), peripheral handbooks (Hew79a) and a large number of other related publications (Zak79, Ell80, Ham80).

The IEEE-488 bus is a bit parallel, byte serial asynchronous bidirectional interface enabling the interconnection of up to 15 instruments at data transmission speeds of up to about 1 Mbyte/sec. There are certain cabling restrictions that limit the separation between devices to a maximum of about 20 metres. In principle the bus is similar to that illustrated in Fig. 6.8 except there are no explicit address lines. Instead, instrument addresses are transmitted over the least significant 5 bits of the data lines. In addition to the eight data lines (labelled DIO1, DIO2, ... DIO8), there are two other groups of signal lines:

(a) Handshake lines - of which there are three,

        DAV   - Data Valid,
        NRFD - Not Ready for Data,
        NDAC - Not Data Accepted, and,

(b) Management lines - of which there are five,

        ATN   - Attention,
        EOI   - End or Identify,
        SRQ   - Service Request,
        IFC   - Interface Clear,
        REN   - Remote Enable.

The attention line (ATN) is used to inform all devices attached to the system that an interface message (for example, an address or other command) is present on the DIO lines. Data transfer cannot take place while ATN is low (logical true). EOI has two functions: it can be used to indicate the end of a multi-byte transfer when it is set during transfer to the last byte of data; alternatively, it can be used as part of the handshaking procedure in parallel polling of devices - polling is the term used to describe a situation in which devices on the bus are invited to indicate their status by placing appropriate signals on the DIO lines. The SRQ line provides a facility whereby individual devices are able to indicate that they require some form of special servicing. IFC is used to reset the interface to a known clear state - usually the idle state. The final management line, REN, can be used with certain devices as a means of instructing the instruments attached to the bus to take their control/operating information off the interface bus rather than from their front control panels.

Devices that are connected to the interface bus can be in any of three states: inactive, receiving or transmitting. Instruments (or devices) that are only able to receive data (for example, printers, paper tape punches, displays or graph plotters) are referred to as 'listeners' while those that are able to transmit data (for example, a tape reader, counter or voltmeter) are called 'talkers'. Some devices can operate as both listener and talker (for example, a digital voltmeter or other progammable instrument). In addition to listeners and talkers the interface may also have 'controllers' attached to it. These are devices that are capable of managing communications over the interface bus such as addressing and sending commands. A calculator or computer with an appropriate I/O interface - such as the Commodore PET (Ham80) - is an example of this type of device. At any one time only one talker can be active but there may be up to fourteen active listeners. Although more than one controller may be attached to the bus, only one can be active at any one time.

Instruments will have individual 5 bit addresses assigned to them by means of a set of switch settings. Devices that can act as both a listener and a talker will have two addresses - a listen address and a talk address. The appearance of the interface connector and the address switches for a typical instrument - a digital multimeter (Hew79) - are illustrated in Fig. 6.11.

Instruments are addressed by the controller when it sets the ATN line low and places the instrument's address on data lines DIO1 through DIO5. While this is happening bits DIO7 and DIO6 are used to specify the function that the device is to play. If the addressed device is to become a talker then DIO7 and DIO6 are set to logic 1 and 0 respectively. If the device is to operate as a listener then the controller sets DIO7 and DIO6 to 0 and 1, respectively. When the ATN line goes high (false) the interface goes into data mode. In this mode data may be transferred between devices that were addressed when the interface was in command mode. Messages that can be transferred in data mode may be programming instructions (for programmable instruments) and data codes. Programming instructions consist of a series of seven bit bytes placed on the DIO lines. The meaning of each byte is device dependent and is selected by the equipment designer. These types of message are usually sent between an interface controller acting as a talker and a single device that has been addressed as a listener. Data codes are also seven bit bytes placed on the data lines. The meaning of each byte is device dependent. For meaningful communication to occur, both the talker and the listener must agree on the meaning of the codes that they use. For the digital multimeter mentioned above (Hew79) the data output format is a fixed length string of 13 characters having the following significance:

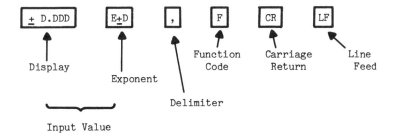

In the above character string sequence the Function Code is a single digit that specifies the mode of operation that the instrument was in when the reading was taken. The allowed values for the digital multimeter are as follows:

Measuring Function	Code
DC volts	1
AC volts	2
DC current	3
AC current	4
ohms	5

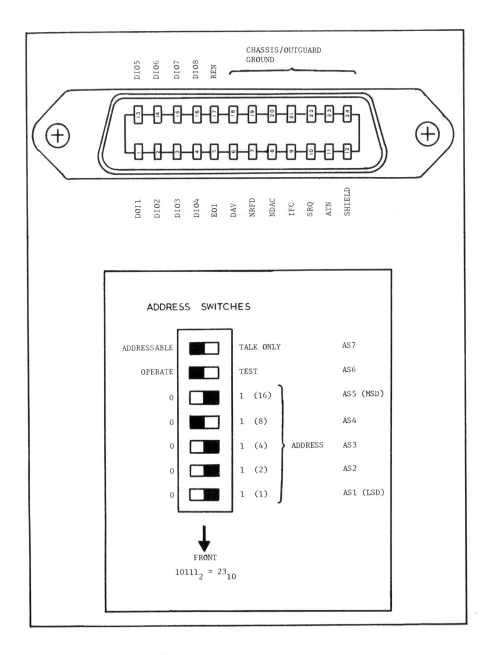

Fig. 6.11 IEEE-488 interface port and address switches that are
used to set up the talk and/or listen addresses of an instrument.

thus, if the reading on the front display of the instrument had been 24.3 volts DC the output data from the instrument would have been

+2.430E+1,1 CR LF

while if the front panel of the instrument was displaying the value 100.1 K ohms the output would have been

+1.001E+5,5 CR LF

Further details of the IEEE-488 interface and Hewlett Packard's implementation of it (HP-IB) will be found in the references cited at the beginning of this section. An example of the use of this interface for interconnecting a PET microcomputer with a digital multimeter will be described later in this chapter.

CAMAC - The IEEE-583 Interface Standard

This standard, like that outlined in the previous section, describes a (word) parallel interface. However, unlike the IEEE-488 interface it is designed to function like a generalised computer Input/Output bus. The name CAMAC is an acronym for Computer Automated Measurement And Control standard. It was introduced in the 1960s to handle interfacing problems encountered in instrumentation for atomic power and nuclear physics. The CAMAC system is based upon the use of card cages called crates. Each of these contains a controller and up to 24 peripheral interfaces (or modules). Different types of functional card slot into the basic cage in order to produce a crate that can thus be tailored to meet the requirements of any particular system. Figure 6.12 (diagram A) shows a typical example of a single crate CAMAC system (Ste75).

Diagram B (of Fig. 6.12) presents a side elevation of a CAMAC module. It illustrates the edge connector by which the card plugs into a common bus. The remaining diagrams contained in Fig. 6.12 depict how individual crates may be interconnected in parallel (sketch C) or serial (sketch D) depending upon the demands of the particular application.

The standard specifications for CAMAC units are rigorously defined. Details are provided for the physical dimensions of each card and containing crate, for connector types, for the power supply, for the mode of connecting crates in serial/parallel form and for the various signal lines (Dataway bus) used to connect the crate controller with its associated modules. There are four relevant standards involved:

IEEE-583-1975   - Modular Instrumentation and Digital Interface
                  System,
IEEE-595-1976   - Serial Highway Interface System,   .
IEEE-596-1976   - Parallel Highway Interface System,
IEEE-683-1976   - Block Transfers in CAMAC Systems,

which are available as a single publication from the Institute of Electrical and Electronic Engineers (IEE77).

The signal lines that constitute the CAMAC bus system and which enable communication and data transfer between a crate controller and individual modules are illustrated in Fig. 6.13. Individual modules are addressed by the crate controller by means of the module address lines. Each module

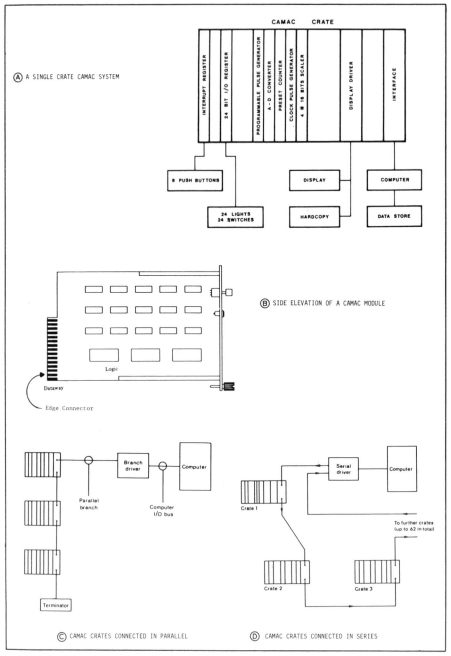

Fig. 6.12   Components of a CAMAC system and the ways in which
these may be interconnected; both parallel and serial combinations
can be constructed.

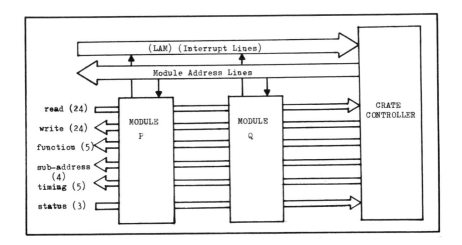

Fig. 6.13 Logical arrangement of signal lines in a CAMAC system.

has an interrupt line (LAM - Look At Me) wired through to the controller. These lines, used in conjunction with the bus timing and status lines, enable basic communication to take place between the crate controller and the modules within the crate. Data is transferred between the two by means of the read and write lines. Each of these has a data width of 24 bits. The overall data transfer rate is 24 Mbit/sec. It is easy to see that CAMAC is thus both a faster and wider bus than the IEEE-488 interface. Furthermore, it is more versatile in that it allows standard sub-addressing and a number of standard functions. Sub-addressing is achieved through the 4 sub-address lines. They permit up to 16 possible sub-sections within a module to be addressed. The 5 function (or command) lines ae used to specify the various operations that are to be performed in a module during a command. The status lines provide feedback information from individual modules to the crate controller, for example, 'command accepted', 'crate busy', and so on.

Further details on the CAMAC system will be found in a variety of sources (Mey80, Ros75, Zak79, Hor76) including the CAMAC bulletin (CAMBU) which is issued three times yearly. A paper by Peatfield (Pea74) leads a whole series of papers devoted to CAMAC; these cover a wide variety of application descriptions. A comparison of the CAMAC and the HP-IB (IEEE-488) interfaces has been presented by Merritt (Mer76) who concluded: CAMAC can support more devices, reach out greater distances and carry more data at higher speeds than can an HP-IB system. However, it is more expensive than HP-IB due to the need for a crate and a controller. Hence, for simple monitoring and data acquisition the HP-IB system is difficult to beat. The advantages of the HP-IB centre around its simple 16 wire interface and uncomplicated command structure.

The EIA RS-232-C and CCITT V.24 Interfaces

These are essentially communications interfaces but can be used for a wide variety of other purposes. They have arisen because of the need to be

able to interconnect data processing equipment (terminals, microcomputers
and computerised instrumentation) to data communications devices. The
standards are concerned with the interface connections between the Data
Communications Equipment (DCE) and the Data Terminal Equipment (DTE) as
shown in Fig. 6.14.

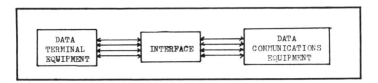

Fig. 6.14   Inter-connection of Data Terminal Equipment (DTE)
and Data Communications Equipment (DCE).

Circuit characteristics for the interface connections and the form of
signals that may appear upon them are standardised in Europe by the
International Consultative Committee for Telephone and Telegraph (CCITT)
and in America by the Electric Industries Association (EIA). The relevant
standards are CCITT V.24 and EIA Standard No. 232-C. They describe the
nature of data exchange operations between any data communications
equipment and any manufacturer's data processing equipment which exchange
serial data at speeds up to 20,000 bits/sec. The standards apply to the
following items:

(1)   Electrical Signal Characteristics   - the definitions
      for the electrical characteristics of the interchange
      signals and the associated circuitry.
(2)   Mechanical Interface Characteristics   - the definition
      of a 25 pin connector between the two inter-connected
      devices.
(3)   Functional Description of Interface Circuits   - the
      specific description of the data timing and control
      circuits for use at this interface.
(4)   Standard Interfaces   - for selected communications
      systems configurations.

Figure 6.15 shows the appearance of the 25-way connector that is used to
connect equipment together. The various interchange circuits involved in
the interface are allocated pin numbers on the connector; the functions of
these are summarised in the tables contained in Fig. 7.22.

The V.24 interface is important because it is widely used in computer
systems for connecting data terminals to computers - either directly or
via a communication network. Many analytical instruments are nowadays
supplied with or utilise an interface of this type. For example, Blank
and Wakefield (Bla79) describe the use of the RS-232-C interface in a
photo-acoustic spectrometer to achieve the storage of spectra in an
external computer. Similarly, Gampp et al (Gam80) have used this same
type of input/output interface in order to store potentiometric data on a
magnetic tape for subsequent analysis using a desk-top computer.

An interesting application of the RS-232 port on a computer has been
outlined by Kuehl and Griffiths (Kue80). They describe a microcomputer
controlled interface between a high performance liquid chromatograph

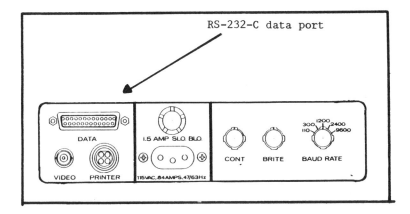

Fig. 6.15   Physical appearance of a typical RS-232-C interface port.

(HPLC) and a diffuse reflectance IR Fourier Transform (FT-IR) spectrometer. A KIM-1 microcomputer, which is based on the 6502 micro-chip
(Zak79a), was interfaced to the HPLC, HPLC/FT-IR interface and the
spectrometer's data system by means of a special Input/Output chip called
a VIA - Versatile Interface Adaptor. All of the software for the
KIM-1 was written in 6502 machine code with the aid of a cross-assembler
and a 6502 simulator running on an IBM 370/158 mainframe computer. The
RS-232 port was used to load the software developed on this system
directly into the KIM microcomputer via a RS-232-C to 20 mA current loop
converter (see next section - Other Standard Interfaces). A similar use
of the RS-232-C interface has been described by O'Driscoll (Dri80) who has
used it to implement a graph plotter interface for serial computer
terminals.

Many laboratory instrument manufacturers are supplying standard RS-232 (or
V.24) interfaces on their equipment. In response to this, suppliers of
data handling systems are making similar facilities available. As an
illustration, the MINC (Modular INstrument Computer) laboratory
system supplied by Digital Equipment Corporation (DEC80) is a good
example. It comes supplied with both a RS-232-C and an IEEE-488
interface. Because it has both of the popular interfaces available, it
thus represents quite a sophisticated laboratory tool.

Further details on the RS-232-C standard will be found in reference
Cam79. The use of this interface in conjunction with communication systems
for the transmission of data will be described in chapter 7.

The S-100 Microcomputer Bus

This interface was originally developed for the Altair microcomputer
system and since its introduction has gained widespread acceptance and has
recently been the subject of IEEE standardisation activities (Lib80,
Elm79). The S-100 bus usually takes the form of a series of 100 pin board
edge sockets wired in parallel via a suitable motherboard. Compatible

devices can then plug straight into the bus. The bus definition includes
16 data lines, 16 address lines, several control lines, three unregulated
DC voltage supply lines (+8v, +16v and -16v) and ground traces. Further
details on each of the pin assignments will be found in Zak79, Mor78 and
Elm79. The papers by Morrow and Elmquist (Mor78, Elm79) outline the
proposed IEEE standard (IEEE-696) for this interface.

An example of the use of this bus in conjunction with an INTEL 8080 microprocessor has been presented by Farley et al (Far80). This paper
describes the construction of a microcomputer controlled microwave optical
spectrometer. The authors claim that the motivation for using this
approach was purely economic since it provided a more cost effective
interface than could be achieved via the use of 'packaged' systems like
CAMAC or the IEEE-488.

The S-100 bus was designed initially for an INTEL 8080 8-bit computer. In
order to upgrade this bus to the levels of performance offered by newer
machines several problems need to be overcome. A series of articles by
Cantrell (Can80a, Can80b, Can80c) describe how some of these problems may
be solved and also illustrate the way in which the INTEL 8088 (16 bit
micro) may be interfaced to the S-100 bus system.

Other Standard Interfaces

A significant number of interfaces have been developed to enable the
various parts of computer based systems to be interconnected with minimal
effort on the part of the user. Many of these relate to the attachment of
peripheral devices to a central computer. Since analytical instruments
are frequently regarded as special purpose computer Input/Output units it
is not unusual to find that instrument producers have tended to adopt
interface standards that have been formulated by computer manufacturers.
The previous parts of this section have outlined some of the popular
standard interfaces. Listed below are some other interesting types of
interface/protocol systems that have been used for interfacing purposes.
Further details on each of these will be found in the reference(s) cited
in the right hand column.

British Standard Interface BS.4421:1969	BSI69, Dav73
Binary Coded Decimal (BCD) Interfaces	Per80
Current Loop Interface	Clu75
Synchronous Data Link Control (SDLC)	Zak79, Don74
High Level Data Link Control (HDLC)	Dav73
IBM 7406 Device Coupler	Col77
IEEE 796 Bus Standard	Bob80

The last of these is extremely useful since it provides both mechanical
and electrical specifications to allow many different manufacturers to
produce varied but compatible microcomputer modules. Functionally, these
may be of many different types such as slaves (memory and/or input/output
boards), masters (CPU and/or controllers) or both. They may be purchased
from a wide variety of different manufacturers and so this new standard
offers the user the flexibility to choose modules that solve microcomputer
problems in the most complete and cost effective manner.

In this section the use of standard interfaces for connecting together
laboratory equipment and computers has been examined. However, the cost
of using these standards may be prohibitive (see Far80) or the approach

too sophisticated. Very often simple interfaces to perform certain types
of tasks can be constructed in the laboratory from readily available
components. Such interfaces can be extremely cost effective. In the
remaining sections of this chapter some examples of simple interfaces will
be given and techniques for their construction described.

INTERFACES - SOME EXAMPLES

The material in this section is restricted to a brief description of the
use of micros in the laboratory for controlling experiments and collecting
data. Interfaces for mainframes and minicomputers are not discussed as
they are adequately described elsewhere (Clu75, Coo77). There are many
micro-systems that could be used in the laboratory. The facilities
available vary from the bare microprocessor chip through single board
systems (containing one or more processors, memory and Input/Output chips)
through to fully packaged desk-top computers such as the Commodore PET and
the range of machines available from Hewlett Packard. In chapter 4 two
examples of micro-systems were given as illustrations of small machines -
the PET and the National Semiconductor SC/MP. These will be used in this
section to describe how typical micro-systems may be interfaced into
laboratory experiments.

Using the SC/MP as a Laboratory Microprocessor

The SC/MP chip is a low cost, easy to use microprocessor. It is supported
by an inexpensive development centre (Nat76a) that enables its user to
develop systems using readily available application cards containing a
processor and memory. Once developed and tested these cards can be
detached from the development centre and 'plugged into' a laboratory
experiment or instrument for control and/or data acquisition purposes.

The configuration of the pins on the SC/MP chip was shown in Fig. 4.13.
Many of these operate at TTL levels (Wit80). TTL is an acroynm for
Transistor-Transistor Logic and refers to a signal convention in which a
low logic state is represented by a voltage in the range 0 through 0.8
volt and a high logic state by a voltage between 2.4 and 5 volts. The
function of each of the pins on the microprocessor chip was explained in
chapter 4 and some schematic diagrams of SC/MP based systems were
presented in Figs. 4.14, 4.18, 4.19 and 4.20. Inspection of Fig. 4.15 (or
4.18) reveals that the basic SC/MP chip permits only serial input through
the SIN pin and serial output via the SOUT pin. This Input/Output is
achieved via the extension register using the serial I/O instruction -
SIO. The output at the SOUT pin is at TTL levels and can be used in this
form or level shifted to RS-232-C levels. Numerous circuits are available
to achieve this, for example, a DS1488 chip (Jer76) or an operational
amplifier. Similarly, input to the SIN pin must be TTL compatible. If
incoming RS-232-C signals are to be processed they can be transformed to
the correct level by an appropriate circuit - for example, a DS1489 chip.

Other ways of interfacing with the chip are via the flag pins (Flag 0,
Flag 1 and Flag 2) which are also TTL compatible and can drive a 1.6 mA
load. The flags are software controlled and can be pulsed in a single or
multiple sequence. They can thus be used for bit serial (or 3 bit
parallel) output but are more often used to implement control switches via

relays that are operated by appropriate driver circuits. Similarly, input
to the processor can be achieved via the sense pins - sense-A (SA) and
sense-B (SB) - which can be used to sense external conditions. They are
TTL compatible and their logic level can be read by the processor but not
altered. SA can serve a special purpose in that it can be used to
generate a processor interrupt provided the interrupt enable mode is in
operation. An interrupt causes the processor to complete the instruction
that it is currently executing and then execute a special interrupt
handling program that is stored in a particular part of its memory space.
Interrupts can be generated by instruments or detectors when they enter
into a state in which they need to have some special operation peformed by
the computer. An instrument can generate an interrupt in the processor
simply by changing the logic level of the SA pin from low to high.

Unfortunately, there is only one interrupt line on the SC/MP which at
first may seem to be rather restrictive. However, the number of interrupt
inputs can be easily extended to 8. All that is needed is some extra
hardware in the form of a 1 to 8 priority encoder - for example, a SN74148
chip (Tex80). The output (EO - pin 15) of the priority encoder goes to
logic high whenever a low signal (logic 0) appears at one of its eight
input pins. The high logic state at pin 15 can be used to drive the
interrupt input (SA) of the SC/MP high causing the CPU to acknowledge the
interrupt request. The BCD outputs of the encoder (pins 6, 7, 9) indicate
which of the inputs is low. This information can be routed to the SC/MP
data bus via three buffers (SN74125 - see Tex80). The 0 input of the
encoder (pin 10) has the highest priority, that is, when this input is
taken low, interrupt requests on any of the other inputs (pins 1, 2, 3, 4,
11, 12, 13) are ignored by the SC/MP. Further details of this arrangement
(including software requirements) will be found in the SC/MP Application
Handbook (Nat76b).

Very often the ease with which a microprocessor can be applied depends to
a large extent upon the power and flexibility of the peripheral support
chips that are available to provide I/O functions. A well supported
microprocessor such as the INTEL 8080 (Can80, Osb78) has dozens of
peripheral interface chips which cater for virtually every interface
problem - parallel I/O, serial (synchronous and asynchronous) input/
output, CRT control, keyboard and display drivers, floppy disc control,
ADC/DAC conversion, stepper motor control and so on. The SC/MP is able to
utilise many of these even though they may originate from different
manufacturers. In addition, it has a wide variety of support chips of its
own that may be utilised in various combinations. For example, there are
integrated circuits (Nat76c, Jer76) to perform serial to parallel
conversion (or vice versa) thereby enabling the SIN, SOUT and Flag outputs
to be utilised for 8 way parallel I/O. Similarly, there are digital
multiplexer and demultiplexer chips to enable the SC/MP to perform serial
I/O operations on more than one input or output device. A useful chip for
use with the SC/MP is the 40 pin RAM I/O device - INS8154 (Nat78, Wil80) -
which provides 128 bytes of RAM storage and a set of 16 signal lines that
can be used for a variety of different input/output options. The output
lines are designated Port A (8 lines) and Port B (8 lines).

The INS8154 support chip incorporates circuitry which provides the user
with a great deal of flexibility with respect to using the 16 I/O lines.
Each line can be separately defined as either an input or an output under
program control. Individual lines can be read as input or set to a high
or low logic level for output. These functions are determined by the
address values used in the software. A further group of usage modes

permit handshake logic (similar to that described at the beginning of the
chapter) to take place in conjunction with 8 bit parallel data transfers
in or out through Port A. When this is done some of the signal lines of
Port B need to be used as control lines. Figure 6.16 illustrates a
typical application of this mode of operation.

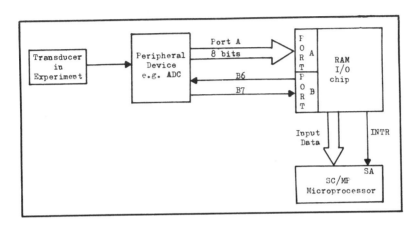

Fig. 6.16   Data acquisition by means of special I/O support chips.

A peripheral device such as the analogue-to-digital converter (ADC) sends
a byte (8 bits) of information to Port A and a low pulse to line B7 of
Port B. This pulse causes the data to be latched into the buffer register
associated with Port A of the RAM I/O chip. At this point, B6 is made
high in order to signal to the peripheral device that the buffer is
occupied and cannot accept any further data. Once the data has been set
in the buffer the INTR pin (provided it has been enabled) goes high to
indicate 'data ready'. Because this is wired to the SC/MP interrupt pin
(SA) this generates a processor interrupt. The interrupt service routine
can read the byte of information from the Port A buffer after which the
RAM I/O chip sets INTR low. In addition, it removes the 'buffer full'
signal from B6 thereby informing the peripheral that the buffer is now
available for the input of new data. The reverse operation, output from
the SC/MP to a peripheral device via the RAM I/O chip, can be accomplished
in a similar way to that outlined above.

The SC/MP can also be used to control laboratory devices through appropriate use of the RAM I/O chip. An arrangement that enables the microprocessor to generate an analogue voltage for control purposes is outlined
in Fig. 6.17 - this time using Port B. In this example, Port B of the
support chip is wired directly to the input pins of a suitable digital-
to-analogue converter chip such as a ZN425E (RSC77). Provided Port B has
been set up for output mode, 8 bit binary values in the accumulator of the
SC/MP can be sent directly to the A/D converter by means of a store
instruction (ST). Further details on analogue-to-digital conversion and
examples of interfacing the SC/MP to other more sophisticated devices will
be found in references Nat76b and Wil80.

The advantages of the SC/MP microprocessor lie in the fact that it is both
easy to use and economical - hence, its name. It is very easily

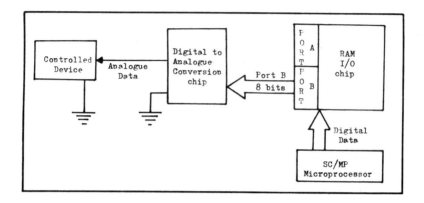

Fig. 6.17 Use of a microprocessor and a peripheral support
chip for instrument control applications.

programmed using machine code, assembler or BASIC. A number of
cross-assemblers are available for it. The chip has been widely used in
many laboratories and an example of its use was described in chapter 4.

Using the Commodore PET as a Laboratory Desk-top Computer

For the majority of laboratory users, desk-top computers will undoubtedly
be much easier to use than the basic microprocessor chips or boards that
were described in the previous section. Consequently, the appearance of
desk-top machines in the laboratory is now quite commonplace. Because of
its simplicity, ease of use and low cost, the Commodore PET has become one
of the most popular small machines for laboratory applications. A brief
description of this machine was given in chapter 4 along with a diagram
depicting the type of devices that can be interfaced to it (Fig. 4.23).
Most of the peripherals that can be attached to the PET plug into a common
board via edge connectors that protrude from the rear - there are two
exceptions, the memory extension port and the second cassette interface
which emerge from opposite sides of the board. The memory extension port
can be used to add on extra memory or additional devices such as a disk
unit. A rear view of the PET showing the location of the interface ports
is presented in Fig. 6.18.

The important connectors from the point of view of interfacing are the J1
and J2 ports. J1 is the edge connector that offers the Commodore implementation of the IEEE-488 programmable instrumentation interface which was
described earlier. It is by means of this port that many of the PET's
peripherals are attached - the printer, floppy disk unit, modem (see
chapter 7) and IEEE-488 compatible instruments such as digital multimeters
(Hew79) and exotic devices such as graph plotters (Hew79a) or digitisers.
As was mentioned earlier the IEEE bus is extremely convenient to use when
appropriate instruments are available. Thus, using this port to interface
an IEEE instrument is a simple matter; it requires only a suitably wired
24-way interface cable and a short program to control the instrument. For
example, suppose it was required to measure the resistance (and hence,

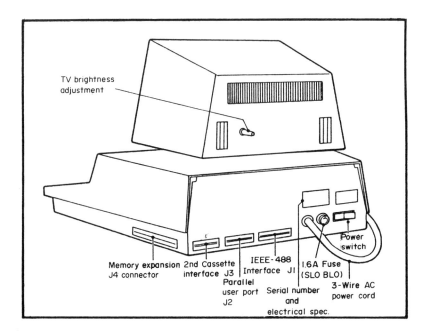

Fig. 6.18   Location of the interface ports on the PET microcomputer.

temperature) of a thermistor probe at regular intervals of time using a
digital multimeter (Hew79). The simple program shown in Table 6.1 is all
that is needed. This would address the device (statement 10) and then
read a value from it every 2 seconds (approximately) - the reading being
printed on the screen of the PET microcomputer. Note that the program
assumes that the multimeter will be appropriately configured (mode setting
and range) prior to its invocation. A more detailed description of the
PET's IEEE-488 interface - its limitations and how some of these may be
overcome - is contained in the book by Fisher and Jensen (Fis80).

TABLE 6.1   BASIC Program for Data Acquisition from a Digital Multimeter

```
10   OPEN 1,5
20   INPUT#1, R$, M
30   PRINT "RESISTANCE= ", R$, "  MODE", M;
40   PRINT TI/60, "SECS"
50   K=TI
60   IF TI<K+100 THEN 60
70   GOTO 20
```

The second port, J2 in Fig. 6.18, is referred to as the User Port. This
port enables the user to attach a wide variety of laboratory instruments
or other devices either for data collection or for control purposes. The

IEEE port may also be used for 'non-standard' applications provided its
operating methods are understood. When used normally, the IEEE bus acts
as a byte serial (8 bit parallel) asynchronous port. The J1 port can be
used for byte parallel or 8 way bit serial operations for a wide variety
of different Input/Output modes. In general, parallel interfaces are used
for local communication. They are extremely useful for transferring large
amounts of data over reasonably short distances at high speed. When used
for non-parallel operations each of the pins function totally
independently. A sketch showing the detailed format of the J1 and J2
connectors is presented in Fig. 6.19.

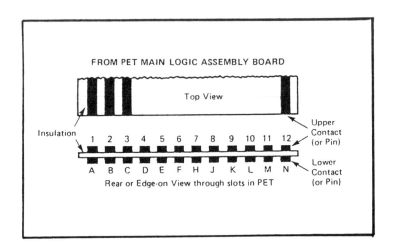

Fig. 6.19 Physical arrangement and designation of the
contact strips that constitute the external interfaces
of the PET parallel user port and the IEEE port.

Each of these connectors consists of a set of 24 contacts arranged in two
parallel rows of 12, located on opposite sides of the printed circuit
board. The function of each of the pins of the IEEE port and of the user
port are listed in Figs. 6.20 and 6.21, respectively.

In the context of applications interfacing the important pins on the User
Port are those labelled PA0 through PA7 (Pins C through L). These are
available for the user's own use. They are TTL compatible and may be set
to logic low or high under program control or by means of external
devices. The logic levels of these pins can be set from within a BASIC
program by means of a special instruction called POKE. Similarly, their
logic state may be read from within a program by means of another
instruction called PEEK. PEEK and POKE are extremely useful functions for
programming data exchange or control of devices through the user port.

Each of the pins PA0 through PA7 may be set individually either for input
or for output. This is achieved by 'poking' appropriate initialisation
values into the Data Direction Register of the I/O chip (a MOS Technology
6522 VIA - Versatile Interface Adapter, see ref. Ham80) that controls the
user port - compare the INS8154 RAM I/O chip described previously.
Further details of the PET's user port and its applications will be found

PET Pin Characters	Standard IEEE Connector Pin Numbers	IEEE Signal Mnemonic	Signal Definition/Label
Upper Pins			
1	1	DIO1	Data input/output line #1
2	2	DIO2	Data input/output line #2
3	3	DIO3	Data input/output line #3
4	4	DIO4	Data input/output line #4
5	5	EOI	End or identify
6	6	DAV	Data valid
7	7	NRFD	Not ready for data
8	8	NDAC	Data not accepted
9	9	IFC	Interface clear
10	10	SRQ	Service request
11	11	ATN	Attention
12	12	GND	Chassis ground and IEEE cable shield drain wire
Lower Pins			
A	13	DIO5	Data input/output line #5
B	14	DIO6	Data input/output line #6
C	15	DIO7	Data input/output line #7
D	16	DIO8	Data input/output line #8
E	17	REN	Remote enable
F	18	GND	DAV ground
H	19	GND	NRFD ground
J	20	GND	NDAC ground
K	21	GND	IFC ground
L	22	GND	SRQ ground
M	23	GND	ATN ground
N	24	GND	Data ground (DIO1-8)

Fig. 6.20   IEEE-488 port identification characters,
associated mnemonics and signal descriptions.

both in the CBM User's Guide (COM79) and in the book by Hampshire (Ham80).

In addition to the PA0-PA7 pins, the lines associated with the CA1 and CB2 contacts are also important during data transfer.   The CA1 pin is used only in conjunction with input operations.   It is used as an interrupt line to the PET system   for latching data currently on the input lines (PA0-PA7) into the input register of the VIA and for handshaking operations.   CB2 can act as a totally independent interrupt, as a peripheral control output, or a serial input or output depending upon how the VIA is programmed.   One of its most important functions, in output mode, is as a handshaking line in conjunction with CA1.

The peripheral input lines (PA0-PA7) can operate in either a latched or unlatched mode.   Reading the input port will transfer the contents of the input register onto the processor data bus.   This is acceptable provided the data on the input lines is not changing.   If it changes during the time it is being read then erroneous results are likely to arise.   In order to avoid this the VIA can be programmed to operate in latched mode. This is shown schematically in Fig. 6.22.

Pin Identification Character	Signal Label	Signal Description
1	Ground	Digital ground.
2	T.V. Video	Video output used for external display, used in diagnostic routine for verifying the video circuit to the display board.
3	IEEE-SRQ	Direct connection to the SRQ signal on the IEEE-488 port. It is used in verifying operation of the SRQ in the diagnostic routine..
4	IEEE-EOI	Direct connection to the EOI signal on the IEEE-488 port. It is used in verifying operation of the EOI in the diagnostic routine.
5	Diagnostic Sense	When this pin is held low during power up the PET software jumps to the diagnostic routine, rather than the BASIC routine.
6	Tape #1 READ	Used with the diagnostic routine to verify cassette tape #1 read function.
7	Tape #2 READ	Used with the diagnostic routine to verify cassette tape #2 read function.
8	Tape Write	Used with the diagnostic routine to verify operation of the WRITE function of both cassette ports.
9	T.V. Vertical	T.V. vertical sync signal verified in diagnostic. May be used for external TV display.
10	T.V. Horizontal	T.V. horizontal signal verified in diagnostic may be used for TV display.
11, 12	GND	Digital ground.
A	GND	Digital ground.
B	CA1	Standard edge sensitive input of 6522VIA.
C	PA0	
D	PA1	Input/output lines to peripherals,
E	PA2	and can be programmed independ-
F	PA3	ently of each other for input
H	PA4	or output.
J	PA5	
K	PA6	
L	PA7	
M	CB2	Special I/O pin of VIA.
N	GND	Digital ground.

Fig. 6.21 User port identification characters, signal names and descriptions.

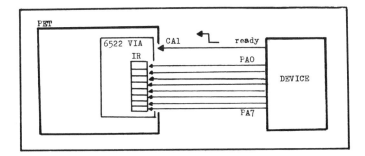

Fig. 6.22 Latched data acquisition via the PET's versatile
interface adapter (VIA).

In this mode, data on the input lines is latched into the input register
(IR) only when there is a transition (high-low or low-high) on the CA1
line. When using a handshaking line such as CA1 to control the latching
of data from an external device it is important to ensure that the data on
the input lines has stabilised prior to an active transition on the
handshake line.

There are various ways of acquiring data from an external device via
PA0-PA7, for example, just reading the input port (either once or
repetitively) or via some form of interrupt service routine. The method
adopted and whether a latched or unlatched input mode is used will depend
upon the particular requirements of the application. Latching is
important in situations where data is changing very rapidly - for example,
a transient data peak could be missed if data input was not being
performed in latched mode.

Although the Input/Output pins of the user port can be programmed in BASIC
using the PEEK and POKE instructions, for many applications programs
written in BASIC will not be fast enough. Thus, the maximum scanning rate
for a test loop waiting for a transition on CA1 during an input operation
is about 40 Hz in BASIC compared with 50 kHz for a similar program written
in machine code. In view of this for applications involving high speed
data acquisition machine code programs are a necessity. This obviously
requires familiarity with the architecture and instruction set of the
microprocessor upon which the PET is based - the MOS 6502 (Zak78).
Production of machine code programs can then be achieved in a variety of
ways, for example,

(a) using an assembler development package for the PET,
(b) using a cross assembler running on another PET, a mini-
     computer such as a PDP-11 or a mainframe such as an IBM
     370 (Kue80), or,
(c) by hand coding a machine code program and 'poking' it into
     the PET's memory using a simple BASIC program.

Method (b) is interesting since it can involve direct inter-connection of
the laboratory PET with a (possibly) remote software development computer,
the software then being transferred to the target machine via a technique
called 'down-line loading'. In this context, the laboratory PET can

itself be used to generate software for and acquire data from other
laboratory microprocessors thereby enabling quite sophisticated networks
of data acquisition/control processors to be set up within the laboratory
environment. Clark et al (Cla80) have described a multi-microprocessor
system of this type for the acquisition of Far Infrared astronomy data.
Their system required the direct interconnection of a Motorola M6800
microprocessor (Hor79) and a PET (via the user port) with subsequent
transmission of the data to a Perkin Elmer 650 printer via the IEEE Port -
using an IEEE/RS-232 conversion interface. A diagram showing the
interfacing arrangements between the M6800, PET and printer is presented
in Fig. 6.23.

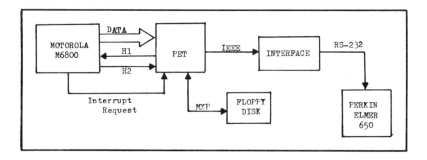

Fig. 6.23 Interfacing the PET microcomputer with a
Motorola microprocessor.

Input to the PET is via the user port using CA1 and CB2 for handshaking
purposes (H1, H2). Interrupts from the Motorola are taken into the PET
via the interrupt request line, IRQ, associated with the memory expansion
port (J4 in Fig. 6.18). This port is labelled MXP in Fig. 6.23. Details
of the interface connections between the PET and Motorola are given in
Table 6.2. The M6800 performs a handshake of 128 bytes of infrared data
with the PET every 8 seconds. Each handshake takes approximately 100
milliseconds. The infrared data is displayed graphically on the PET
screen, partially processed and then stored on floppy disk connected to
the PET via the memory expansion port. The programs for performing the
handshaking are written in assembler. Further details of the programs,
assembler listings and handshake protocol are given in the original paper
(Cla80).

TABLE 6.2 Interface Connections for PET-Motorola Interconnection

Function	Motorola Line(s)	PET Line(s)
DATA	PB0-PB7	PA0-PA7
H1	CB2	CA1
H2	CB1	CB2
Interrupt Request	PA1	IRQ

INTERFACE DESIGN

There are several standard interface systems available which offer an easy
to use solution to the problem of interfacing. Despite this, situations
often arise in which an interface has to be constructed from basic
components (Est80, Dri80, Rod79). Instrument interfacing is not a
difficult problem (Des73, Lev73) provided those involved have some basic
background in electronics and software design. In this section a brief
overview of the steps involved in designing and constructing interfaces
will be presented.

Principles Involved

When approaching the problem of interfacing there is a standard set of
design questions that need to be asked. The most important of these are
listed below:

   (1)   What types of devices need to communicate?
   (2)   How many devices are to be considered?
   (3)   What distances are involved?
   (4)   What messages need to be carried?
   (5)   What data rates are required?
   (6)   What error checking, debugging and diagnostic facilities
         should be included?
   (7)   What resources are available?

The answer to each of these questions will significantly influence both
the design and final implementation of the interface. In all probability
the interface will be bus orientated and so answers to questions 1, 2, 4
(and possibly 5) will determine the bus width and the way in which signals
will be partitioned between the functions of data transportation,
addressing and control. Questions 1 and 4 are also used to determine the
directionality of the interface, that is, the direction of data flow and
whether it is unidirectional or bidirectional. The answer to question 3
will help the designer to decide upon the types of power loss and signal
distortion that may be expected. Ways of overcoming these problems will
need to be considered. The data rates that are involved is an extremely
important factor as this will strongly influence the design of the
circuits involved and the choice of components for constructing them - for
example, the use of optical fibre links, high speed IC chips, special
coding techniques, etc. Another important consideration when designing
complex interfaces is that of error checking and debugging. In situations
such as this it is imperative that facilities are designed into the
interface so that it can easily be checked and tested during construction,
routinely when in operation, and, when an error condition arises.
Techniques such as signature analysis (Kel80) are often used for this
purpose. The incorporation of appropriate diagnostic indicators and fail-
safe procedures is critical when the interface is being used to control
potentially dangerous equipment. During the design stage it is often
possible to use computer simulation techniques to test out particular
types of interface ideas and then evaluate the performance and merits of
candidate designs.

After all the design decisions have been made the next stage is the actual
construction of the interface. More often than not the interface will
have been designed to within some previously specified cost restraints
involving both labour and materials. In addition, the previously made

decisions will, of necessity, have taken into account the types of
components, tools and techniques required for the construction of the
interface. Once it has been constructed the interface has to be
exhaustively tested in order to locate any 'bugs' that may reside in the
hardware (for example, faulty chips, resistors, capacitors, etc) or the
software (for example, program or data errors).

A basic introduction to the principles involved in interfacing will be
found in the books by Zissos et al (Zis73) and Zaks (Zak79). Design of
the interface could involve both digital and non-digital techniques.
Typical digital interfacing techniques include level shifting, buffering,
bus inter-connection, serial to parallel conversion, parallel to serial
conversion and synchronisation. In contrast, transducer selection,
amplifier design, signal conditioning, analogue-to-digital and digital-
to-analogue conversion are examples of non-digital interfacing problems.
Many interfaces will consist of a combination of hardware, software and
firmware. Certain parts will necessarily be implemented in hardware,
other parts will optimally be constructed in hardware or software.
Deciding upon the optimal combination of hardware and software is one of
the major design decisions. Often parts of the interface will be implemented in firmware. This term is used to describe software that has been
permanently stored in Read-Only-Memory (ROM). There are considerable
advantages to the firmware approach with respect to speed, security and
reliability of the interface.

In a large number of laboratory applications the simplest type of
interface is one which inter-connects two devices. A simple model for
such an interface was presented in Fig. 6.4. In this instance the design
steps involved can be made much simpler than the general case described
above. The following steps could form the basis of a design procedure:

(1) Describe how the interface is to be controlled.
(2) Look at the characteristics of Device-1.
(3) Examine the characteristics of Device-2.
(4) Formulate the differences/compatibilities that will
    influence their inter-connection.
(5) Design the interface to overcome the incompatibilities
    and utilise the compatibilities.
(6) Optimise the hardware/software tradeoffs - that is,
    decide which parts should be built in hardware, which
    in software and which in firmware.

Examples illustrating these techniques will be found in papers by Estler
(Est80), O'Driscoll (Dri80), Goeringer (Geo79), Hordeski (Hor78) and
Rogrigues (Rod79).

Basic Components and Tools

The practical realisation of interface construction will depend critically
upon the types of components available. Similarly, the ease with which
the interface can be fabricated and tested will relate directly to the
types of tools and test equipment that is to hand. In this section a few
brief comments on the practical aspects of interfacing will be presented.

(1) Integrated Circuits and Other Electrical Components

Integrated circuit chips - along with the other familiar electronic
circuit components such as transistors, resistors and capacitors - are the

fundamental building blocks from which most modern interfaces are
constructed. A familiarity with the different types of IC chips available
and the functions they perform is a necessary pre-requisite for successful
design. Some commonly available classes of integrated circuit are listed
below:

- microprocessors of different speeds and architectures,
- memory chips of various types,
- microprocessor support chips,
- communications chips (such as UARTS, parity bit
  generators, etc),
- character/sound generators,
- stepper motor controllers,
- timers, latches, triggers, ADC/DAC converters,
  registers, etc,
- basic gates such as AND, OR, NAND, NOR, EOR, and others.

This range is continually expanding with respects to both functional
capability and variety within type. When choosing chips several factors
have to be considered - the chip family (for example, TTL, CMOS, etc),
speed, power requirements, signal levels, physical size, pin connections,
drive capability and so on. Much of this information is contained in data
books (such as Tex80) or manufacturer's data sheets (for example, RSC77).
Most interface designers build up a library of manufacturer's data sheets
appropriate to their needs.

In addition to integrated circuits the interface constructor will need to
have available a reasonable stock of other items including resistors,
capacitors, switches, various types of relays, amplifiers, wires and
cables, connectors, light emitting diodes (LEDs) and stocks of copper clad
board for constructing printed circuit boards.

(2) Tools

A certain minimal set of tools is required for interfacing work. Amongst
these are included: soldering iron, wire wrapping gun, several analogue
multimeters, digital multimeter, logic probe, logic pulser, logic clip, an
oscilloscope and appropriate power supplies. The task of interfacing can
be made much easier if, in addition, to the standard requirements,
additional tools such as current tracers, logic comparators and
temperature probes are available. While the previously listed facilities
will suffice for the construction of simple interfaces, if complex computer based systems are to be constructed then a more powerful range of
tools will be required. Consequently, the above list might now be
extended to include:

                (a) a desk-top computer,
                (b) a logic state analyser,
                (c) a logic pattern generator, and,
                (d) a signature analyser.

A desk-top computer (such as the Commodore PET) is useful as a tool for
testing and helping to debug interfaces. It can be programmed to collect
data from various points in the interface circuits or generate test
signals to simulate likely interface conditions. These functions can
often be performed more conveniently using logic state analysers and logic
pattern generators. The state analyser is capable of collecting state and

timing information from many different parts of the interface circuit
simultaneously. At a later stage it can replay the results on an oscilloscope screen. The logic pattern generator is used to generate control
signals, addresses and data. This enables interfaces to be tested
independently of other parts of the system. Another useful hardware
debugging tool is the signature analyser. This detects and displays
digital signatures unique to the bit streams present at data nodes of a
circuit under test. By comparing these actual signatures to the correct
ones tabulated when the system was working correctly it is possible to
locate a faulty component in a circuit and then replace it.

(3) Prototyping

Once an interface circuit has been designed it is important that it is
tested and its performance evaluated before all the components are finally
assembled and soldered together. In other words, a prototype must be
built. There are many different ways to approach the prototyping of
circuits. One method is to use a circuit board or patch-board consisting
of a matrix of connection points into which can be inserted discrete
components (such as resistors, capacitors, transistors, etc) and connecting wires to facilitate their rapid interconnection. A variety of
different types of boards are available. The most useful are those which
enable the insertion of digital integrated circuit chips. A typical
example is the 'Bimboard' type (Dun79). These are available in many
different sizes all of which are compatible with each other so that they
can be connected together and interlocked to form large work areas. Some
systems come supplied with built-in power units suitable for driving a
wide variety of IC chips.

During the initial construction of the prototype the use of a patch-board
such as that outlined above greatly facilitates testing of the interface
since attaching test probes at appropriate points in the circuit is an
easy matter. When the circuits are functioning correctly the circuit
parameters can be recorded as an aid to subsequent trouble-shooting using
signature analysis techniqus similar to those outlined above.

(4) From Prototype to Production

As soon as an interface circuit has been finalised, the production version
will need to be constructed. This will involve transferring the prototype
from the environment of the design bench into the more permanent setting
provided by an 'interface box'. To achieve this several different stages
are likely to be required, for example, the design and construction of
suitable printed circuit boards, fitting and soldering (or wire-wrapping)
the components together, testing the board(s) and housing them in a
suitable container fitted with appropriate connecting plugs/sockets. If
the interface involves the use of a microprocessor then the associated
control software/data may need to be transferred from the temporary RAM
storage of a development centre into some form of permanent ROM storage
that can be housed within the interface. Finally, once the production
version of the interface has been constructed and checked out, appropriate
documentation should be prepared. This will usually contain the details
of how to use the interface, a specification of its limitations and
detailed descriptions for trouble-shooting in the event of errors arising.

OTHER CONSIDERATIONS

The interfacing of one instrument to another or of an instrument to a computer system can be an involved and complex procedure which may often require participation by experts from several different disciplines if it is to be achieved successfully. In this chapter total emphasis has been given to the different aspects of digital (and analogue) electronic interfacing. However, although this is one of the most important considerations - particularly if computers are involved - there are many others.

A more generalised approach to interfacing would cover a much wider variety of factors - electrical, mechanical, hydraulic, ergonomic and many more. In chapter 3 the problems of interlinking various types of instrument were outlined. A brief description was given of combinations of instruments such as the GC/MS, the LC/MS, the GC/IR, the GC/MS/IR and several others. Instrument combinations of this type require much more than just digital signal interfacing: appropriate sample handling facilities are required for passing the effluent from one machine into another, suitable instrument control panels need to be designed along with many other facilities to achieve total integration of the system. This integration must result in an instrument that is 'smooth' running, easy to use and safe.

As in illustration of the many different aspects involved in interfacing within sophisticated machines one needs only to mention the Technicon SMAC biochemical analyser. A description of this highly automated computer based system was given in chapter 3 in the section that dealt with 'sophisticated instruments'. The construction of the SMAC reflects many of the different aspects of interfacing outlined above - ergonomic, involving the design of suitable work areas and modes of interaction for the human operators; digital, involving the design of appropriate interfaces between the computer and data/acquisition/control units; optical, requiring the design of high quality optical systems based upon the use of optical fibres for transmitting light from a central source to the many visible/ultra-violet detectors; fluidic, involving the leak free interconnection of flow lines between the reagent reservoirs and the continuous flow reaction vessels; mechanical, involving the provision of suitable pumps to aid the smooth flow of materials through the system. These are just a few of the many different types of interface involved. Each of these is extremely important since the instrument as a whole could not function correctly if any one of them was incorrectly designed or implemented. The amount of space devoted to these aspects of interfacing should not in any way be taken as indicative of their lack of importance. Unfortunately, a detailed discussion of these other considerations is outside the scope of this book.

CONCLUSION

Interfacing is the term used to describe the interconnection of devices such as laboratory instruments and computers. Various standard interfaces are available (for example, CAMAC, IEEE-488, S-100, RS-232) which enable instruments to be connected to local computer systems - micros, minis or mainframes. Undoubtedly, the most popular are the IEEE-488 and CAMAC each of which have their advantages and particular areas of application.

However, not all problems require (or can be handled) by this type of interface. When unusual or non-standard applications arise the experimenter is forced into the situation of having to design and construct the required interface - or employ someone else to do it. In view of this, some of the basic principles/techniques involved in designing and constructing interfaces have been outlined.

The emphasis in this chapter has been on the linking of instruments to a local computer system. Not all interfacing falls into this category. Sometimes instruments have to be linked to a remote computer via some form of communication link. In the next chapter a description will be given of some of the communication links that are available and how these may be used to connect instruments to remote computers.

REFERENCES

Bla79   Blank, R.E. and Wakefield, T., Double Beam Photoacoustic
           Spectrometer of Use in the Ultraviolet, Visible and Near
           Infrared Regions, Analytical Chemistry, Volume 51, No. 1,
           50-54, January 1979.

Bob80   Boberg, R.W., Proposed Microcomputer System 796 Bus Standard
           (Task IEEE P796/D2), IEEE Computer, Volume 13, No. 10, 89-105,
           October 1980.

BSI69   British Standards Institution BS.4421:1969, A Digital Input/
           Output Interface for Data Collection Systems, British Standards
           Institution, London, UK, 1979.

CAMBU   The CAMAC Bulletin, available from CEC, D.G. XIII, 28 rue
           Aldringen, Luxembourg.

Cam79   Camp, R.C., Smay, T.A. and Triska, C.J. Microprocessor Systems
           Engineering, Matrix Publishers, Inc., Portland, Oregon, ISBN:
           0-916460-26-6, 1979.

Can80a   Cantrell, T.W., An 8088 Processor for the S-100 Bus - Part (I),
           BYTE: The Small Systems Journal, Volume 5, No. 9, 46-64,
           September 1980.

Can80b   Cantrell, T.W., An 8088 Processor for the S-100 Bus - Part (II),
           BYTE: The Small Systems Journal, Volume 5, No. 10, 62-88,
           October 1980.

Can80c   Cantrell, T.W., An 8088 Processor for the S-100 Bus - Part (III)
           BYTE: The Small Systems Journal, Volume 5, No. 11, 340-360,
           November 1980.

Cla80   Clark, A.R., Smith, C.D. and Kirk, I., Far Infrared Astronomy
           Ground Station - Using the PET with Interrupts, CPUCN
           (Commodore PET Users Club Newsletter), Volume 2, Issue 7,
           21-23, 1980.

Clu75   Cluley, J.C., Computer Interfacing and On-line Operation,
           Edward Arnold Ltd., ISBN: 0-7131-2504-7, 1975.

Col77   Cole, H. and Guido, A.A., The IBM 5100 and the Research Device
           Coupler - A Personal Laboratory Automation System, IBM Systems
           Journal, Volume 16, No. 1, 41-53, 1977.

COM79   CBM User Manual, Model 2001-16, 16N, 32, 32N, Publication No.
           320856-3, 1979. Available from: Commodore Business Machines,
           3330 Scott Blvd., Santa Clara, California 95051.

Coo77   Cooper, J.W., The Minicomputer in the Laboratory: With
           Examples Using the PDP-11, John Wiley, ISBN: 0-471-01883-X,
           1977.

Dav73   Davies, D.W. and Barber, D.L.A., Communications Networks for
           Computers, John Wiley, ISBN: 0-471-19874-9, 1973.

DEC80   Digital Equipment Corporation, Maynard, Massachusetts, USA., A
           Closer Look at the MINC System, MINC-11 Product Specification,
           1980.

Des73   Dessey, R.E. and Titus, J.A., Computer Interfacing, Analytical
           Chemistry, Volume 45, 124A, 1973.

Don74   Donnan, R.A. and Kersey, J.R., Synchronous Data Link Control: A
           Perspective, IBM Systems Journal, Volume 13, No. 2, 140-162,
           1974.

Dri80   O'Driscoll, R.C.O., A Plotter Interface for Serial Computer
           Terminals, J. Phys. E: Sci. Instrum., Volume 13, No. 9,
           935-937, September 1980.

Dun79   Duncan, T., Adventures with Micro-Electronics, John Murray
           (Publishers) Ltd., ISBN: 0-7195-3671-5, 1979.

Ell80   Ellefsen, P.R., IEEE Bus Standard, Wireless World, Volume 86,
           No. 1534, 75-78, June/July 1980.

Elm79   Elmquist, K.A., Fullmer, H., Gustavson, D.B. and Morrow, G.,
           Standard Specification for S-100 Bus Interface Devices (IEEE
           Task 696.1/D2), IEEE Computer, Volume 12, No. 7, 28-52, July
           1979.

Est80   Estler, R.C., Data Acquisition and Control System Based Upon the
           Rockwell AIM-65 Microcomputer, Rev. Sci. Instrum., Volume 51,
           No. 10, 1428-1430, October 1980.

Far80   Farley, J.W., Johnson, A.H. and Wing, W.H., Microcomputer
           Controlled Microwave-Optical Spectrometer, J. Phys, E: Sci.
           Instrum., Volume 13, 848-856, 1980.

Fis80   Fisher, E. and Jensen, C.W., PET and the IEEE-488 Bus (GPIB),
           Osborne/McGraw-Hill, ISBN: 0-931988-31-4, 1980.

Gam80   Gampp, H., Maeder, M., Zuberbühler, A.D. and Kaden, T.A., Micro-
           processor Controlled System for Automatic Acquisition of
           Potentiometric Data and their non-linear Least Squares Fit in
           Equilibrium Studies, Talanta, Volume 27, 513-518, 1980.

Goe79   Goeringer, D.E. and Pardue, H.L., Time Resolved Phosphorescence
           Spectrometry with a Silicon Intensified Target Vidicon and
           Regression Analysis Methods, Analytical Chemistry, Volume 51,
           No. 7, 1054-1060, June 1979.

Ham80   Hampshire, N., The PET Revealed, (Second Edition), January
           1980. Available from: Computabits Ltd., P.O. Box 13, Yeovil,
           Somerset, UK.

Hew79   Hewlett Packard, Digital Multimeter Model 3438A - Operating and
           Service Manual, Publication Number: 0438-90002, May 1979.
           Available from: Hewlett Packard, 16399 W. Bernardo Drive, San
           Diego, California 92127, USA.

Hew79a   Hewlett Packard, 9872B and 9872S Graphics Plotters Operating
           and Programming Manual - Using the HP-GL Instructions,
           Publication Number: 09872-90008, August, 1979.

Hor76   Horelick, D. and Larsen, R.E., CAMAC: A Modular Standard, IEEE
           Spectrum, Volume 13, 50-55, April 1976.

Hor78   Hordeski, M., Interfacing Microcomputers in Control Systems,
           Instruments & Control Systems, Volume 51, Number 11, 59-62,
           November 1978.

Hor79   Hordeski, M.F., Microprocessor Cookbook, Tab Books, Blue Ridge
           Summit, PA 17214, USA., ISBN: 0-8306-1053-7, May 1979.

IEE75   Institute of Electrical and Electronics Engineers Inc., IEEE
           Standard 488 - 1975 - IEEE Standard Digital Interface for Pro-
           grammable Instrumentation, available from IEEE Inc., 345 East
           47 Street, New York, NY 10017.

IEE77   Institute of Electrical and Electronic Engineers, Inc., CAMAC
           Instrumentation and Interface Standards, John Wiley, ISBN:
           0-471-02307-8, April 1977.

Jer76   Jermyn Group, The Jermyn Book - Catalogue and Specification of
           Electronic Components, 1976.

Kel80   Kelly, K., Tracing Faults in Digital Systems with Bit Streams,
           Control & Instrumentation, Volume 12, No. 9, 39-40, September
           1980.

Kno75   Knoblock, D.E., Loughry, D.C. and Vissers, C.A., Insight into
           Interfacing, IEEE Spectrum, Volume 12, 50-57, May 1975.

Kue80   Kuehl, D.T. and Griffiths, P.R., Microcomputer Controlled
           Interface between a High Performance Liquid Chromatograph and a
           Diffuse Reflectance Infrared Fourier Transform Spectrometer,
           Analytical Chemistry, Volume 52, No.9, 1394-1399, August 1980.

Lev73   Levine, S.T., Instrument Interfacing can be Done But It Takes a
           Bit of Doing, Electronic Design, Volume 24, 74-79, November
           1973.

Lib80   Libes, S. and Garetz, M., Interfacing to S-100 Micro-
           computers, Osborne/McGraw Hill, ISBN: 0-931988-37-3, 1980.

Lou72   Loughry, D.C., A Common Digital Interface for Programmable
           Instruments: the evolution of a System, Hewlett-Packard
           Journal, Volume 24, 8-11, October 1972.

Lou74   Loughry, D.C., What Makes a Good Interface?, IEEE Spectrum,
           Volume 11, 52-57, November 1974.

Mer76   Merritt, R., Universal Process Interfaces - CAMAC vs HP-IB,
           Instrument Technology, Volume 23, No. 8, 29-36, August 1976.

Mey80   Meyer, H. (Ed.), Real-Time Data Handling and Process Control,
           Proceedings of the First European Symposium on Real-time Data
           Handling and Process Control, Berlin (West), 23-25 October 1979,
           North-Holland Publishing Company, ISBN: 0-444-85468-1, 1980.

Mor78   Morrow, G. and Fullmer, H., Proposed Standard for the S-100 Bus
           (IEEE Task 696.1/D2), IEEE Computer, Volume 11, No. 5, 84-90,
           May 1978.

Nat76a   National Semiconductor, SC/MP Low Cost Development System
           User's Manual, Publication No. 4200105A, November 1976.
           Available from: National Semiconductor Corporation, 2900 Semi-
           conductor Drive, Santa Clara, California 95051.

Nat76b   National Semiconductor, SC/MP Microprocessor Applications
           Handbook, Publication No. 420305239-001A, October 1976.
           Available from: National Semiconductor Corporation, 2900 Semi-
           conductor Drive, Santa Clara, California 95051.

Nat76c   National Semiconductor, SC/MP Technical Description, Publi-
           cation No. 4200079B, September 1976. Available from: National
           Semiconductor Corporation, 2900 Semiconductor Drive, Santa
           Clara, California 95051.

Nat78   National Semiconductor (UK) Ltd., 301 Harpur Centre, Hone Lane,
           Bedford, MK40 1TR, INS8154N - Channel 128 x 8 Bit RAM I/O Chip
           Data Sheet, DA-B15M48, April 1978.

Nel72   Nelson, G.E. and Ricci, D.W., A Practical Interface System for
           Electronic Instruments, Hewlett-Packard Journal, Volume 24,
           2-7, October 1972.

Osb78   Kane, J. and Osborne, A., An Introduction to Microcomputers -
           Volume 3: Some Real Support Devices, Osborne and Associates,
           Inc., ISBN: 0-931988-18-7, September 1978.

Pea74   Peatfield, A.C., Spurling, K. and Zacharov, B., CAMAC as a
           Computer Peripheral Interface System, IEEE Trans. Nucl. Sci.,
           NS-21(1), 867-869, 1974.

Per80   Person, J.C. and Nicole, P.P., Interface for Rapid Data Transfer
           from NIM Counters to Small Computers, Rev. Sci. Instrum.,
           Volume 51, No. 10, 1425-1426, October 1980.

Ric74   Ricci, D.W. Nelson, D.W., 1974 Standard Instrument Interface
           Simplifies System Design, Electronics, Volume 47, No. 23,
           95-106, November 14th, 1974.

Rod79   Rodrigues, A.R.D. and Siddons, D.P., Inexpensive Computer
          Controlled Experimentation, J. Phys. E: Sci. Instrum., Volume
          12, 403-408, May 1979.

Ros75   Rosner, R.A., Penny, B.K. and Clout, P.N. (Eds), On-line
          Computing in the Laboratory, Proceedings of a Conference
          entitled, 'On-line Computers for Laboratory Use', held at
          Imperial College, London, 11-12th September 1975, ISBN:
          0-903012-33-2, 1975.

RSC77   R.S. Components Ltd., Data Sheet R/2911, 8 Bit D to A/A to D
          Converter IC., March 1977. Available from: R.S. Components
          Ltd., 13-17 Epworth Street, London, EC2P 2HA, UK.

Ste75   Stephens, C.L., On-Line Computing in Astronomy, 305-319 in On-
          line Computing in the Laboratory, Proceedings of a Conference
          entitled, 'On-line Computers for Laboratory Use', held at
          Imperial College, London, 11-12th September 1975, (Ed: Rosner,
          R.A., Penny, B.K. and Clout, P.N.), ISBN: 0-903012-33-2, 1975.

Tex80   Texas Instruments, The TTL Data Book for Design Engineers,
          ISBN: 0-904047-27-X, Publication No. LCC4112, Fourth Edition
          1980.

Wil80   Williamson, I. and Dale, R., Understanding Microprocessors With
          the Mk14, MacMillan Press Ltd., ISBN: 0-333-31075-6, 1980.

Wit80   Witten, I.H., Communicating with Microcomputers, Academic
          Press, ISBN: 0-12-760752-8, 1980.

Zak78   Zaks, R., Programming the 6502, Sybex Inc., ISBN: 0-89588-
          099-1, 1978.

Zak79   Zaks, R. and Lesea, A., Microprocessor Interfacing Techniques,
          Sybex Inc., ISBN: 0-89588-029-6, 1979.

Zak79a   Zaks, R., 6502 Applications Book, Sybex Inc., ISBN: 0-89588-
          015-6, 1979.

Zis73   Zissos, D. and Duncan, F.G., Digital Interface Design, Oxford
          University Press, ISBN: 0-198-517-149, 1973.

7

Communication Channels

INTRODUCTION

Analytical science is concerned with the acquisition of scientific data and information from one or more inter-related processes. The act of observation inherent in the operations involved in data collection forms the basis of the 'process-observer' model that was introduced in chapter 1. Both data and information are derived from a more fundamental class of entity, namely, signals. Our ideas, obviously, now need to accommodate these. A more detailed analysis of the observation process will reveal that there are three basic items involved in the handling and propagation of signals: generators, consumers and carriers. The relationship between these three is depicted in Fig. 7.1.

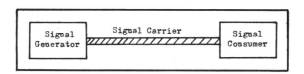

```
┌─────────────────────────────────────────────────────┐
│  ┌──────────┐   Signal Carrier    ┌──────────┐       │
│  │ Signal   ├─//////////////////─┤ Signal   │       │
│  │ Generator│                     │ Consumer │       │
│  └──────────┘                     └──────────┘       │
└─────────────────────────────────────────────────────┘
```

Fig. 7.1 Entities involved in signal generation, propagation and detection.

Signals are generated as a result of observations made on a process; these may originate from a transducer or a detector of some sort. The consumer converts signals into data or information that may be either stored or used for decision making. The function of the carrier is to transmit/transport the signals from the physical locality of the generator to that of the consumer. It should do this with minimal time delay, signal distortion, attenuation and cost while maintaining maximum security and reliability of the signal. Often, the signal carrier is referred to by the synonymous term 'communication channel'. Based upon what has been said above this term may now be defined quite generally as a medium for the transmission of signals, data or information from a source to a

destination. Correspondingly, communication may be thought of as the
transmission of signals between two or more points via a suitable
communication channel.

When a signal is transmitted from one location to another it is subject to
a wide variety of distorting effects. These may change or modify the
signal in many different ways. The changes produced in a square-edged
digital pulse as a result of transmitting it over a length of copper wire
(Heb75, Mar72b) are often used to illustrate the combined effects of the
various perturbing influences. In an ideal communication link the signal
pulse should be received in a form identical to that in which it was
transmitted - as shown in Fig. 7.2 (case A). However, in most practical
communication channels signals are likely to be subject to the effects of
attenuation and distortion. In addition, the influences of noise and
finite propagation delays will undoubtedly cause the received signal to
differ considerably from that which was originally transmitted. A comparison of the two will probably reveal a relationship between transmitted
and received signals similar to that depicted in case B of Fig. 7.2. The
effects caused by the transmission channel (and the environment through
which it passes) can be minimised by the addition of suitable signal conditioning units. These can be used to compensate for unwanted influences
of signal distortion and attenuation. However, their use will increase
the cost of the overall transmission facility.

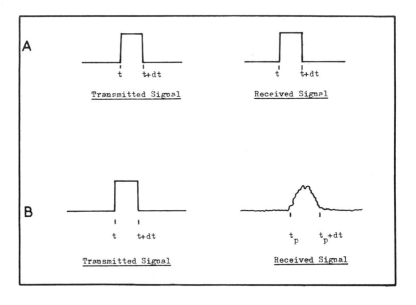

Fig. 7.2 Transmission of a square wave pulse along a communication
link. Case A illustrates communication via an ideal data link;
case B shows the effects that various perturbing forces are likely
to have on the transmitted signal.

When data is transmitted from one location to another it must be transmitted reliably. In other words, the signals presented to the receiver

must be an exact representation of those which originated from the sender.
If this does not happen then in situations where the signals are used for
decision making (either automatic of manual), it is likely that incorrect
decisions could be made. For this reason it is imperative that if the
communication channel is subject to effects which may cause transmission
errors, appropriate techniques are made available to enable these errors
to be detected and, where possible, corrected. There are many well
established error detecting and correcting codes available for this
purpose (Mar70).

Security of the transmitted data is another area of significant concern.
In addition to transmission errors arising from natural environmental
causes, it is possible that errors can be introduced as a result of deliberate attempts to corrupt the message. Problems of this type can arise
either due to the action of vandals or as a consequence of industrial
espionage. Thus, when the data that is being transported is of a
confidential nature (for example, the results of medical or industrial
analyses), it is important to ensure that signals are transported in such
a way that they are inaccessible to those for whom they are not intended.
A variety of techniques are used in order to be certain that this
requirement is met. Perhaps the most well known, in the context of
computer based communication systems, is that of data encryption using
either hardware or software techniques. These will be described in more
detail in a later section of this chapter.

There will usually be a wide variety of different types of communication
channel: wire links, optical fibres, microwave and infrared are examples
of some of those which are in common use. Each will vary in its basic
properties and suitability for particular types of application. The links
may be used for the transmission of data or for remote control of laboratory or process control instruments via optical or ultrasonic transducers.
In the next section some of these different types of communication link
will be described, while in a later chapter a more detailed discussion of
communication networks will be presented.

TYPES OF CHANNEL

Bearing in mind the discussion that was presented in the previous section,
it is likely that there will be many different kinds of data communication
channel in use for the transmission of scientific data between geographically separate locations. Often it is possible to choose between different
alternatives depending upon the speed, reliability and security that is
required for the dispatch of particular categories of data. A few of the
commonly used data links will be outlined in this section.

(1) Conventional Data Links

The use of couriers. Couriers are probably one of the oldest means of
transmitting data from one location to another. They may be used both for
the local movement of data and for its global migration from one location
to another. The courier technique is still widely used in two chief forms:

        (a) Private Couriers, and,
        (b) Postal Services.

Private couriers are often used for the in-house distribution of data. This method ensures data authenticity and is often used as a back-up facility to support the more sophisticated methods of transfer such as the telephone and computer based systems. If the latter types of link become inoperable for any reason then a 'runner' can be used - provided the distances involved are not too great. The courier approach is still used extensively in many process control applications for the communication of results between the analytical laboratory and a central control room or plant locations that cannot be accessed by telephone. Private couriers can also be used to ensure the confidentiality of information in situations where results cannot be disclosed to anyone other than the intended recipient.

Postal services offer another means of dispatching data between organisations. As the coverage of postal systems is world-wide it is feasible to use this system for sending data between installations anywhere on the globe. This, of course, assumes that both the sender and recipient each have local access points to a postal service. Should this not be the case then a private courier may be required to supplement the postal facility. Thus, when a research centre is located at some geographically remote site it will be necessary for a courier to transport data to/from the nearest postal access point.

When using either private courier or a postal service, the data to be transmitted may be recorded in a variety of forms. It may be recorded as conventional written reports/documents that have been hand-written or type-written. Alternatively, it is possible to send the data on magnetic tape, paper tape, magnetic disk, etc. When computer based magnetic media are used care must be taken to ensure that the data does not become corrupted in any way - as a consequence of environmental effects caused by electrical or magnetic fields. If small amounts of data are involved, cassette tapes and flexible disks offer an inexpensive means of data transfer - provided they are packaged in a way that prevents them from becoming damaged.

The use of telephones. Like the postal service, the telephone network(s) offered by different countries provide a world wide means of transmitting data from one location to another. Although telephone systems were originally designed for the transmission of voice signals, they can also be used for sending other types of information. Particularly important in this context is data produced by machines, analytical instruments and computers. Of course, suitable conversion techniques are needed in order to transform the data into a form that is acceptable to the telecommunications carrier. Some of the methods that are used will be outlined later in this chapter.

There are many ways in which a telephone system can be used to communicate the results of analyses from a laboratory or measuring instrument to those who are waiting to receive them. Probably the simplest technique involves the analyst manually dialling the telephone number of the intended recipient and then 'talking' the results down the line. If the person concerned is not available for receipt of the data the results might be recorded on an automatic answering machine - otherwise, the analyst might try to establish contact again later. At the other extreme, some form of total automation might be involved. For example, at some particular instant in time a microprocessor based instrument takes some measurements; it then uses an auto-dialling facility (Heb75) in order to dial the telephone number of a remote control station. Once the number has been

obtained, the data to be transmitted is converted to sonic form using a
voice synthesis system (Phi80, Wit80). The results are then 'spoken' over
the telphone just as though a human analyst was reporting them. Alternatively, the results could be deposited at the recipient control station
using an 'electronic mailbox' facility. This approach will be described
in more detail in Chapter 12.

(2) Wired Links

Various types of wire and cable may be used for the transmission of
signals from one location to another. Typical examples include:

        (i)   twisted pairs,
       (ii)   multi-core cables,
     (iii)   co-axial cable, and,
     (iv)   ribbon cables.

Each of these will vary in their voltage/current ratings, resistance,
covering, cost and immunity to corrosion and noise. These latter factors
are important in situations where safety standards have to be complied
with. A comprehensive review of the properties of different types of
cable has been given by Hickey (Hic80).

Twisted pairs are probably the most well established type of cable used
for the transmission of analogue signals. They are still widely used in
both shielded and unshielded form. Sometimes many shielded pairs are
included in a common cable to form a multi-channel link; 8, 12 and 24
shielded pairs are common. Multi-core cables consist of a large number of
individual colour coded conductors in one bunch that has a common shield.
Co-axial cable is used for carrying video signals to remote VDU screens.
It is also used where high frequency digital information is being transmitted. There are three common impedence standards: 50, 75 and 93 ohms of
which 75 ohms is the most popular. Ribbon cables are constructed from
flat or round copper conductors that run side by side. These are embedded
within some form of insulating material. They are often used for inter-
connecting printed circuit boards and constructing parallel bus systems.

Obviously, cables carrying signals can pick up noise and can themselves
produce interference in other conductors. It is imperative that cables
used to transmit measurement signals are protected from all sources of
noise. Various techniques are available to ensure the minimisation of the
unwanted influences of noise (Kli77, Nal77). The most common of these is
the use of braid or foil shielded cables (Kin80).

The use of wired links for the transmission of data enables a degree of
permanency to be built into the data communication system. Figure 7.3
(case A) illustrates the simplest type of wired link that could be used
for data transfer. It consists of a single conductor linking the locations
between which data is to be transmitted. Changes of voltage on the wire
could be used to represent the data items being transmitted. For example,
the binary message sequence 011010 might be encoded as voltage changes
similar to those depicted in the waveform presented in Fig. 7.4. Provided
a suitable coding scheme was designed to enable actual data values (such
as pH=4.76, Cu concentration=3.56 g/l) to be represented as appropriate
sequences of voltage values, the data could be transmitted over limited
distances at fairly slow speeds. However, any attempt to use such a
technique at high speeds over substantial distances would result in

significant distortion and attenuation effects similar to those that were
illustrated in Fig. 7.2B.

If two wires were used to connect the locations involved in data transfer
- as indicated in case B of Fig. 7.3 - then a more versatile system could
be set up. One of the wires could carry the common low voltage for logic
0 and the other could then be used for signalling relative to the low
logic level. Alternatively, one of the wires could be used to transmit
data from P to Q while data was simultaneously being transmitted from Q to
P on the other. This mode of usage is referred to as full-duplex transmission. The more restricted method of communication in which data
travels only in one direction at a time (as might be the situation in case
A of Fig. 7.3) is called half-duplex operation. Conventional telephone
systems (at least in the United Kingdom) are two wire circuits. This means
that after a subscriber has dialled a telephone number a two-wire circuit
will connect the local telephone to that of the other subscriber whose
number was called. This facility offers significant potential for data
transmission. However, because it is not permitted to attach scientific
instruments and equipment directly to the public telephone circuits,
special devices called modems are required. These serve to electrically
isolate the subscriber's devices from those of the telecommunications
authority. Wired circuits, as used in telephone systems, are primarily
designed for the transmission of speech signals. Consequently, the design
constraints associated with their use for this purpose impose severe
limitations on their applicability for the transfer of digital data. The
limitations are manifest in the speed of error free data transmission that
can be achieved. In order to improve signalling speed it is possible to
make special arrangements with the telephone agencies which involve their
supplying permanent four-wire links consisting of two 2-wire pairs. Using
this type of approach, improved transfer speeds can be obtained.

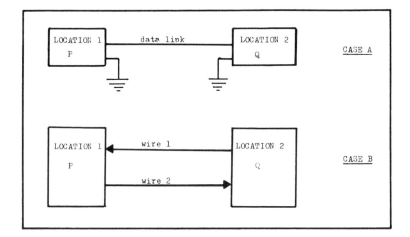

Fig. 7.3 Possible approaches to the use of wire links for
data communication.

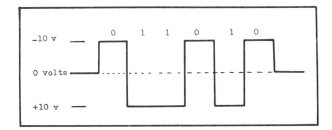

Fig. 7.4 Data transmission by means of square wave pulses.

When wires are used for data communication links between devices there are
various ways in which connections may be made. The simplest approach is
similar to those described in Fig. 7.3 in which the two devices are
connected by a single link. This is called point-to-point. Because of
the expense of wired links - particularly where long distances are
involved - techniques for data link sharing are often used. One of the
commonest methods of achieving line sharing is through the use of multi-
point (or multi-drop) connections in which several devices share the same
line. This approach is illustrated in Fig. 7.5.

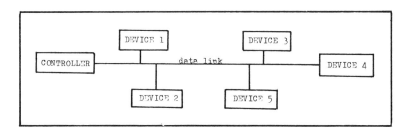

Fig. 7.5 A multi-point data link.

In multi-point operation, one of the devices on the link is always
designated as the controller whose function is to control data transfer by
means of a technique called polling (Heb75). This involves asking each
device in turn if it has any data to transmit or if it is ready to receive
data. Using this mode of inter-linking, wire and cable connections can be
used to construct quite complex data communication networks - a subject
which is dealt with in more detail in Chapter 12.

Another commonly used method of line sharing may be achieved through the
technique of multiplexing. Figure 7.6 indicates the basic principles
involved. A multiplexer is a communication tool that enables the simultaneous sharing of a data link by several devices. Because of the way in
which a multiplexer operates a corresponding device called a demultiplexer
is required at the opposite end of the data link in order to regenerate
the original data. Two techniques of multiplexing are in common use
(Heb75) based upon either frequency or time division of the communication

bandwidth. In time division multiplexing, each device that uses the data
link has access to it for a fixed portion of time according to a well-
defined cyclic pattern. When frequency division multiplexing is used, the
available frequency range (or bandwidth) that the data link will support
is divided up into sections and each of these allocated to a particular
device. Thus, in frequency division multiplexing there is parallel sharing
of the data link between devices compared with the serial access
necessitated by the time division approach. Further details on each of
these techniques will be found in books by Hebditch (Heb75), Martin
(Mar70, Mar72a, Mar76), Smol et al (Smo76) and Johns et al (Joh72).

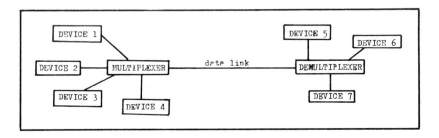

Fig. 7.6 A Multiplexed data link.

(3) **Optical Fibres**

The use of optical fibres for data links in digital data transmission is
relatively new and growing in popularity. Transmitting data by light
through optical glass or plastic fibres is potentially far superior to
communicating information over conventional electrical wires. Such fibres
offer several advantages, including wide signal bandwidth, electrical
isolation, no crosstalk, immunity from interference and lightweight, low
volume cabling. A typical optical fibre data communication system will
consist of at least one transmitter, a length of fibre optic cabling (with
appropriate terminal and in-line connectors) and a receiver. Fibre cables
come in several different forms. Two of these, the single strand and the
fibre bundle are illustrated in Fig. 7.7.

In the early days of fibre optic development, bundled fibres were
considered necessary for reliability because breakage of one or more
fibres could be tolerated without loss of signal transmission. Also, the
large diameter of the fibre bundle allowed greater tolerance with respect
to connector alignment when linking cables together. However, the popularity of fibre bundles has decreased because the single fibre cable
durability is much greater than had been anticipated. Also, connectors
are now available which are capable of providing the precise alignment
required for low coupling loss with small diameter fibres.

An optical fibre is essentially a cylinder made of dielectric materials.
A central region (called the core) is surrounded by one or more cladding
regions and the whole structure is usually protected by a jacket. Figure
7.8 shows a schematic cross section of a single strand fibre.

The optical characteristics of the fibre are determined by its refractive
index distribution which is usually circularly symmetric and depends only

on the radial coordinate, r. Two basic types of fibre commonly used:

    (a)   step-index, and
    (b)   graded index.

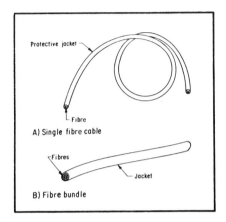

Fig. 7.7   Types of optical fibre; A: single fibre
cable; B: fibre bundle.

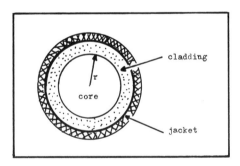

Fig. 7.8   Cross-section of a single strand optical
fibre cable of radius r.

The step-index fibre consists of a core of uniform refractive index made
either from a highly transparent solid material such as high quality
silica glass, multicomponent glass or a low loss liquid such a tetra-
chloroethylene. The cladding that surrounds the core is a dielectric of
slightly smaller index of refraction and is made of silica glass, multi-
component glass or plastic. The appearance of a typical step-index fibre
optic cable (Hew78a) is illustrated in Fig. 7.9.

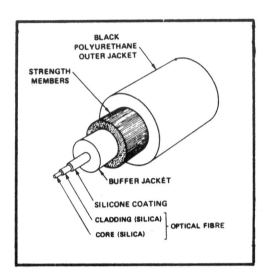

Fig. 7.9   Appearance of a step index fibre optic cable.

In graded-index fibres the refractive index of the core gradually
decreases from the centre towards the core/cladding interface.   Either
high-silica content or multi-component glasses are used to make these
guides.   The graphs presented in Fig. 7.10 show the distribution of
refractive index (RI) with radius for each of the two types of fibre.

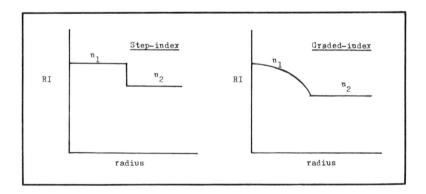

Fig. 7.10   Variation of refractive index (RI) with radius
for step-index and graded index optical fibres; n_1 and
n_2 represent the refractive index of the core and
cladding, respectively.

Graded-index fibres are more costly than those of the step-index variety.
They are used mainly in applications requiring transmission over many
kilometres using high bandwidths. For shorter distances and/or lower
bandwidths, a variety of less expensive step-index fibre is available.

All dielectric fibres guide light because of total internal reflection.
Light flux coupled into an optical fibre is largely prevented from
escaping through the wall by being re-directed towards the centre of the
strand. The basis of this re-direction is the index of refraction of the
core (n_1) relative to that of the cladding (n_2). The total internal
reflection mechanisms that are involved in re-directing light in a step-
fibre are illustrated in Fig. 7.11.

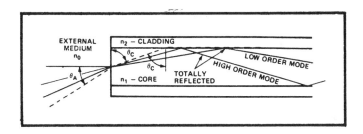

Fig. 7.11 Total internal reflection in an optical fibre.

Rays may propagate through the fibre at various angles. Those propagating
at small angles with respect to the fibre axis are called low order modes
and those propagating at larger angles are called high order modes. These
modes do not occur as a continuum. At any given wavelength, there are a
number of discrete angles where propagation occurs. Single mode fibres
can be made in which only one mode can propagate.

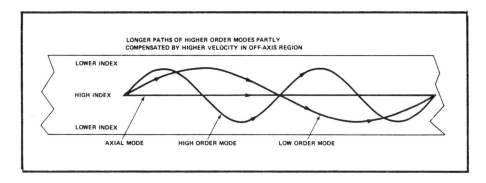

Fig. 7.12 Gradual bending of light in a graded-index
optical fibre.

Total internal reflection does not take place exactly at the boundary
between the core and the cladding. There is some small penetration of the

ray into the cladding which causes a loss of light. To reduce such reflection losses it is possible to make the rays turn less sharply by reducing the index of refraction gradually, from core to cladding. This is the basis of the graded-index fibres which have much lower transmission loss. Light propagation in such a fibre is depicted schematically in Fig. 7.12.

Fibre optics are emerging as a practical cost effective technology for data communication applications. The transmission of information over optical fibres offers many advantages - some of which are not available with any other communication technology. A comparison of some of the important properties of optical and conventional cables (Wil76) is presented in Table 7.1.

TABLE 7.1 Comparison of Fibre Optic, Coaxial and Twisted Pair Cables

Parameter	Fibre Optic	Coaxial	Twisted Pair
Low level of cross-talk	●	●	
No measurable cross-talk	●		
Immunity to RFI/EMI/EMP	●		
Total electrical isolation	●		
No sparks or fire hazards	●		
No short circuits or loading	●		
No contact discontinuity	●		
Will withstand temperatures up to 300°C	●	●	●
Will withstand temperatures up to 1000°C	●		
Signal bandwidth to 1 MHz (over 300 metres)	●	●	●
Signal bandwidth to 20 MHz (over 300 metres)	●	●	
Signal bandwidth > 200 MHz (over 300 metres)	●		
Lightweight materials	●		●
Low cost	●	●	●
Vibration tolerant	●	●	●

As Table 7.1. reveals, fibre optic cabling scores (●) under every comparison parameter listed. This is not the case with the other cables.

Optical fibres, being made of dielectric materials, provide optimum immunity to radio-frequency and electromagnetic interference (RFI/EMI). Since they neither pick up nor radiate signal information such fibres offer greatly improved electromagnetic compatibility over wire cable systems. Because of their dielectric properties they are also immune to electro-magnetic pulses (EMP). Fibre optic transmission systems provide total electrical isolation between the sending and receiving terminals so that many problems associated with common ground connections are eliminated. Similarly, such cables present no fire hazards when their fibres are damaged. In addition, no local secondary damage can occur because fibre cables neither produce sparks nor dissipate heat. Damage to a wire cable, however, can do considerable harm to the terminal circuits that they connect by short circuiting or grounding them. In contrast, short circuits or circuit loading produced by shorting a fibre optic cable does not reflect back to the terminal equipment being connected.

The signal attenuation of typical fibres ranges from 2 dB/km (decibels per kilometre) to 1000 dB/km. For a 300 metre fibre which has 50 dB/km of attenuation and the proper light transmission characteristics, the signal bandwidth is about 200 MHz. This limit is primarily a function of the intensity modulation rate limit of the light source used to carry the data. The bandwidth of coaxial cable, independent of signal processing electronics, is limited to 20 MHz for the equivalent diameter and length of cable, while twisted pair wire has a 2 MHz bandwidth limit.

The type of equipment needed to support a fibre optic data transmission system is shown in Fig. 7.13.

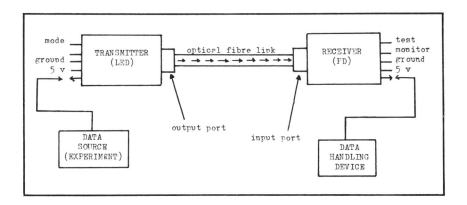

Fig. 7.13 Basic components of a fibre optic data transmission system.

The transmitter consists of a light emitting diode (LED) as the light source. The intensity of the light that it emits will vary in accordance with the modulating signal produced by the data source. The light generated by the LED enters the optic fibre cable and propagates through it until it reaches the light sensitive photodiode contained within the receiver unit. The receiver converts the light intensity variations back into digital data pulses. Transmitter and receiver units are available commercially from a number of different sources (RSC80, Hew78a, Hew79). The detailed construction of the transmitter and receiver circuits for a typical optical fibre transmission system (Hew78b, Hew78c, Hew78d) are shown in the diagrams presented in Fig. 7.14.

Each of the units shown in these diagrams are TTL compatible both for data input and output. They can be used for data rates up to 10 Mbits/sec over distances up to 100 metres with a bit error rate of less than 10^{-9}. The bandwidth can be increased by using several fibres in parallel with each other. The transmitter generates optical signals in either of two externally selectable modes. The internally coded mode produces a 3-level coded optical signal which is a digital replica of the data input waveform.

The simplest type of data transmission that could be achieved using an optical fibre link is unidirectional or simplex flow between a transmitter (T) and a receiver (R). A full-duplex transmission system would require two transmitter/receiver pairs and two optical cables (single fibre). These arrangements are illustrated in the diagrams shown in Fig. 7.15.

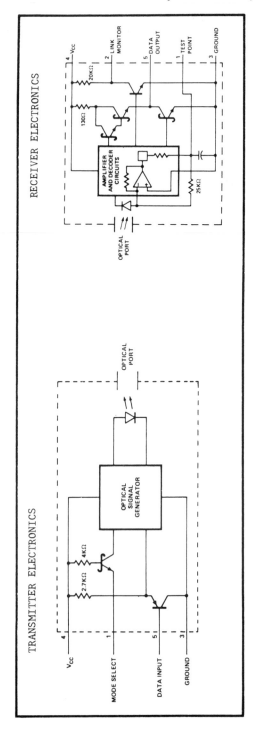

Fig. 7.14 Electronic circuits to support data transmission via optical fibres.

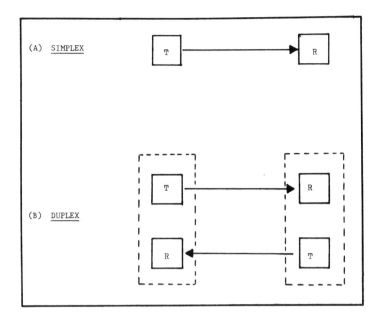

Fig. 7.15   Arrangement of transmitters (T) and receivers(R)
to support simplex and full-duplex transmission via optical
fibre data links.

When using wire cabling, half-duplex operation (a means of allowing two
stations to alternately use the same transmission medium) is commonly and
easily implemented.   However, with fibre optic systems this is not a
common practice since it is often easier and more desirable to set up
full-duplex links.   Various techniques are available to enable data
interchange between a large number of stations by means of multiplexing
(Hew78a).

Three types of multiplexing can be considered for fibre optic signals.
Electronic parallel-to-serial (or time division multiplexing) may be used
to transmit many signals in a serial format.   Alternatively, each fibre
optic bundle may be subdivided to make each group of fibres a separate
channel for parallel data transmission.   Indeed, a single fibre may be
used as one channel if the redundancy provided by a group of fibres is not
required.   A further possibility is to use several light sources which
have different wavelength characteristics or colours.   This not only
achieves multiplexing but also increases the data capacity of the cable.

It is likely that the decreasing cost of fibre optic components coupled
with their highly desirable data transmission properties will make this
type of communication link very popular for laboratory and process control
applications.   Coupled with a standard interface system such as the
IEEE-488, optical fibre links can provide an easy to use tool for data
collection, control of laboratory experiments and control of plant
equipment.   Such an application has been described by Grady (Gra79) who
has outlined the use of an optical fibre link to interconnect a desktop

computer and a process control computer using a 20 kilobyte/second
IEEE-488 interface. Further details on the use of optical fibres for data
transmission will be found in the books by Miller and Chynoweth (Mil79),
Midwinter (Mid79), Sandbank (San79) and Howes et al (How80).

(3) Infrared Data Links

One of the major problems associated with 'hard-wiring' terminals and
instruments to particular locations within an office or laboratory is the
high cost that is usually involved. In the past copper coaxial cables
have most often been used. These are generally quite expensive;
installing them is even more expensive and once they are installed it is
difficult to move any of the devices connected to them. Although the use
of optical fibre cables may reduce the material costs, the cabling and
mobility problems still exist. Each time a new device is installed a new
cable has to be laid; if a machine is moved, appropriate modification of
the cabling is necessary. A different method of data transmission that
might prove a feasible alternative approach to hard-wiring a laboratory
has been investigated by International Business Machines in their Zurich
Research laboratories (Gfe79). The techniques that they have developed
would enable computers, terminals and other electronic devices to communicate by broadcasting at infrared wavelengths.

In principle a computer and several remote terminals could be linked by
radio-frequencey signals. Each terminal would be equipped with a transmitter, a receiver and an antenna - the basic hardware for communication.
Unfortunately, it is not possible to control the range of radio-frequency
signals precisely so that this would mean that neighbouring computer
networks might interfere with one another. Of course, there are many
other sources of interference including radiation from the terminals or
other instruments that might be in use. Of more concern is the fact that
some part of the radio-frequency spectrum would have to be set aside for
this type of communication. This would not be a simple matter since the
spectrum is already crowded. The use of infrared radiation does not
penetrate most materials and so the signals could readily be confined to a
single room thus enabling nearby networks to operate independently. The
system would also be immune to most other electromagnetic emissions.

The transmitter for an infrared data communications link would be a light
emitting diode (LED) - a semiconductor device that emits infrared
radiation in response to an applied voltage. The emitted radiation could
be modulated directly by turning the diode off and on to represent the 0's
and 1's of binary data; alternatively, the diode could be switched at a
fixed frequency to generate a carrier wave which could then be modulated
by any of several methods, such as shifting the phase of certain pulses.
The receiver is another semiconductor device, a photodiode, that gives
rise to a voltage when it absorbs light or infrared radiation. Amplifying
and detecting circuits would recover the original signal from these
voltages. An essential feature of the infrared data link is that it does
not depend on a direct beam to connect two devices. The radiation is
diffuse, it is reflected by walls and the ceiling and ideally it permeates
an entire room like light from an electric light bulb suspended centrally
from the ceiling. As there is no need to maintain a clear line of sight
between a transmitter and a receiver, terminals and other devices could be
placed anywhere in the room and in any orientation. Signals from the
central computer would be broadcast by a large array of light-emitting
diodes near the centre of the room; each signal would be prefaced by an

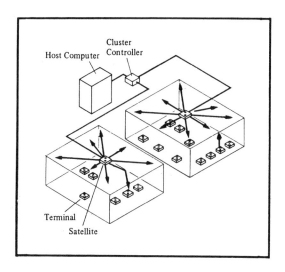

Fig. 7.16 Arrangement of transmitters and receivers
for data transmission via infrared data links.

address designating a particular terminal and would be ignored by all
other terminals. Signals in the other direction would be transmitted by
light-emitting diodes mounted on the terminals or instruments and would be
received by photodiodes in the central array. Figure 7.16 shows a typical
arrangement of transmitters and receivers as they might be located in a
typical laboratory environment.

Figure 7.16 illustrates how an infrared data communication system might
inter-connect terminals to a common controller via satellite stations that
are mounted above the working areas. These satellite stations poll each
terminal in turn asking for data. The maximum speed with which information can be transmitted is determined in part by the size of the room.
The interval between successive pulses must be long enough so that a pulse
reflected from a distant wall reaches the receiver before the next pulse
gets there by direct transmission. In practice a lower speed limit is set
by noise introduced into the photodiodes by background illumination. The
maximum practical speed is probably less than a million bits per second.
Prototype senders and receivers built at the Zurich laboratory operate at
64,000 and 125,000 bits per second.

The performance of such links in an analytical laboratory environment has
yet to be investigated. However, initial experiments conducted by
International Business Machines have used a variety of potential sources
of interference. For example, data was transmitted in the proximity of
arc welding - a powerful source of both heat and electrical static.

(4) **Microwave links**

There has been an enormous increase in demand for communication facilities
over the last few years. This demand has led to the development of
various techniques for transmitting high volumes of information simultaneously between two points. One aspect of the problem of transmitting
large amounts of information from one point to another over a single
communication link is the provision of channels capable of passing high
enough frequencies, that is, channels of large enough bandwidth. This is
particularly true in the case of television picture transmission - an area
of increasing importance for many scientific experiments. Thus, because a
TV picture requires signal frequencies up to several MHz, it was not
possible to transmit trans-Atlantic television signals before the advent
of communication satellites - submarine cables did not provide sufficient
bandwidth.

In order to transmit information requiring large bandwidths the frequencies of channels carrying the information have had to become considerably
higher. Often frequencies above 1 GHz are employed. Electromagnetic
radiation of this frequency is referred to as microwaves. Microwave radio
is one of the chief means of transmitting long distance telephone calls
and television pictures. Radio relay systems using frequency modulation
(Heb75) and operating in the 4GHz and 6GHz bands are currently in use.
Such systems may carry many radio channels each consisting of as many as
2700 telephone links with modulating signals of the order of 12 MHz.
Microwave systems of this type can be built with highly directive aerials
thus concentrating the transmitted radiation into relatively narrow beams,
the propagation being essentially in straight lines. Transmitting and
receiving aerials carrying microwave dishes or horn antennae can often be
seen on the roof-tops of buildings in many cities. Circuits may consist
of many stations, each some 30 miles apart, relaying large amounts of
information from tower to tower and thereby around the curvature of the
earth.

Special types of electronic circuit (such as cavity magnetrons, klystron
oscillators, etc) are required in order to generate microwave signals. In
transmission circuits the waves must be guided with very small leakage
from one point to another. The guiding of the waves is achieved by the
use of two-wire or coaxial cable - the latter being used at high frequency
because of the smaller leakage and more controllable performance.
However, to reduce power loss waveguides are more often used (Sta61,
Mar71). A waveguide is, in essence, a hollow metal tube down which radio
waves of high frequency travel. Power loss in waveguides (typically, 0.04
dB/metre) is usually much lower than that of cables (0.6 dB/metre).
Various designs of waveguide have been used - rectangular, circular,
elliptical and helical are the most common. Rectangular waveguides are
often used to feed the signals between microwave antennae and their
associated electronic equipment. They are not used for long distance
communication and are rarely employed for distances in excess of a few
thousand feet.

Currently there is much interest in the possibility of using optical waveguides in the form of glass fibres which could carry modulated light
signals. Light is yet another form of electromagnetic radiation, and
infrared light at frequencies of 100 THz (1 THz = 10^{12} Hz) could
probably be used to make vast bandwidths available. Lasers are another
important area of current interest for communication applications because
of their potential for high speed signalling. Indeed there has been much

interest shown in recent years in the use of lasers in conjunction with
waveguides constructed from optical fibres of the type described in the
last section.

(5) Satellite Links

Communication satellites have been in use for a variety of purposes since
about 1958 when the USA launched their SCORE system. Since then many
others have been sent into orbit - for example, TELSTAR (1962), MOLNIYA
(1965), EARLY BIRD (1965), INTELSAT IV (1971), WESTAR (1974) and INTELSAT
V (1980). The major functions of satellite systems fall into five broad
categories:

        (a)   earth observation,
        (b)   civilian communication,
        (c)   distributed broadcasting,
        (d)   military communication, and,
        (e)   anti-satellite activities.

Civilian communication satellites are currently being used for the interchange of live television pictures, news and telephone services to over 50
countries. Because of their high bandwidth, large geographical coverage
and high data transmission speeds, civilian communication satellites are
being used increasingly to transmit scientific and business data in such a
way as to provide two-way linking of information between computer systems
anywhere on the earth's surface. Such a satellite based communication
system has recently been set up by an IBM based company called Satellite
Business Systems or SBS (Mar78).

In principle, the use of satellites for data communications is quite
simple. Basically the technique involves transmitting the desired signal
from an earth station to an orbiting satellite. The equipment on board
the satellite receives the signals, amplifies them and then rebroadcasts
them to another earth station thereby providing a point-to-point data
link. Communication is achieved using a narrow beam of microwave carrier
in the 4-6 GHz range. In many satellite systems the transmission up to
the satellite (the up-link frequency) operates in the 6 GHz band. The
signal is then returned to earth on a different frequency (the down-link
frequency) usually in the 4 GHz band. Because of their widespread use for
telecommunication applications the capacity of a satellite is often
described in terms of the number of voice channels it will support in
relation to its physical size. INTELSAT IV (Mar78) which was launched in
1971 provided 4000 voice channels and weighed 700 kilograms. INTELSAT V
which was launched in 1980 provides 12,500 voice channels and weighs 950
kilograms. The relative smallness of the latter is due to the extensive
use of microtechnology. Usually, a satellite has to support a large
number of earth stations and so multiplexing is used (both time and
frequency division) in order to achieve appropriate sharing of resources.
Quite often a technique called frequency division multiple access (or
demand access) is also used (Mar78). In this technique the available
channels are not permanently allocated to particular stations. Instead,
channels remain part of a common pool which are allocated when needed.
When transmission has terminated, channels are returned to the pool.

Most satellites are launched into a geosynchronous or geostationary orbit
at a distance of about 22,500 miles from the centre of the earth in a
plane that includes the equator. In such a position they will (in theory)

follow the rotation of the earth and always make the same footprint (area covered by the satellite beam) which works out at about 40% of the earth's surface - using earth coverage beams. Thus, three satellites operating in this mode could achieve total coverage of the earth. Earth coverage beams radiate from the satellite at an angle of about 17.5 degrees. If this angle is reduced, thereby narrowing the beam, spot beams are produced which have a much narrower footprint. Thus a 1° beam produces a coverage of about 500 miles diameter - depending upon the altitude of the satellite. Satellites that provide multiple spot beams can be used to implement distributed broadcasting or distributed data communication. Further details on the use of satellites for data communication will be found in the book on communication satellite systems by Martin (Mar78).

BASIC TELECOMMUNICATIONS

In the previous section some of the commonly used methods of implementing data links between terminals, computers, measuring instruments and control equipment were outlined. These may be used to construct two broad types of data communication facility according to whether the overall system is (a) totally in-house, or (b) requires the (additional) use of services provided by special agencies (called 'carriers') such as the national and regional telecommunication authorities. As far as users of this latter type of facility are concerned, the kind of data link used to provide the communication capability will be totally transparent to them. Often, the facilities that are provided will appear to be based upon existing telegraph (Ren74) or telephone circuits - even though the latter may involve the use of optical fibre, microwave or satellite links. As telephone circuits are more extensively used than telegraphic systems, they will form the basis of the discussion that follows.

As can be seen from Fig 7.17, the basic ideas underlying the use of a telephone connection for the transmission of data is quite simple.

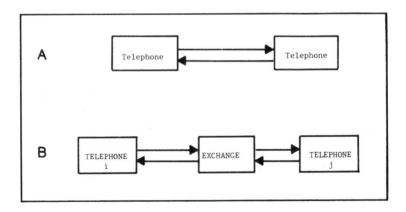

Fig. 7.17   Simple approaches to data transmission using dedicated (case A) and switched (case B) telephone links.

The easiest situation to envisage is one in which two particular telephones are linked by a pair of wires (case A in Fig. 7.17). An arrangement
such as this can be set up on an in-house basis or via the telecommunication authority when it involves using public telephone ciruits. In the
latter case the arrangement is often referred to as a leased line.

If some form of manual/automatic switching exchange is introduced so that
the telephones in the previous example are not dedicated to each other as
a result of permanent connections then a more general purpose system
results - as is illustrated in case B of Fig. 7.17. The exchange may be a
private one if the system is in-house or a public exchange if it involves
the telecommunication agencies. In such a case the exchange probably
forms part of the Public Switched Telephone Network (PSTN) a section of
which is depicted in Fig. 7.18.

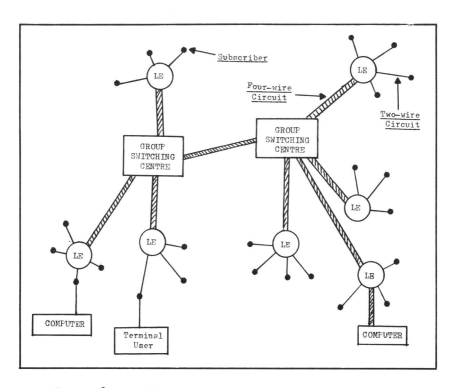

Fig. 7.18 A section of a public switched telephone network
LE: subscriber's local exchanges.

In the type of system shown in Fig. 7.18 each subscriber is connected to a
local exchange (LE) which handles dialled calls between users in that
local area (Smo76). Calls involving distant subscribers require the use
of group switching centres. These are connected to local exchanges (and
each other) by 4-wire circuits. In these 4-wire trunk circuits one pair
of wires is used for transmitting and the other for receiving information.
An arrangement such as this is used because the long transmission
distances involved require the use of amplifiers which operate only in one
direction. Note that with non-switched least lines it is also possible to

have four-wire connections brought out from a local exchange. This then
permits full-duplex operation by using two wires for the transmission and
two for the reception of data.

Most of the older telephone circuits were based upon the use of analogue
signals associated with the transmission of voice data - signals in the
frequency range 300 - 3400 Hz. Consequently, if the telephone system is
to be used for digital data transmission some additional hardware is
required - a modem (or modulator - demodulator). Its relationship to
the other items of equipment is illustrated in Fig. 7.19.

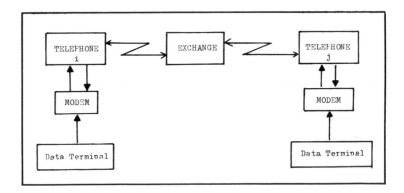

Fig. 7.19 Modems used for data communication over a telephone network.

The function of the modem is to convert digital data into signals that can
be transmitted over the telephone network. This transformation can be
achieved by a process called modulation in which digital data levels
leaving the terminal device are used to modulate a sine-wave carrier. The
function of the carrier is simply to provide a means of 'carrying' the
data over the telephone line to its destination. At the receiving end the
carrier is demodulated using a second modem. The details of the
connections for these modems on two-wire and four-wire links are shown in
the diagrams contained in Fig. 7.20.

The connection between the data terminal (or other device such as a
scientific instrument or computer) and the modem is made by means of a
standard interface - the RS-232-C convention (or its European CCITT V.24
equivalent) which was discussed in chapter 6. Physically, this consists
of a 25-way connector (see Fig. 6.15) with each of its pins allocated to a
specific function. The significance of each of the interface connections
is shown in Fig. 7.22. Not all of these connections need to be used as
most modems will operate using an appropriate subset.

A particularly useful and inexpensive type of modem for use with data
speeds up to about 300 bits/second is the acoustically coupled modem
(Cia80, Amo80, Tra79a, Tra79b). This type of modem uses a loudspeaker and
microphone for output and input of audible frequencies. Connection to the
telephone network is accompolished by plugging an ordinary telephone
handset into a special mounting cradle contained within a sound-proof box.
A typical arrangement of equipment is depicted in Fig. 7.21.

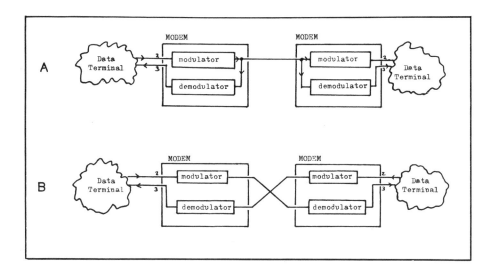

Fig. 7.20   Connections for two-wire (case A) and four-wire
(case B) modems.   The numbers associated with the modems
indicate pin connections on an RS-232-C interface system.

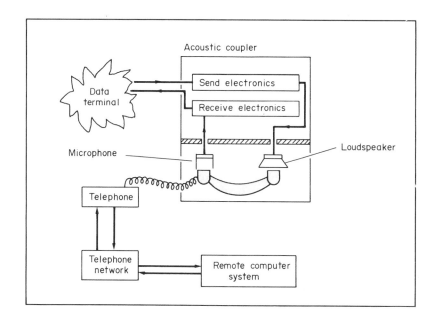

Fig. 7.21   Use of an acoustic coupler for data transmission.

INTERFACE CONNECTIONS

PIN NO.	FUNCTION	CCITT V24 DESIGNATION	DESCRIPTION OF FUNCTION
1	Protective Earth	101	The Protective Earth is connected to the a.c. power earth and the modem frame.
2	Transmitted Data	103	Data from the Terminal Equipment that is to be transmitted.
3	Received Data	104	Received data from the modem to the Terminal Equipment.
4	Request to Send (RTS)·	105	The Terminal Equipment requests the modem to assume a Transmit condition by placing the 'Request to Send' line in an 'ON' condition.
5	Clear to Send (CTS)	106	The modem provides an 'ON' condition on this line to indicate to the Terminal Equipment that it is ready to transmit data after receiving a 'Request to Send' signal.
6	Data Set Ready	107	The modem provides an 'ON' condition on this line to indicate that it is connected to line and all operate switches are in the correct mode to Transmit or Receive.
7	Signal Earth	102	Signal Earth is usually connected to Protective Earth inside the modem power supply.
8	Data Channel Received Line Signal Detector (Also referred to as Data Carrier Detector)	109	An 'ON' condition indicates that the data carrier is being received.
9	Unassigned		
10	Unassigned		
11	Select Transmit Frequency	126	An ON condition indicates that the Backward Channel carrier is being received.
12	* Backward Channel Received Line Signal Carrier	122	The modem provides an ON condition when the Backward Channel is ready to transmit data
13	* Backward Channel Clear to Send	121	The modem provides an 'ON' condition to indicate that it is ready to send back Backward Channel Data.

Fig. 7.22 Details of the signals and circuits used for data transmission via the RS-232-C (CCITT V.24) interface.

PIN NO	FUNCTION	CCITT V24	DESCRIPTION OF FUNCTION
14	✳Backward Channel Transmitted Data	118	Data to be transmitted on the Backward Channel.
15	Transmitter Signal Element Timing	114	The modem provides a square wave signal at the clock frequency for synchronisation of the data to be transmitted from the data terminal
16	✳Backward Channel Received Data	119	Received Data on the Backward Channel
17	Received Signal Element Timing	115	This signal is a square wave at the bit frequency which is synchronised with the receiver
18	Unassigned		
19	Backward Channel Request to Send	120	The terminal equipment requests the modem to assume a Transmit condition on the Backward Channel
20	Data Terminal Ready	108	An 'ON' condition allows the modem to operate in the data mode when it receives some other stimulus
21	Unassigned		
22	Calling Indicator	125	An 'ON' condition indicates receipt of a call signal
23	Data Signalling Rate Selector	111	A Signal from the Terminal indicating the data rate to be transmitted.
24	Unassigned		
25	Unassigned		

DIFFERENCE BETWEEN PIN ALLOCATIONS IN RS232 AND EUROPEAN STANDARDS
Pin 11 is unassigned in RS 232 but allocated in Europe to 'Select Transmit Frequency'.
Pin 21 is unassigned in Europe but allocated in RS232 to a 'Signal Quality Detector Signal'.
Pin 24 is also unassigned in Europe but allocated in RS232 to 'External Transmitter Clock'.

✳NOTE: A backward channel is incorporated in some modems to enable
low data rate signals (e.g. 150 bps) to be transmitted
simultaneously with the main data stream.

Fig. 7.22 (continued) Details of the signals and
circuits used for data transmission via the RS-232-C
(CCITT V.24) interface.

The acoustic coupler converts the digital inputs into speech frequencies which can then be transmitted on normal dialled telephone lines. At the receiving end a similar device detects the audible tones issuing from the telephone earpiece and transforms them back into serial data pulses. A typical set of frequencies for modem operation might be (Cia80),

Originate Mode	Logic 0	1070 Hz
	Logic 1	1270 Hz
Answer Mode	Logic 0	2025 Hz
	Logic 1	2225 Hz

One set of tones (1070 Hz and 1270 Hz) is used by the originating terminal and another set (2025 Hz and 2225 Hz) is used by the answering terminal. Modems of this type may be originate only, answer only or originate/ answer. For most applications in which data is transmitted from a terminal or scientific instrument to a remote computer, simple originate only modems may be used at the data source. Modems that are capable of answering will only be required for certain types of specialised applications in which the computer 'calls' the remote equipment. The main advantage of an acoustic coupler is its light weight and small size which enables it to be moved from one location to another. Consequently, data transmission can take place from any location that has a telephone. A wide variety of other types of modem exist. They cover a broad range of signalling speeds and modulation techniques. Obviously, modems do not need to be used in conjunction with telephone systems, they can be used with any 'wired' data system. The telecommunications network, however, offers a useful and versatile national/international 'wired' data network. Such a system used in conjunction with automatic call and automatic answering facilities (Heb75) provides the means to enable quite sophisticated data acquisition systems to be constructed.

DATA ENCRYPTION

The increasing use of computers for the storage of scientific, technical, business and personal information has caused considerable research and development effort to be devoted to the problems of security, privacy and integrity of the stored data. Similarly, because of the expanding use of data communications facilities for the collection and exchange of data, it is imperative to ensure the secrecy of the transmitted data. This is particularly true where business information, 'industrial secrets' or other highly confidential information is being transmitted. Several authors (Kat73, Mar73, Pri79, IBM77, IBM79a) have considered these problems in some detail. The discussion that follows is an attempt to provide an overview of the major principles involved.

Data encryption or cryptography refers to techniques for converting data from its plain or clear text form into a cipher form (either for storage or transmission) and back again when required. The process of converting plain or cipher text is called encryption. The reverse process is then referred to as decryption. Cryptography is an important counter-measure to the threat of accidental (or deliberate) disclosure of confidential data to persons who are not authorised to see, handle or use it. Since data transmission often involves movement of information in localities remote to the organisation that holds it, the problems of secrecy become

more critical than when just in-house storage is involved. A typical
system that might be used for encryption of data that is transmitted over
a communication link is illustrated in Fig. 7.23 (Dif76).

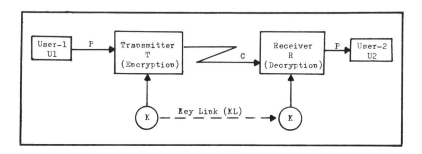

Fig. 7.23 Hardware arrangement to implement data encryption.
U1, U2: users; P: plain text; C: ciphered text; K: key.

In this system the user (U1) enters confidential data into the system and
wishes to transmit it over a communication link (C) to another user (U2).
The plain text (P) entered by user U1 is encrypted at the transmitter
using the same (or a different) key (K) as is used by the receiver to
decrypt the cipher text (C). As a result of the encryption the plain text
is protected during transmission because an intruder who manages to get
access to the cipher text is unable to decrypt it without knowledge of the
key. However, there is still a problem, the secrecy of the key itself.
It must be securely protected both at the transmission (T) and receiving
(R) devices where it is used and also during its transmission over the key
transport link (KL) that transports it between the two locations. The key
transport link is also used to transmit new keys or key changes between
the two (or more) users. It is feasible to use the same circuit for the
key link as is used for the transmission of the cipher text. However,
there are advantages to using a different circuit or an entirely different
type of data link - for example, a courier or postal service.

Various techniques have been used to achieve encryption of data - both
hardware and software. Software approaches depend upon the use of an
appropriate algorithm (set of rules) for manipulating the plain text in
various ways - usually substitution, transposition and concealment. The
use of a fixed algorithm, however, means that if an intruder determines
the algorithm he/she then has access to all the information. For this
reason algorithms with a variable part (the key) are used. The security
of the data then rests with the key which can be changed as often as
desirable - perhaps once or more each day - depending upon the system
design. If it is decided to change the key on a daily basis, the key and
algorithm should be selected in such a way that the time needed for an
intruder to decipher the key would exceed the frequency with which it
changes. Thus, if an intruder was able to find out the enciphering
algorithm, to obtain even a small key by trial and error could take an
extremely long time (Gar77). For example, suppose a key was constructed
by choosing 12 alphabetic characters from the English alphabet, the time
taken to deduce it using a fast computer would be about 1500 years.
Similarly, a twenty character key could take as long as 3×10^{11} years
to decipher. This type of approach to data encryption is illustrated in
the algorithm listed in Fig. 7.24 (Mar73).

```
 1     SECRET: PROCEDURE OPTIONS(MAIN);
 2     DECLARE
 3       (TEXT, MESS, ALPH, KEY)     CHAR(48),
 4        M(48)                      FIXED BIN,
 5        MODE                       CHAR(30) VARYING;
 6     GET LIST (KEY);
 7     GET LIST(MODE);
 8     ALPH='ABCDEFGHIJKLMNOPQRSTUVWXYZ1234567890-.,)(:;"=+ ?';
 9     DO K=1 TO 48;
10       M(K)=INDEX(KEY,SUBSTR(ALPH,K,1));
11     END;
12     IF MODE='TRANSMIT'
13       THEN DO;
14             GET EDIT(TEXT) (COL(1),A(48));
15             DO K=1 TO 48;
16               N=INDEX(KEY,SUBSTR(TEXT,K,1));
17               SUBSTR(MESS,M(K),1)=SUBSTR(ALPH,N,1);
18             END;
19             PUT SKIP EDIT(MESS) (A);
20           END;
21       ELSE DO;
22             GET EDIT(MESS) (COL(1),A(48));
23             DO K=1 TO 48;
24               N=INDEX(ALPH,SUBSTR(MESS,M(K),1));
25               SUBSTR(TEXT,K,1)=SUBSTR(KEY,N,1);
26             END;
27             PUT EDIT(TEXT) (A(48));
28           END;
29     END SECRET;
```

Fig. 7.24 Software implementation of a typical
data encryption algorithm.

The algorithm is used at both the transmitter and receiver stations. It
uses the same key both for encryption and for decryption. This is entered
into the program via the statement embodied in line 6 of the listing. The
code for the encryption process is contained in lines 12 through 20. When
presented with the following three lines:

```
T1    'NY+BTVG FCED18WS:)X564AQZ,.-OLOPIKJM=U9732?"(;HR'
T2    'TRANSMIT'
T3    FLUORIDE IN TOOTHPASTE IS TO LOW AT 361 PPM  ..
```

the output produced is:

```
3PH4PKL17.?3EWHW6H73041 HMU(3HHEHA7EHEEHH31H6OK6
```

which is the result of encrypting the message contained in line T3. At
the receiving end of the data transmission system, presenting the
following three items to the program:

```
R1    'NY+BTVG FCED18WS:)X564AQZ,.-OLOPIKJM=U9732?"(;HR'
R2    'RECEIVE'
R3    3PH4PKL17.?3EWHW6H73041 HMU(3HHEHA7EHEEHH31H6OK6
```

would result in the orginal message (FLUORIDE IN TOOTHPASTE IS TO LOW AT
361 PPM ..) to be produced as a result of the decryption process embodied
in lines 22 through 28 of the program.

There are many different algorithms that could be used for data encryption. One of particular importance is that embodied in the Data Encryption Standard (DES) formulated by the United States Bureau of Standards (NBS77). This utilises both substitution and transposition techniques in order to encipher the plain text. The DES operates with a 64-bit key upon a 64-bit input block of data to produce a 64-bit output block. The detailed operations are highly suitable for hardware implementation and many DES chips and devices incorporating these chips (or equivalent hardware) are available - such as the IBM 3845 and 3848 Encryption Devices (IBM79, IBM79b). Each of these units contain a hardware implementation of the US Federal Data Encryption Standard. The IBM 3845 is a desktop device employing cipher text feedback. It has a detachable hand-held Personalisation/Key Entry Unit that permits the entry of initialising data, the key and other data to match the 3845 to specific data communication line characteristics.

Of the 64 bits used in the DES standard 8 of them are parity bits that are generated by the system. This means that only 56 of them can be chosen independently. Because of this, researchers in cryptography have claimed that the DES provides an inadequate level of security since the key is short enough to be determined by an exhaustive search. Diffie and Hellman (Dif77) suggest that it would be possible to build a machine that would be capable of breaking the code in about 12 hours computation time. Although the cost of constructing such a machine was high at the time the original estimates were prepared (1977), it was claimed that within ten years the diminishing cost of hardware components would enable such a machine to be built relatively cheaply. The NBS standard would then become totally insecure. However, changing the length of the key to 128 or 256 bits would increase the time and cost of deducing it to limits beyond the capabiltiy of any foreseeable technological advances. Indeed, it is claimed (Dif77) that both quantum mechanical and thermodynamic arguments rule out the feasiblity of exhaustive searches on keys of several hundred bits. The evidence presented by Diffie and Hellman seem quite convincing, however, despite this the assertions that have been made are not accepted by the NBS. Data encryption is thus an area which is of considerable controversial concern.

CONCLUSION

Data and the information that it produces are commodities that are of vital importance to the scientist, technologist and manager. Many situations arise in which the information required for accurate decision making needs to be collected in real-time. Often, it has to be derived from data produced by scientific instruments and other measuring devices that are located at sites which are remote to the centre at which it will be processed. Arrangements that enable these types of activity to take place will require the implementation of appropriate communication links capable of transmitting the data from its source to its required destination. In this chapter some of the various types of data links have been described: couriers, cables, optical fibres and satellites. Particular emphasis has been given to the use of the telephone. The enormous distribution web associated with the telecommunication networks of the world enable the facile transmission of data between instruments/equipment that is located virtually anywhere on the earth's surface. Because computers can be easily interlinked via the use of telecommunication facilities,

distributed computer networks are a logical next step. This topic is
dealt with in more detail in Chapter 12.

In storing data and transporting it from one location to another there are
three important factors that need to be considered: its security, its
accuracy and its privacy. Those aspects of these topics that are relevant
to data communication have been briefly outlined. The methods and
problems of data encryption have been discussed since the use of cryptographic techniques is vital if the confidentiality of data (either during
transmission or storage) is to be ensured.

In this and previous chapters the subjects of computers, interfaces,
instruments, and communication have each been discussed in considerable
detail. These basic components have to be put together in appropriate
ways to form an integrated whole that forms some useful function to the
analytical scientist. In the next chapter the topic of laboratory automation will be considered in order to illustrate how technology is
influencing laboratory activities.

REFERENCES

Amo80   Amor, K., Acoustically Coupled Telephone Modem, Practical
           Electronics, Volume 16, No. 2, 39-44, February 1980.

Cia80   Ciarcia, S., Build-it-Yourself Modem for Under $50, BYTE: The
           Small Systems Journal, Volume 5, No. 8, 22-38, August 1980.

Dif76   Diffie, W. and Hellman, H.E., New Directions in Cryptography,
           IEEE Transactions in Information Theory, Volume IT-22, No. 6,
           644-654, November 1976.

Dif77   Diffie, W. and Hellman, H.E., Exhaustive Cryptoanalysis of the
           NBS Data Encryption Standard, IEEE Computer, Volume 10, No. 6,
           74-84, June 1977.

Gar77   Gardiner, M., A New Kind of Cipher that Would Take Millions of
           Years to Break, Scientific American, Volume 237, No. 2, 120-
           124, August 1977.

Gfe79   Gfeller, F.R. and Bapst, U., Wireless In-house Data Communi-
           cation via Diffuse Infrared Radiation, Proceedings of the
           IEEE, Volume 67, No. 11, 1474-1486, November 1979.

Gra79   Grady, R.B., High Speed Fibre Optic Link provides Reliable Real-
           Time HP-IB Extension, Hewlett-Packard Journal, Volume 30, No.
           12, 3-9, December 1979.

Heb75   Hebditch, D.L., Data Communications - An Introductory Guide,
           Elek Science, ISBN: 0-236-31098-4, 1975.

Hew78a   Hewlett Packard, Digitial Data Transmission with the HP Fibre
           Optic System, Application Note 1000, Publication No. 5953-0391,
           November 1978.

Hew78b   Hewlett Packard, Fibre Optic 100 Metre Digital Transmitter
          HFBR-1001, Publication No. 593-0387, November 1978.

Hew78c   Hewlett Packard, Fibre Optic Digital Receiver HFBR-2001,
          Publication No. 5953-0388, November 1978.

Hew78d   Hewlett Packard, Fibre Optic Connector/Cable Assemblies,
          Publication No. 5953-0389, November 1978.

Hew79   Hewlett Packard, Fibre Optic HP-IB Link Products: 12050A Fibre
          Optic HP-IB Link; 39200 Series Cable, Publication No. 5953-
          4227, June 1979.

Hic80   Hickey, J., Wire and Cable - What's Happening, Instruments and
          Control Systems, Volume 53, No. 8, 39-42, August 1980.

How80   Howes, M.J. and Morgan, V.V., Optical Fibre Communications:
          Devices, Circuits and Systems, John Wiley, ISBN: 0471-27611-1,
          1980.

IBM77   IBM Corporation, Data Security through Cryptography, Form No:
          GC22-9062-0, October 1977.

IBM79   IBM Corporation, IBM 3845 Data Encryption Device, IBM 3846 Data
          Encryption Device, General Information Manual, Form No: GA27-
          2865, September 1979.

IBM79a   IBM Corporation, IBM Cryptographic Subsystems - Concepts and
          Facilities, Form No: GC22-9063-3, November 1979.

IBM79b   IBM Corporation, IBM Cryptographic Subsystem featuring the IBM
          3848 Cryptographic Unit, Form No: G520-3233-1, December 1979.

Joh72   Johns, P.B. and Rowbotham, T.R., Communication System
          Analysis, Butterworths, ISBN: 0-408-70197-8, 1972.

Kat73   Katzan, H., Computer Data Security, Van Nostrand Reinhold,
          ISBN: 0-442-24258-1, 1973.

Kin80   Kincaid, J., Consider Foil-Shielding Cables for Data Trans-
          mission, Instruments and Control Systems, Volume 53, No. 4,
          63-65, April 1980.

Kli77   Klipec, B.E., How to Avoid Noise Pick Up on Wire and Cable,
          Instruments and Control Systems, Volume 50, No.12, 27-30,
          December 1977.

Mar70   Martin, J., Teleprocessing Network Organisation, Prentice-
          Hall, ISBN: 0-13-902452-2, 1970.

Mar71   Martin, J., Future Developments in Telecommunications,
          Prentice-Hall, ISBN: 13-345868-7, 1971.

Mar72a   Martin, J., Systems Analysis for Data Transmission, Prentice-
          Hall, ISBN: 0-13-881300-0, 1972.

Mar72b   Martin, J., Introduction to Teleprocessing, Prentice-Hall,
          ISBN: 0-13-499814-6, 1972.

Mar73   Martin, J., Security, Accuracy, and Privacy in Computer
           Systems, Prentice-Hall, ISBN: 0-13-798991-1, 1973.

Mar76   Martin, J., Telecommunications and the Computer, Prentice-Hall,
           ISBN: 0-13-902494-8, 1976.

Mar78   Martin, J., Communications Satellite Systems, Prentice-Hall,
           ISBN: 0-13-153163-8, 1978.

Mid79   Midwinter, J.E., Optical Fibres for Transmission, John Wiley,
           ISBN: 0471-60240-X, 1979.

Mil79   Miller, S.E. and Chynoweth, A.G., Optical Fibre Telecommuni-
           cations, Academic Press, ISBN: 0-12-497350, 1979.

Nal77   Nalle, D.H., Kill Signals that Attack Measurement Data,
           Instruments and Control Systems, Volume 50, No. 2, 35-39,
           February 1977.

NBS77   National Bureau of Standards, U.S. Department of Commerce, Data
           Encryption Standard, FIPS PUB46, January 1977.

Phi80   Philpotts, P., English as She is Spoke ... Nearly, Control and
           Instrumentation, Volume 12, No. 7, 50-51, July 1980.

Pri79   Pritchard, J.A.T., Security in On-line Systems, National
           Computing Centre, UK., ISBN: 0-85012-211-2, 1979.

Ren74   Renton, R.N., The International TELEX Service, Pitman
           Publishing, ISBN: 0-273-31774-1, 1974.

RSC80   R.S. Components Ltd., 13-17 Epworth Street, London, EC2P 2HA,
           UK., Data Sheet No. R/4030, Optical Fibres, July 1980.

San79   Sandbank, C.P. (Ed.), Optical Fibre Communication Systems, John
           Wiley, ISBN: 0471-276677, 1980.

Smo76   Smol, G., Hamer, M.P.R. and Hills, M.T., Telecommunications: A
           Systems Approach, George Allen & Unwin Ltd., ISBN:0-04-621022-9,
           1976.

Sta61   Starr, A.T., Microwave Techniques, Chapter 10, 351-390, in
           Telecommunications, Pitman and Sons, 1961.

Tra79a   Transdata Ltd., 11 Garrick Street, London WC2E 9AR, UK., Data
           Sheet for the Model 307 Acoustic Data Modem, 1979.

Tra79b   Transdata Ltd., 11 Garrick Street, London WC2E 9AR, UK., Data
           Sheet for the Model 307A Originate Acoustic Modem, 1979.

Wil76   Williams, D.N., Fibre Optics for Data Transmission, Instrumenta-
           tion Technology, Volume 23, No. 9, 61-66, September 1976.

Wit80   Witten, I.H., Communicating with Microcomputers, Academic
           Press, ISBN: 0-12-760752-8, 1980.

8

Automation in the Laboratory

INTRODUCTION

Some years ago in his book on automation Sir Leon Bagrit (Bag64) suggested that the directions of modern science and technology pointed towards the creation of a series of machine-systems based upon Man as a model. To Bagrit it appeared as though Man was engaged in creating an extension of himself - 'He is extending his eyes with radar; his tongue and his ear through telecommunications; his muscle and body structure through mechanisation. He is extending his own energies by the generation and transmission of power and his nervous system and his thinking and decision-making faculties through automation'.

Automation is the means by which a machine-system is able to operate with minimal human intervention and maximum efficiency. It achieves these objectives as a consequence of making appropriate measurements and observations and then controlling its behaviour accordingly. In order to achieve automation of a system it is necessary to obtain a detailed and continuous knowledge of the functioning of the system so that the best corrective actions (to achieve or maintain desired goal states) can be applied as soon as they become necessary. Fundamental to the process of automation are the three concepts of communication, computation and control - all of which have been briefly outlined in previous sections of this book.

Historically, automation has developed as a natural evolutionary step that follows the process of mechanisation. Mechanisation is the term used to describe those situations in which machines are employed either to simplify or to ease the tasks that humans might otherwise have to perform. Mechanisation does not usually involve replacement of people by machines. Indeed, human operators are still required in order to maintain control over and operate the machines. However, in automation, the machines control themselves automatically and no direct intervention by humans is required. A basic introduction to the concepts of mechanisation and automation and an outline of the fundamental differences between the two will be found in the book by Handel (Han67).

The essential transitions involved in moving from a highly labour
intensive system to one which is largely automatic in operation can be
deduced from a comparison of the diagrams presented in Fig. 8.1. They
illustrate three approaches to analysis: manual, mechanised and
automated.

Each of the diagrams reflects the role of analytical science as a process
control tool. In these simple models the symbol Ⓗ is used to denote
some form of intervention by a human operator or link. It is easy to see
that mechanisation should reduce the effort required on the part of those
involved in manual analysis (e.g. titrimetry, colorimetry, gas analysis,
etc). The driving force for mechanisation is reduction in cost, removal
of tedium on the part of human operators, increased performance (results
available more quickly), reliability and accuracy. The introduction of
machines as a result of mechanisation does not necessarily imply a substantial depletion of the man-power involved even though the level of
skills might be substantially less (Sim72). The third of the diagrams (C)
in Fig. 8.1 shows - as a result of automation - that the man-power
involved in the control cycle is minimal. Human operatives are still
required to maintain the machines and perhaps override decisions made by
the control software but they are not a necessary part of the fundamental
control loop.

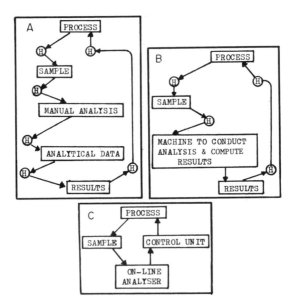

Fig. 8.1 Three approaches to chemical analysis; A: manual;
B: mechanised; C: automatic. Ⓗdenotes intervention by a
human operative.

When considering automation within an analytical laboratory it is
important to distinguish between two different approaches: instrument
automation, and, laboratory automation. The major difference between
these is illustrated in Fig. 8.2. In the diagram on the left (case A),
instrument automation can be seen as an application of computer technology
to particular instruments or experiments within a laboratory. This may

involve the use of 'hidden' microprocessors embedded in an instrument,
external microcomputer systems, desk-top computers or minicomputers.

In contrast to this, laboratory automation (case B of Fig. 8.2) is
regarded as being the total integration of the instruments/experimental
rigs within a laboratory to form a unified whole. As will be discussed in
subsequent sections of this chapter, there are a variety of reasons why
this second approach is useful. Sometimes instrument automation forms the
basis of the initial steps towards laboratory automation - especially
where cost may be a limiting factor. However, where this is not the case,
design for total automation is often a more desirable way to proceed.

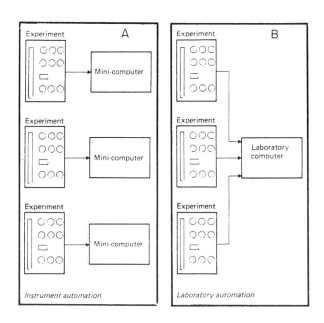

Fig. 8.2   Approaches to analytical automation;
A: instrument automation; B: laboratory automation.

When mainframe computers were the only source of computational resources,
the costs involved could not be made to justify their widespread use for
automation applications. However, during the late 1960s and early 1970s
there was a spate of interest in laboratory and instrument automation.
This arose as a result of the availability of relatively inexpensive
minicomputer systems such as the DEC PDP-11 series, IBM's System/7 and a
variety of computers manufactured by Hewlett Packard. A series of
articles in the IBM Journal of Research and Development (IBM69) serve to
indicate the scope and potential offered by the laboratory and instrument
automation. Typical of the advantages cited are: automatic data
acquisition, open and closed loop control, overall system integration,
greater work throughput, new types of analyses, cost reduction per
analysis, greater productivity, higher precision, better calibration
and drift control of instruments, broader ranges of activities, more data
available, new information extracted from the raw data and real-time data

analysis through the use of such devices as interactive graphics terminals.

Nuclear physicists were probably the first major users of computers for automation in their experiments. Their pioneering work set the scene for future developments in instrument automation in a variety of other areas. Medical applications (IER70), chemistry and other physical sciences then soon followed suit. Perone (Per71) has listed a large number of digital computer applications in the chemistry laboratory. These directly relate to the area of instrument automation using minicomputers. Cole (Col74) has described a hierarchical laboratory automation system based upon the use of a series of minicomputers and a telecommunications link to a remote mainframe system. A similar but more comprehensive description of laboratory automation in a research environment has been described by the Royal Dutch/Shell Company (IBM75). Overall development of this system necessitated the interconnection of eight gas chromatographs, two mass spectrometers and a number of balances/rotameters to a small minicomputer which was itself attached to a larger machine. In addition to the dual minicomputer system housed within the laboratory, there was also access to a remote mainframe computer by means of a communication terminal.

Each of the above examples illustrate laboratory automation as is applicable within a research environment. Often, more demanding applications of automation are to be found in process control laboratories having high throughput requirements (number of samples per unit time) and which involve multi-component analyses (perhaps 20-30 measurements/components per sample). Situations of this type often arise in clinical laboratories (health screening), the pharmaceutical industry (drug screening) and other high volume manufacturing industries requiring the application of analytical science as a process control tool. Descriptions of some of the problems involved in the automation of laboratories for these latter types of application have been described by Sims (Sim72), Haan (Haa80), Arndt (Arn80), Horsely (Hor80) and others (IER70).

During the late 1970s and early 1980s further interest in laboratory automation has been promoted by the common availability of inexpensive microcomputer systems that enable the facile construction of hierarchical data collection and control systems. Another significant factor that has influenced the further development of automation in this area is the ease with which integrated data base management systems can now be constructed. Further details on data bases will be presented later. However, for the moment, an integrated laboratory data base may be thought of as being a collection of control programs and data suitable for use in an automated laboratory. Together these items are responsible for running and controlling experiments and checking the flow of samples through the laboratory. In addition they provide the means to enable archival of results and appropriate techniques for result reporting and checking.

AUTOMATING ANALYTICAL PROCEDURES

An analytical control laboratory is concerned with providng a variety of analytical services. Fundamentally, the laboratory will be involved in processes that deal with the input of samples/specimens, their analysis and the return of analytical results to those who require them. In order to operate effectively and efficiently a variety of organisational and managerial procedures need to be observed if a smooth running operation is

to be maintained. Thus, in most laboratory environments there will
usually be a routine procedure for handling incoming samples presented for
analysis. Typically, samples along with their analysis request forms
(ARFs) are submitted to a sample reception area. This is usually designed
to cater for sample registration, allocation of a unique job (or
accession) number and preparation of a work flow sheet that specifies the
route that a sample should take through the laboratory in order to obtain
the various analyses that are required. Sample and work sheet are then
dispatched to an appropriate section of the laboratory for analysis.
Figure 8.3 presents a simple model that summarises this approach.

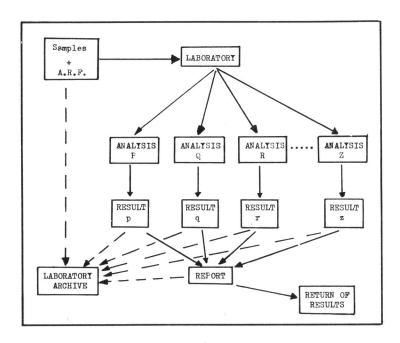

Fig. 8.3 Data and material flow involved in a typical
analytical laboratory.

As each analysis is performed an associated set of results is produced.
These are entered onto the sample's work sheet and the sample then
dispatched to its next port of call within the laboratory. Ultimately,
when all the required analyses have been performed the results are
collated and a report produced which is then sent to the person/section
which submitted the original sample for analysis. Copies of the final
report and individual analytical results are filed away along with the
original request form for future reference - this may be a legal requirement. Some laboratories also archive the samples - where this is feasible.

In a scheme similar to that outlined above automation can be applied in a
variety of ways:

(a) to the record keeping and reporting procedures,
(b) to the analytical function of the laboratory as a whole, and
(c) to individual analytical techniques.

The introduction of automation into a laboratory that is orientated towards conventional manual analytical techniques will not proceed without problems - particularly managerial ones. Many of the problems likely to be encountered by laboratory managers have been outlined in a recent paper by Foreman (For80) - to which the reader is referred. Some of the different aspects of automating record keeping procedures will be discussed in the next section, following this, a section on the use of turnkey systems as a means of integrating many laboratory functions within a single 'analytical machine' will be outlined. This present section deals with some of the fundamental steps involved in automating individual analytical techniques. The demand for such a requirement could arise from two sources. Firstly, a high sample throughput requirement - thereby necessitating the design of a machine that can be run for extended periods of time (perhaps overnight, etc) without any form of manual intervention. Secondly, the need to integrate newly developed analytical techniques into an existing automated system often arises.

Essentially, there are two well defined approaches to automating chemical analyses - the discrete approach (as in centrifugal analysis) and the continuous flow method (as in flow-injection analysis). In the discrete approach each sample is maintained as a separate entity throughout the duration of the analysis and is tranported to individual stations for operations like dilution, reagent addition, incubation and measurement. This category is ideally suited for the application of robotic techniques as will be described later. In the continuous flow method each sample is introduced into a flowing stream and reactions are carried out by merging streams, the final measurement being made in a flow-through unit of some sort. The choice of approach taken will depend upon a number of factors such as complexity of the analysis, the sample throughput and the relative capital and maintenance costs of the equipment involved. Often hybrid techniques consisting of combinations of each of these two extremes are employed depending upon the particular analytical objectives to be achieved. A summary of some of the different types of instruments available for chemical assay has been presented by Mitchell (Mit80); his findings are summarised in Table 8.1.

TABLE 8.1 Types of Instruments available for Automated Analysis

```
(1)    Continuous flow
(2)    Discrete
(3)    Discretionary
(4)    Non-discretionary
(5)    Large multi-channel
(6)    Smaller multi-channel instruments (sometimes assembled by
       multiplexing single channel machines to suit the operator)
(7)    Single channel fast analysers operating in both continous
       and batch mode
(8)    Single channel continuously operating relatively slow
       instruments
(9)    Instruments operating with completely pre-packaged reagents
(10)   Stand-alone automatic instruments for measuring one, or a
       small number of analytes, e.g. glucose, urea or electrolyte
       analysers, with the minimum of operator involvement
(11)   Automatic colorimeters reading out in concentration units
       but with no automation of chemistry
```

Once a decision has been made to look at the possibilities of introducing
automation into an analytical procedure there are several fairly well
defined steps involved. These are listed in Table 8.2.

TABLE 8.2 Steps Involved in Automating an Analytical Procedure

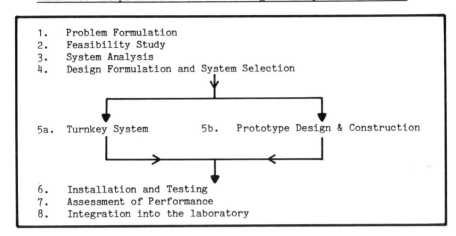

1. Problem Formulation
2. Feasibility Study
3. System Analysis
4. Design Formulation and System Selection

5a. Turnkey System   5b. Prototype Design & Construction

6. Installation and Testing
7. Assessment of Performance
8. Integration into the laboratory

When an attempt is made to automate individual analytical techniques within a laboratory the first decision that has to be made relates to whether
automation is feasible. Many analytical procedures used in standard
laboratory testing may not be suitable for direct mechanisation or automation. In situations such as this it becomes necessary to devise and
design new analytical methods that are capable of being automated to
various degrees depending upon the exact requirements of the situation to
which they are to be applied. Once a positive decision has been made
regarding the feasibility of the project, the systems analysis stage
commences. During this phase a detailed analysis of resource requirements is made along with appropriate procedural recommendations for system
implementation. Next, system designs are suggested and drawn up. This
may involve the use of readily available equipment (turnkey system) or the
detailed design and in-house construction of a suitable analyser. As soon
as the automated equipment is available in the operational laboratory it
has to be installed, tested and assessed in various ways. Procedures for
interfacing the new automated method into the existing laboratory system
then have to be implemented. The complete process of specifying,
designing and fabricating an automatic analytical unit and also implementing it in a routine environment is a complex task involving many stages
and a number of interactions between many staff.

In the initial stages of the automation study it is useful to have available a suitable questionnaire that can be used to determine the nature of
the automation process. This can also be used to gain some estimate of
the likely resource requirements for the operation. A typical example of
such a questionnaire is shown in Fig. 8.4 (IBM75).

Based upon an analysis of questionnaires of the type shown in Fig. 8.4
(and interviews with laboratory staff) it is possible to get an indication
of the type of automatic system that is required and whether the computational demands will require the use of a microcomputer, a minicomputer or

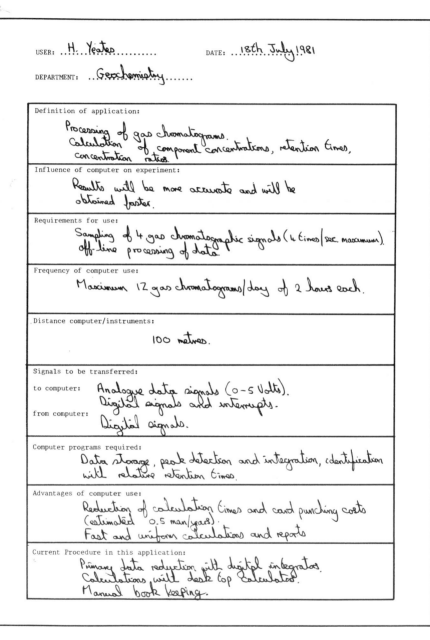

USER: ..H..Yeates.......... DATE: ...18th July 1981

DEPARTMENT: ..Geochemistry.......

Definition of application:

Processing of gas chromatograms.
Calculation of component concentrations, retention times,
concentration ratios.

Influence of computer on experiment:

Results will be more accurate and will be
obtained faster.

Requirements for use:

Sampling of 4 gas chromatographic signals (4 times/sec. maximum).
Off-line processing of data.

Frequency of computer use:

Maximum 12 gas chromatograms/day of 2 hours each.

Distance computer/instruments:

100 metres.

Signals to be transferred:

to computer: Analogue data signals (0-5 Volts).
Digital signals and interrupts.

from computer: Digital signals.

Computer programs required:

Data storage, peak detection and integration, identification
with relative retention times.

Advantages of computer use:

Reduction of calculation times and card punching costs
(estimated 0.5 man/year).
Fast and uniform calculations and reports

Current Procedure in this application:

Primary data reduction with digital integrator.
Calculations with desk top calculator.
Manual book keeping.

Fig. 8.4 Questionnaire used to assess user's requirements
during the early stages of planning for laboratory automation.

both. When the automation is to take place within an existing automated
laboratory the possibility of integrating the new experiment into a
currently operational system will need to be considered. In addition to
providing some insight into the type of system that is required, the
initial survey will also give some indication of the nature of the data
acquisition, storage, retrieval and processing requirements of the application. These considerations will also strongly influence the choice of
basic system that is to be employed as a basis for the automation.

Questionnaire design and interviewing techniques form an important part of
the systems analysis phase of the automation study. During this step the
analyst will usually use some simple model to provide a basis for the
initial data gathering investigations. A typical example is illustrated
in the diagram shown in Fig. 8.5. This attempts to subdivide the analysis
and design into six basic categories ranging from hardware requirements
(on the left) through to procedural requirements (on the right). Analysis
can then proceed in such a fashion that enables information to be gathered
for each individual consideration: hardware, software, interfaces, and so
on. Using the results of the systems analysis stage some preliminary
design possibilities can be formulated. In doing this, there are certain
preferred directions in which the design should proceed in order to
produce a system that is of optimum utility. Some important considerations are presented below:

(1) the design should be modular, to permit processing,
     acquisition or storage techniques to be modified in
     the light of new knowledge gained through research
     and experience,

(2) the person running the experiment should be able to
     communicate with and control the computer and thus,
     if necessary, over-ride any actions it may take,

(3) the system must be capable of growth without redesign,

(4) data acquisition and storage should be separate to
     allow the maximum opportunity for processing information,

(5) the system should be designed in such a way that it has
     high reliablity,

(6) where possible, it should be able to easily adapt to
     the needs of different operators, and,

(7) if a centralised system is involved then, where possible,
     the advantages of digital data networks (see chapter 12)
     should be used to minimise the number of unique connections
     to the computer (see also Fig. 7.16)

A detailed discussion on the in-house design and construction of automatic
analysers for laboratory use has been presented by Porter and Stockwell
(Por80).

Some interesting examples of automation applied to individual analytical
procedures have been described by Dobson (Dob80) and Luft (Luf80). Dobson
has described the design and construction of an auto-titrator based upon
the use of a PET desktop computer. In this system the computer displays
appropriate instructions to the operator then, at the press of a button,
it performs a titration, locates the end point and calculates the results
in the required units. During the titration the computer controls the
stirrer, adds titrant at appropriate time intervals, monitors the pH and
determines the equivalence volume. Two applications of the titrator are
outlined. In the first it is claimed that the auto-titrator is able to
increase the number of daily titrations by over 100%. In the second, an

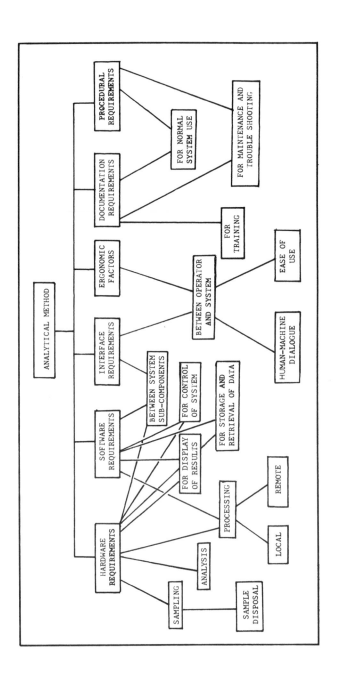

Fig. 8.5   Simple design model for automation of analytical methods.

analysis which would take 1.25 hours by manual techniques can be conducted
totally automatically under computer control.

Luft's system is similar to that of Dobson but is based upon the use of
gravimetric titrations. Figure 8.6 shows the arrangement of equipment
that is used.

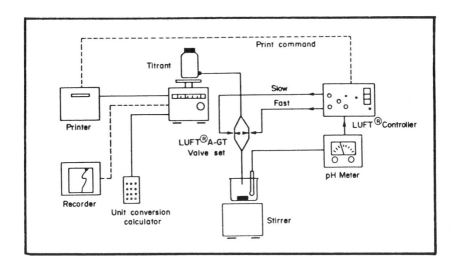

Fig. 8.6 Automated system for performing gravimetric titrations.

In this system gravimetric titrations are performed by a highly modular
automated system consisting entirely of readily available components - a
digital top loading balance, a pH meter, a titration controller, a
titration valve assembly and other related accessories. The system
permits semi-automatic or automatic titrations to be conducted with substantial increases in accuracy and speed. Thus, a typical acid-base
titration can be completed in 10-20 seconds with an endpoint precision of
0.01 pH units. Because the system does not require time to fill a burette
(the electronic tare performing the same function more quickly and
reliably), overall cycle times can be substantially reduced with respect
to conventional titrimetry. Further details on the implementation of
titrators and techniques for automating individual analyses in wet
chemistry laboratories will be found in the paper by Arndt and Werder
(Arn79).

A useful application of automation in the laboratory - in order to achieve
reproducibility of performance - has been described by Stockwell and
Copeland (Sto79) in their experiments with a smoking machine for the
analysis of cigarette smoke by infrared techniques. The apparatus that
was used is illustrated schematically in the set of diagrams shown in Fig.
8.7. There are four basic stages involved in the operation of the
machine. In the first (A - pre-smoking) the apparatus is initialised in
preparation for the experiment. The next step (B) involves the use of a
piston to draw air through a burning cigarette, leaving particulate matter
in the Cambridge Filter (CF) which allows the vapour phase to pass into a

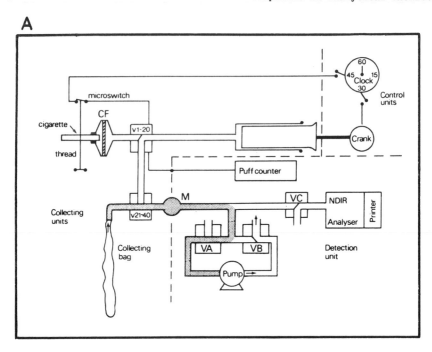

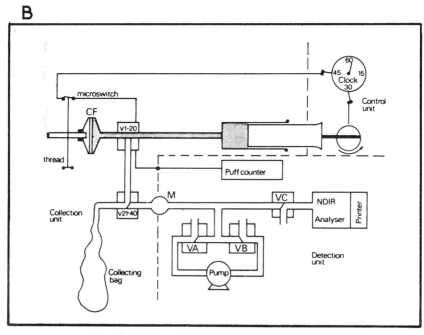

Fig. 8.7 Cigarette smoke analysis using a
smoking machine; stages A and B.

C

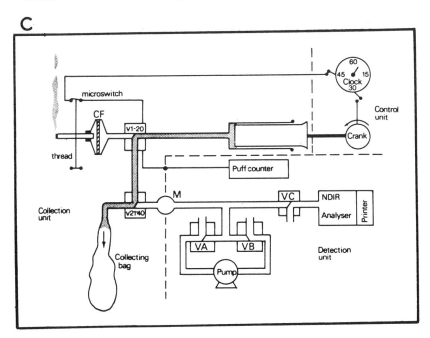

D

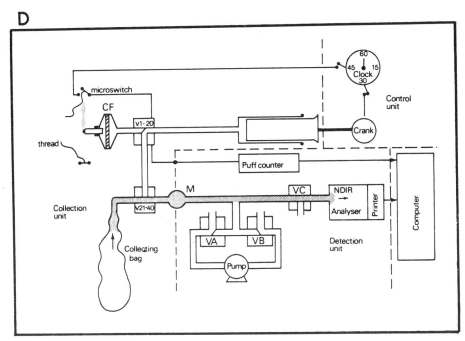

Fig. 8.7 (continued). Cigarette smoke analysis
using a smoking machine; stages C and D.

piston. In the third step (C) the vapour in the piston is passed to a collection bag. Finally, in the last step (D - called post-smoking) the material in the collection bag is transferred to the infrared analyser which is used to measure the carbon monoxide content. This value, along with the puff-count (number of piston strokes), is passed across to the computer which processes and stores the results. The machine used in the experiments is a Filtrona 300 20-port machine consisting of 20 independent suction lines all connected to a common control and drive unit. Each suction line has its own discrete termination device (a microswitch controlled by a weighted cotton thread) that causes smoking to cease when the thread is burnt.

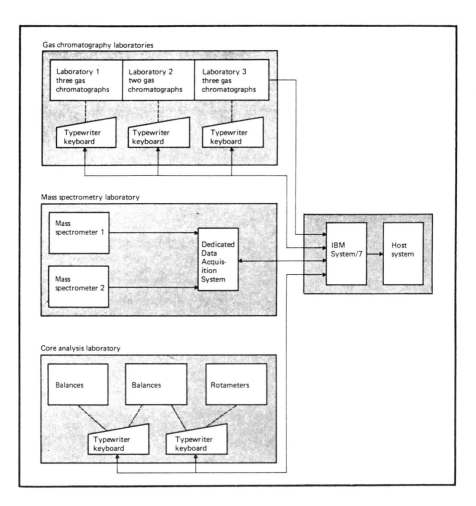

Fig. 8.8 A typical approach to total laboratory automation.

Each of the previous examples of automation have involved single experiments or items of equipment that have been automated in isolation

from any other laboratory activities. They are thus examples of
instrument automation - as depicted in case A of Fig. 8.2. The other
approach to automation, involving integration of the whole laboratory, has
been described by a variety of authors (IBM75, Arn80, Por80). One typical
example of total laboratory automation is illustrated in Fig. 8.8.

In this geo-chemistry laboratory (IBM75) gas chromatography and mass
spectrometry are both used to analyse the hydrocarbon content of crude oil
samples from a variety of sources. The core analysis laboratory is used
to measure the properties of cores produced from test drilling operations
associated with oil exploration. Typical analyses on cores might include
water content, oil content, porosity, permeability to air and density.
Many of these measurements involve some form of weighing before and after
some operation has been performed on them. Permeability and porosity
experiments require data from rotameters to be recorded. For a given oil
well about 300 samples might be taken, each requiring about ten measurements to be made. Because the measurements for one well could take many
months to complete, and several wells have to be processed at one time,
the data handling is both extensive and complicated. The computer system
is able to offer considerable assistance through the availability of a
core analysis data base and appropriately designed human-machine dialogues
to enable the facile entry of experimental data by analysts via typewriter
keyboards as they conduct their analyses. The simple dialogue shown on
the left of Table 8.3 illustrates the type of human-machine interaction
that takes place between the analyst and the computer.

TABLE 8.3 Human-Machine Dialogue Involved in Automated Analysis

```
                          ┌──────── computer asks for oil-well number
WELL    305/AA  ◄─────────┘               (analyst replies)
CODE    DWS     ◄──────────── computer asks for analysis code
1,35.145        ┐                         (analyst replies - DWS)
20,36.472       │
H41,32,452      ├── entry of results
S2,34.567       ┘
CODE    VWC     ◄──────────── entry of analysis code - VWC
1,10.6          ┐
31,9.45         ├── entry of results
S2,10.4         ┘
END             ◄──────────── dialogue terminated
```

```
┌─────────────────────────────────────┐
│ Legend:                             │
│ DWS = Dry Weight of Sample          │
│ VWC = Volume of Water Collected     │
└─────────────────────────────────────┘
```

In the gas chromatography laboratories the computer is used to perform
data acquisition with subsequent data reduction of the acquired data -
measurement of peak positions and the corresponding areas. For many
analyses automated procedures are used - particularly when only a few
well-defined results are needed. These pre-defined methods can be stored
in the data base associated with the central computer and re-called when

needed. Connection of the chromatographs to the computer is via three
cables each containing five twisted pairs and each wire separately
shielded. These cables terminate in a plug-board panel in each of the
three laboratory rooms. Chromatographs can then be plugged into
appropriate points on the plug-board. The analysts communicate with the
computer by means of a typewriter keyboard, one in each laboratory. The
computer has space for 500,000 data points (corresponding to about 30
chromatograms) and the sampling rate used is usually two sample points per
second. A typical human-machine dialogue that might take place during the
running of the GC analysis is illustrated in Table 8.4. The trace on the
left (case A) is an example of a pre-analysis dialogue that is used to
prepare the computer for an analysis on a particular chromatograph using a
specific pre-stored method. When the computer has retrieved the
appropriate programs from its memory it informs the analyst (using the OK
RUN message) that the analysis can proceed. The analyst then injects the
sample for analysis into the chromatograph and signals the completion of
this activity to the computer by means of an appropriate event marker (a
foot switch or pressure sensitive pad, etc). When the analysis has been
completed the analyst requests the computer to stop the data monitoring
for the GC that has been used - as shown in dialogue B or Table 8.4.

TABLE 8.4 Human-Machine Dialogue Trace during a GC Analysis

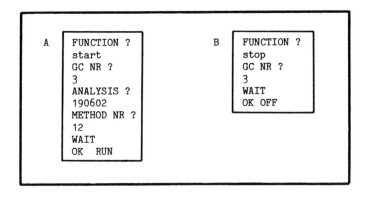

Computer assistance is also used in the mass spectrometry laboratory.
From the point of view of data acquisition the two major characteristics
of mass spectrometry are:

(a) the time taken for the analysis is extremely short
     compared with the time needed for interpreting the data;
     several mass spectra can be obtained in a few seconds,
     however, each spectrum may require several hours for its
     interpretation, and,

(b) large amounts of data are produced - one component may
     give rise to a mass spectrum with more than 100 peaks.

Because of the special problems of mass spectrometric data handling, a
small dedicated system is used to provide data acquisition, correlation
and presentation facilities. The power of the dedicated system is
enhanced by connecting it to the central laboratory computer. This inter-
connection utilises direct communication lines between the two systems and

consists of two 16-bit parallel data links and a series of interrupt/
acknowledge lines. Figure 8.9 illustrates the mode of operation of the
interfaces.

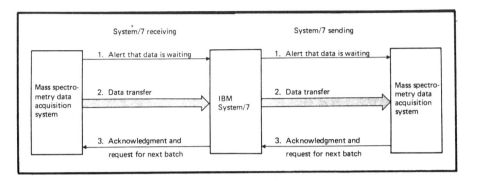

Fig. 8.9 Details of the computer-instrument interface
involved in the laboratory automation example presented
in Fig. 8.8.

The system is designed in such a way that it is interrupt driven - the
data acquisition system generates an interrupt when it requires service
from the central computer. In normal use, the operator of the mass
spectrometry data acquisition system loads a primary program and then
supplies it with details of the type of transaction involved, the names of
the program or the amount of data required. Communication between the two
systems then proceeds automatically in accordance with the protocol
depicted in the above diagram.

Considerable benefits are claimed to accrue as a direct consequence of the
automation of the geo-chemistry laboratory using a central computer. In
addition to the financial benefits gained by the laboratory as a whole,
there are substantial advantages realised by each of the application areas
involved. Some of these are listed in Table 8.5. One major advantage not
listed in the table is the potential ease of expansion of the system.
Because it was designed and implemented in a strictly modular fashion,
future expansion of the system should be a relatively simple matter.

In this section consideration has been given to the automation of
individual experiments and to the laboratory as a whole. In the latter
case, when this is done using equipment that is developed on an in-house
basis considerable effort is required - particularly in the areas of
interface design and software development. The use of turnkey (ready
made) systems can be used to eliminate some of the many problems
associated with laboratory automation. Turnkey systems will be discussed
later in this chapter.

TABLE 8.5 Benefits of Laboratory Automation

(A) In the Gas Chromatography Laboratory
  • Administrative effort for routine analyses reduced
  • Uniform calculation methods make it easier to correlate
    results from different analyses
  • The use of digital integrators was discontinued - these
    are awkward to use and difficult to maintain
  • New routine analyses are easier to set up
  • Greater work throughput for the laboratory
  • Data acquisition and processing were separated; this
    allows more elaborate experiments to be conducted

(B) In the Core Analysis Laboratory
  • The procedure for filing results was simplified
  • Costs were reduced by on-line data entry
  • Fewer data input errors resulting in less overhead
    in calculation times
  • Measurement procedure was made easier by keyboard input
  • Tasks completed more quickly

(C) In the Mass Spectrometry Laboratory
  • Instrument efficiency is higher
  • Data correlation can be enhanced by the extra data
    storage facilities

AUTOMATING RECORD KEEPING

Nowadays, in both large and small laboratories, computers are being used
in order to help automate record keeping procedures. The form of computer
that is used varies from situation to situation. Sometimes a dedicated
desk-top machine is employed while on other occasions a larger centralised
laboratory automation computer is used for this purpose. Alternatively,
there may be some communication link to a company mainframe located at the
central administrative offices. Time-sharing bureaus may also be used.
Techniques for automating record keeping fall into three broad categories:
data input, data/result output and storage.

In the past data input to the computer has involved the use of card/paper
tape facilities. Visual display units and other typewriter keyboard
devices are now quite commonplace within the laboratory and enable the
analyst to enter data directly into the computer without the need for the
intermediate storage of results on punched cards or paper tape. The more
sophisticated types of analytical instruments have a built-in keyboard for
data entry, an on-board computer for local data processing and an
interface to enable their connection to external computer systems. These
advances in data entry techniques are supported by many other types of
device that enable quicker and easier data entry by the analyst. Two of
these, the bar-code scanner and pressure sensitive writing pad will be
described later - because of their ease of use and growing popularity.

The use of digital bar-code readers is increasing substantially as a means
of data capture. It permits data entry at a rate of about four times that

which may be attained by the use of a conventional keypad. In addition, it overcomes the problem of incorrect data entry arising as a result of keystroke errors. It is feasible that this approach to routine data capture could have considerable value within the analytical laboratory for identifying samples and analytical reagents. Identification is achieved through the use of labels containing special bar-codes. The mode of operation of a wand and some examples of code tags are shown in Figs. 8.10 and 8.11, respectively.

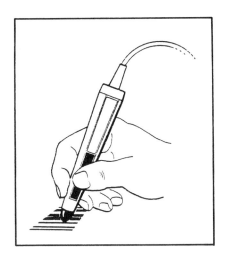

Fig. 8.10   Method of using a bar-code scanner.

Fig. 8.11   Examples of bar-codes; A: standard test tag for checking equipment performance; B: typical European article number bar-code.

312   Computers in Analytical Chemistry

The components used to construct a wand are indicated in the schematic
circuit diagram illustrated in Fig. 8.12 - sketch A. It consists of a
precision optical sensor, an analogue amplifier, a digitising circuit and
an output transistor. These elements are used to provide a TTL compatible
output from a single voltage supply V_S. A non-reflecting black bar
results in a logic high (1) level while a reflecting white space will
cause a logic low (0) at the V_O connection. A push-to-read switch is
used to energise the LED emitter (700nm) and electronic circuitry. When
the switch is initially depressed, its contact bounce may cause a series
of random pulses to appear at the output terminal. This pulse train
normally settles to a final value within 0.5 ms. The details listed above
are for the Hewlett Packard HEDS-3000 wand (Ueb81). The logic interface
for this device is depicted in diagram B of Fig. 8.12.

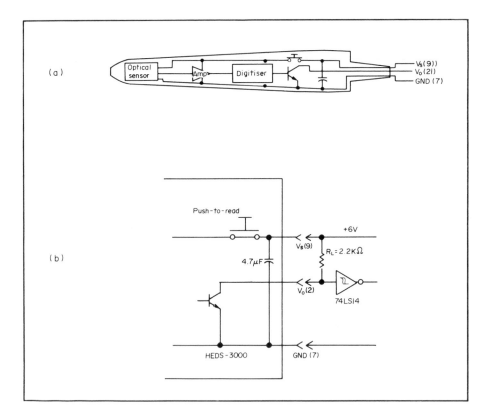

Fig. 8.12 Constructional details of a bar-code scanner;
A: cross-section through the scanner showing the components
involved; B: details of the interface circuit for connection
to the computer.

The 74LS14 integrated circuit is an inverting Hex Schmitt Trigger - a low
input produces a high output, a high input produces a low output. This IC
is used to latch the signal. Unlike a normal TTL gate, the inputs possess
a snap action used to condition slowly changing or noisy inputs. The

input impedance of the circuit is about 6K. The trip point for a positive
going signal is 1.7 volts and for a negative-going signal it is about 0.9
volt. The snap action or hysteresis range is about 0.8 volt. The
propagation delay of the gate is typically 17 ns.

In order to use bar-code labels in the laboratory a source of printed
labels is required. Special machines are available for producing these.
They enable labels to be printed in a wide variety of formats using a
number of different encoding techniques. In Europe the one most often
used is the European Article Numbering Convention (see Fig. 8.11 - example
B). The bar-codes shown in Fig. 8.13 were produced on a printer capable
of printing the required bar-code, an interpretation line and one line of
free text. The larger of the two labels (see Fig. 8.13) permits a total
of 25 data characters to be entered while the smaller label can accommodate
only 10. The bar-code reader for processing labels of this type is easily
interfaced to a desk-top computer. The system is then easy to use -
provided it is equipped with suitable software to interpret the incoming
bit streams produced by the reader. Labels similar to those shown in Fig.
8.13 are being used increasingly for sample and specimen identification
within a large number of laboratories - particularly those that have
significant numbers of samples to handle each day. Some types of optical
wands (Con81) are able to read colour coded bars - a facility that can
often be put to use in an analytical laboratory. Of course, in addition
to their use for reading identification labels it is also possible to use
this type of device for reading computer programs into bench calculators
or desk-top computers. Short programs for processing experimental results
can easily be encoded using bar-codes. These can then be stored in the
analyst's laboratory notebook or along with the documentation for the
analytical method to which the program applies. They can be read into the
computer very rapidly when they are required.

Fig. 8.13 Examples of bar-code labels that are designed
in such a way that they can be read by both a scanner and
a human reader.

Another useful device for automating data input and generating result
reports is the hand print data terminal. Several different types are
available. One of the more popular of these is illustrated in Fig. 8.14
(Mic80a). The hand print terminal is a peripheral device that can act as
a local or remote terminal. It enables hand written alphanumeric data to
be captured at the time of writing. The main components from which it is
constructed are a pressure sensitive writing surface, a microprocessor and
an integral 40 character line display. Data is written using an ordinary
ball-point pen or pencil on special documents that are designed by
individual users to suit their particular applications (some illustrations

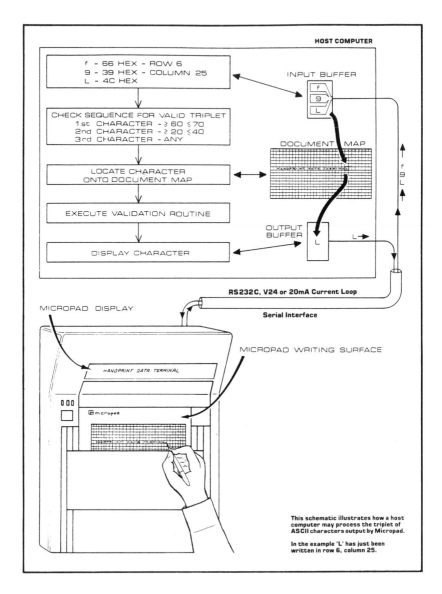

Fig. 8.14 Mode of operation of the MICROPAD hand
print terminal.

are given in Fig. 8.15). As data is written it is recognised and converted, character by character, into standard ASCII code. Each character
is then transmitted to a host computer for processing. In addition to the
code for the written character the device also transmits two additional
characters which identify its position on the document. The data entered

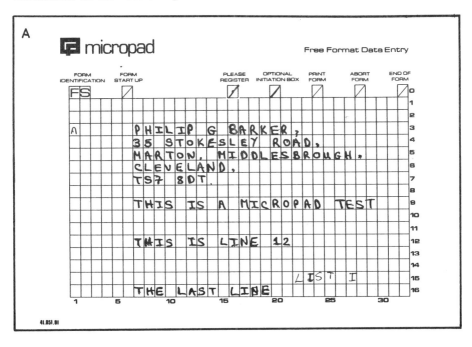

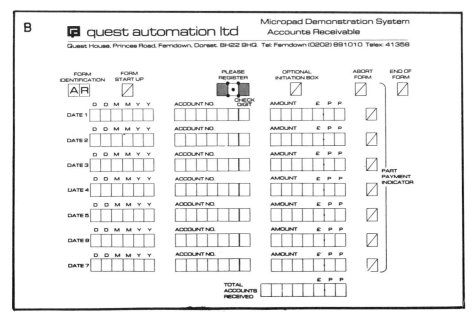

Fig. 8.15 Examples of data entry forms for use with
a hand print terminal; A: free format data capture;
B: formatted numeric data entry.

via the input pad, or responses generated by an application program as a
result of input from the pad, can be displayed on the single line display
of the device or on a associated CRT screen of a visual display unit.
Interfacing the device to a computer is easily achieved be means of the
standard RS-232-C or current loop interfaces (see chapter 6) which the
system provides. The diagram presented in Fig. 8.14 summarises the way in
which the computer software for processing the input would need to respond
to the characters written by the user. In this example a document map is
used as a means of deducing the context of the written characters. Once
the position and validity of a character has been checked it is 'echoed'
back to the MICROPAD for presentation on the display. A more detailed
description of some end-user applications of the hand print terminal are
given elsewhere (Bar81).

The illustrations contained in Fig. 8.15 show two examples of input
documents (Mic80b). That on the top is typical of the type used for free
format data entry while that on the bottom is characteristic of the design
used for highly formatted data input. As can be seen from the upper
diagram the MICROPAD makes available a matrix of rectangles (17 rows by 32
columns) into which may be entered any of the allowed alphanumeric
characters (Mic80c). Some of the boxes in row 0 are reserved for special
purposes, for example, document registration, form identification,
initialisation, etc. In the free format input form the user can write in
any of the boxes in rows 1 through 16 - the boxes shown in line 0 being
used to initiate the indicated special control functions. The program to
support free format data entry must contain a document map in the form of
a 16 x 32 character array in order to store the individual characters it
receives from the MICROPAD. There is a special touch sensitive strip down
the left hand side of the printed form that enables any of the lines of
the internally stored matrix to be retrieved and displayed on the MICROPAD
display; they may then be modified if the need arises.

Each special purpose document (for example, form AR in Fig. 8.15) will
have a corresponding data acquisition program (and internal map) that is
invoked when the document is registered and initialised. In the example
shown, data will be recorded by the computer only if it is entered within
certain pre-defined areas of the form (within the small rectangles),
characters written anywhere else will be ignored. Documents can be
designed for virtually any application - particularly manual data entry of
analytical results in a laboratory environment. Some of the analysis
report forms presented in chapter 1 would be ideally suited for automation
applications using a MICROPAD since for most of them minimal re-design
would be required.

In addition to the use of devices based upon pressure sensitive pads, a
wide variety of other techniques can be used to automate data entry or
control activity within the laboratory environment. Touch sensitive keyboards are now commonly employed in many instruments because they contain
no moving parts and are therfore much more reliable and robust than keyboards based upon the use of mechanical switching. Similarly, many
different techniques of CRT screen interaction are being employed to
facilitate interaction with a laboratory computer. Typical of these are
the use of light pens (Bar78,JEO80) and touch sensitive screens (Heb75,
TSD80). Using such devices it becomes possible to program an automated
instrument by means of a series of simple 'pointing' operations. One
simply points to a screen in order to implement a menu selection technique
which specifies the particular analytical options required. Further menus
then enable specification of the mathematical data processing techniques

that are to be applied to the data and the way in which the results are to be reported.

Many instruments provide a number of facilities for automating result reporting. Some contain built-in printers (that have the ability to plot graphs) while others just provide a standard interface (typically RS-232-C) which enables the attachment of an external printer. In both cases it is usually possible to use a wide variety of stationery - either blank or pre-printed, continuous or individually loaded, coloured or white - to meet the requirements of the application. Facilities of this type are important where conventional result reporting procedures must be used - production of a printed result form which has contractual or legal implications. In situations where results have to be reported to centres remote to the laboratory at which the analysis was conducted, instruments can be equipped with auto-dialling facilities so that the analytical results they produce can be automatically dispatched over a telecommunication network to those for whom they are intended. Research and progress in this area is rapid. Currently, there is much interest in the application of speech input/output techniques for instrument control and result reporting in the laboratory. A recent paper by Levinson and Shipley (Lev80) illustrates the present capabilities of such systems. It describes a sophisticated conversational information system based upon the use of speech input and output. The unit is capable of handling 19 different semantic categories. Within each of these there are several alternative and equivalent syntactic structures permitted. The vocabulary of the speech recogniser is 127 words and that of the voice response unit is 191 words. Speech input to the machine is in the form of a sentence with brief pauses (100 ms) between each word. The accuracy of the speech recognition part of the system is claimed to be over 96%. Response time is about five times real time and could easily be reduced. Obviously, input/output techniques such as these will have considerable impact on the future of laboratory automation.

THE USE OF TURNKEY SYSTEMS

The construction of automated laboratory systems using in-house techniques can be both an expensive and timely exercise. Consequently, many laboratories are turning increasingly towards the use of turnkey systems for the solution of their laboratory automation problems. A turnkey system is essentially one which has been totally designed, constructed, installed and tested by a specialist equipment manufacturer or systems house. The supplier of such a system thus provides a complete operational solution to the laboratory automation problem posed by the customer. In doing this the supplier either (1) builds a system to meet the exact requirements of the customer's laboratory, or (2) modifies the characteristics of some readily available item of equipment to match those of the problem to be solved. The provision of a turnkey package usually involves providing appropriate hardware, software, operational procedures, suppliers for consumable items, operator training and suitable equipment maintenance agreements. Various types of turnkey system are available that are designed to cater for a wide spectrum of user needs. Often they will vary both in the degree to which they meet a user's automation requirements and in how closely they satisfy the turnkey system criteria outlined above. Applications of simple systems (for example, Spectra Physics SP4100 system) have been described by Henshall (Hen78), Phillips (Phi80) and Leuenberger (Leu80) while more advanced turnkey systems (such

as the Technicon SMAC, LKB systems, Coulter, JEOL, etc) have been
described by Roy (Roy79) and Haan (Haa79). Two typical systems are
illustrated in Figs. 8.16 and 8.17. Each represents a different level of
approach to the provision of a turnkey package.

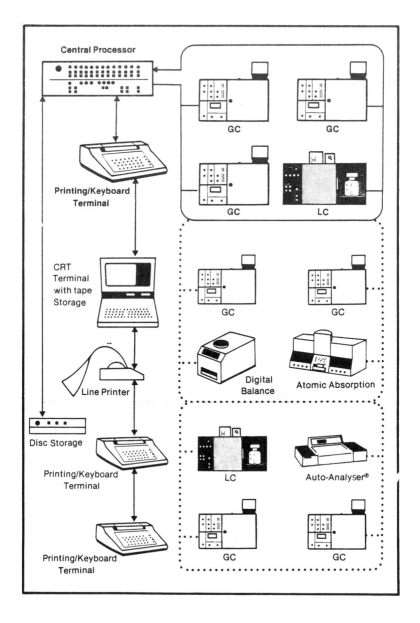

Fig. 8.16   Hardware and instrumentation associated
with a turnkey laboratory automation system.

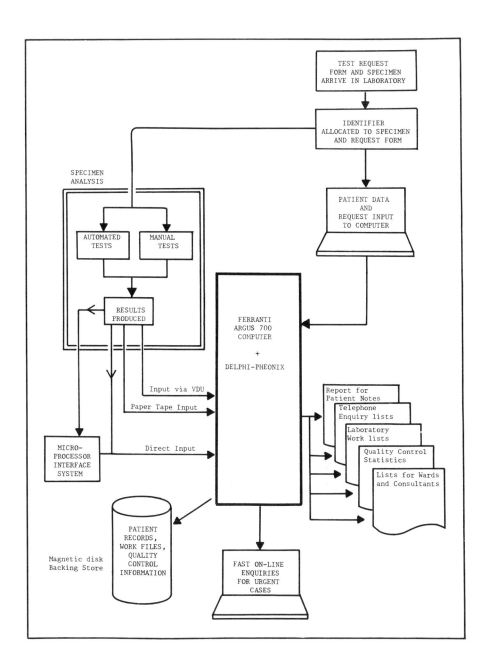

Fig. 8.17   Example of the data and material flow that takes place in a typical turnkey laboratory automation system.

Figure 8.16 (Hewlett Packard HP3350 series - see Hew80a, Hew80b) shows the type of equipment involved and illustrates the ease with which the computer hardware and analytical instruments can be inter-connected by means of a simple interface loop. This is represented by the solid line at the top of the right-hand side of the diagram to which are attached three gas chromatographs and one liquid chromatograph. The dotted lines in this illustration represent possible future directions of expansion, that is, instruments that are likely to be added to the laboratory in future years. This sketch thus illustrates an important point - when installing a laboratory automation system it is important to allow for future growth. Although it is not apparent from the diagram, the system comes ready equipped with a wide range of processing software for data acquisition, control and storage of results. No special knowledge of computers or programming is required in order to use the system. Once installed, the software contained within it enables laboratory personnel to set up operating parameters and control settings by means of simple interactive dialogues conducted via typewriter keyboards attached to hardcopy or CRT screen devices. A large variety of software for processing results is usually available with turnkey packages of this type. For those who wish to extend the software capability of the system an easy to use programming language (most often BASIC) is provided. Once processed, both the raw data from experiments/analyses and the derived results can be archived onto a user specified storage medium - disk store, cassette tape and others. They can then be used to produce many different types of report to suit the requirements of the various user population. A description of some of the advantages and problems of using this particular system has been given by Wadsworth (Wad80).

The second example of a turnkey package illustrated in Fig. 8.17 (Ferranti Delphi-Phoenix system) is designed for data handling in the pathology laboratory (Fer80). The system is based upon the use of a central mini-computer (Argus 700G) and a microprocessor subsystem which can be used for the purpose of interfacing on-line instruments to the system. Editing and pre-processing of results may also be achieved by means of the micro-processor facility. The data handling package receives and processes information from a wide range of complex clinical chemistry instruments - both automated and manual. The throughput of the system is claimed to be in excess of one million (blood test) samples annually. As can be seen from the diagram, the results that are produced can be archived onto magnetic disk backing store. From there they may be interrogated directly by medical staff (or analysts) using on-line terminals located anywhere throughout the hospital. As well as providing an on-line interrogation facility, software is provided to enable the production of a wide selection of reports to aid the smooth running of the laboratory and those other sections of the organisation with which it interfaces.

Because of the importance of clinical chemistry in health screening investigations (both diagnosis and prognosis), medical laboratories face a continual demand for chemical analyses at frequencies and throughputs not often met in many other areas of analytical science. Usually, large numbers of samples have to be processed - each requiring multi-component analysis. In all probability, it is for this reason that most of the major advances in laboratory automation have often been made in areas associated with chemical assay of clinical samples. The variety of automatic analysers available from the Technicon company (the Technicon SMAC was outlined in Chapter 3 - sophisticated instruments) and others that have been mentioned briefly in this chapter are evidence of this. Indeed, nowadays on a world-wide basis there is, in total, a significant

number of different turnkey systems available both in the area of clinical
chemistry (specifically) and in the other more general areas of automated
chemical analysis. In view of the complexity of turnkey systems and
because of the many various facilities that they provide product selection
is a difficult task. Indeed, it becomes a substantial problem when it is
required to compare and assess the different product offerings from the
various manufacturers who are active in this area. Some of the ways of
approaching the problem of equipment selection have been outlined in a
series of papers (under the general title, "Design Criteria for the
Selection of Analytical Instruments Used in Clinical Chemistry") by
Haeckel (Hae80), Mitchell (Mit80), Buttner (But80), Hjelm and Geary
(Hje80), Sandblad (San80) and Craig (Cra80). Taken together these papers
cover the majority of aspects that need to be considered when evaluating
the types of instruments that are available. Some of the non-monetary
decision criteria (taken from Buttner's paper) are listed in Table 8.6.

In addition to making decisions with respect to evaluating the
suitability, capability and performance of the candidate equipment
available for purchase (or lease), there is also a requirement to be able
to accurately assess both the present and future automation needs of a
laboratory. This problem is addressed in the paper by Sandblad (San80)
who has described the use of computer based simulation experiments for
investigating different proposed laboratory functions and configurations.
An important factor likely to strongly influence automation applications
is cost effectiveness. Details of techniques for the cost evaluation of
automation analysers have been discussed in considerable depth by Craig
(Cra80). He uses a computer based model (containing 20 different cost
equations) which produces a cost/benefit analysis report from the data
input to it. The data is collected from a variety of sources but is
mainly based upon interviews with laboratory personnel and administrators.
Although the details and discussion in these papers relates to the
particular area of clinical chemistry, there is no reason why the
principles involved could not be generally applied to automatic analysis
in other areas.

For a variety of reasons - economic, staff availability, volume of
samples, sample throughput and frequency, etc - applications of automation
in the analytical laboratory are becoming increasingly popular. This
popularity is likely to increase significantly as turnkey systems of
various types become more readily available and the cost of employing
conventional analytical methods increases. Automation, in general, often
requires few people and in many cases (as was pointed out in the introduction) can involve a considerable reduction in skills. If machines are
to perform analytical procedures on behalf of the human, what will happen
to the analytical chemists that the machines will replace? If machines
can automatically perform titrations, calculate results and write reports,
what is left for the analyst to do? Some possible answers to these
questions will be discussed in the next section.

THE CHANGING ROLE OF THE ANALYST

The history of analytical chemistry has been described and reviewed by a
variety of authors (Sza66, Bel80, Sza77). Although Man's interest in the
composition of matter is probably as old as Man himself, it is only in the
last 100 years or so that analytical science has transcended from a

TABLE 8.6   Decision Criteria for Instrument Selection

Class A - Basic Considerations
1. Single or Multiple Testing
   - fixed program
   - possibility of selecting tests according to request
   - possibility of selecting clinically orientated groups
     of tests
2. Analytical Mode
   - analytical principle (photometric, others)
   - endpoint or reaction rate measurement
3. Sample size (Volume)
   - use for paediatric departments
4. Degree of Mechanisation
   - expected number of patient specimens
   - fully mechanised (or automated) system needed
   - electronic data processing needed
5. Possibility of Performing Statistical Emergency Tests

Class B - Instrument Performance Considerations
1. Analytical Reliability of the Measuring Part of the
   Instrument
   - precision
   - accuracy
   - specificity
   - detection limit
2. Speed Criteria
   - specimens/samples per unit time (frequency)
   - analysis time
   - throughput time
3. Correction of Interferences
   - automated detection and correction of interferences
     such as two wave lengths' turbidity
4. Contamination
   - carry over between samples
5. Stability
   - temperature
   - measurement process (drift, noise)

Class C - Instrument Practicality
1. Space required
2. Energy Consumption, other services required
3. Ease of Operation - operation by unskilled personnel
4. Adaptability and changeover of analysis methods
5. Possibility of introducing urgent samples into routine
    procedures
6. No restrictions in terms of selection of reagents that
    can be used
7. Safe operation
8. Fault detection and signalling, eg. using microprocessors
9. Easy maintenance
10. Operating instructions
11. Safety for the operating personnel
12. Environmental Aspects, eg. production of effluent, waste

somewhat 'hit and miss' art to the important science of measurement that
it is today. Undoubtedly, in many modern analytical situations the rigour
that the analyst is able to bring to bear on a problem would not be
possible without the use of the sophisticated machines that are often used
- least of all the computer and its many derivatives. Over the years, the
use of instruments and mechanised techniques in analytical chemistry has
lead to the development of a new rapidly expanding area of interest
referred to as Automatic Analysis. Much of the (relatively) modern
history of the development of this subject is covered in the book by
Foreman and Stockwell (For75).

As was suggested at the end of the last section there are many reasons why
automatic methods in the laboratory are increasing. Betteridge (Bet80)
summarises some of these as follows:

> 'Social, political and economic pressures on analysts require a
> large number of analyses to be performed daily and reliably.
> The problems of coping with such large numbers pose new problems
> for the analyst'

and cites an example of 12 scientists annually having to analyse between
them 30,000 samples each requiring 11 chemical determinations and 10,000
samples each needing 3 microbiological tests. Although details of their
laboratory hours were not provided, it is easy to see that their work
throughput (about 1 sample/hour) is nowhere near that of an automatic
instrument such as the Technicon SMAC (200 samples/hour). Betteridge
suggests that it is not only the sheer number of analyses that creates a
demand for automated methods. Indeed, the detection limits involved and
the need for the analyst to 'measure more and more in less and less'
nowadays often provides the motivation for machine involvement. Of
course, as more analyses are performed, so, there is a parallel increase
in the amount of accompanying paperwork, data to be processed and results
to be stored. Automation of this aspect of analysis is also of
significant importance.

With the increasing use of automation - especially in industrial and
commercial laboratories, Deens (Dee77) has posed the question 'Where are
we going?' in the context of the future role of the analytical chemist.
He suggests that:

> 'In the past the analyst has had a number of roles to play. One
> of these roles which the introduction of automation is tending
> to remove is that of expert manipulator of apparatus and expert
> interpreter of readings and indications'

and feels that as certain of these skills become displaced, so the analyst
must be encouraged to develop and exercise the other more important skills
- such as knowledge of the capabilities and error characteristics of the
various analytical methods available. Thus, automatic analysis is
unlikely to displace analysts but merely change their role.

As demand for automatic analysis increases, there is likely to be an
increasing need for expert analytical chemists to become part of multi-
disciplinary teams whose function is to design and build the automatic
instruments required to solve particular analytical problems. The
composition and structure of such groups is likely to be similar to that
described by Stockwell and Foreman (Sto79). The interactions between the
members of the multi-disciplinary teams involved is depicted in Fig. 8.18.

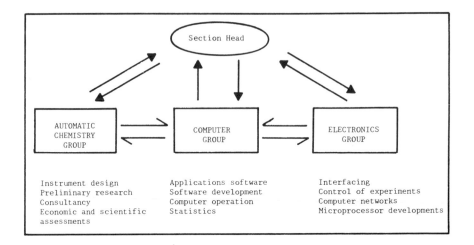

Fig. 8.18   Group interactions in a multi-disciplinary
team involved in the design and construction of automated
laboratory equipment.

Here the expertise of the analyst compliments the skills offered by other
members of the group to form a problem solving team capable of applying a
wide range of knowledge to the solution of problems arising in automatic
chemical analysis.

Automatic analysis and the growing involvement of computer based data
archives in this area is certainly changing the role of the analyst.   This
view is supported by the simple (but not untypical) model of laboratory
automation shown in Fig. 8.19 (Hew80b).

In Fig. 8.19 the laboratory function is seen as part of a much larger
global management information system (MIS).   Samples come into the
laboratory, are analysed (wherever possible and appropriate) by automatic
instruments which produce results that go forward automatically to produce
reports for management decision making.   Data archival in a suitably
designed computer data base (highly structured non-redundant collection of
data that can easily be retrieved and processed in various ways) is an
integral and central part of this model. The reliability and accuracy of
much of this data is the responsibility of the analyst in that the values
that are ultimately stored will have been derived from measurement
techniques and processing methods that he/she will have devised, tested
and, hopefully, proved.

Despite what has been said in this chapter on automation in the laboratory
it is imperative to remember that this tool, like any other, should only
be applied to those areas where,

(1)   it produces cost effective solutions to a problem,
(2)   service requirements demand its introduction, and
(3)   the processes concerned are capable of automation.

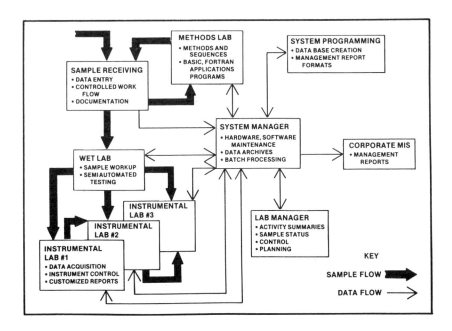

Fig. 8.19   A summary model for an integrated management
information/laboratory automation system.

Whatever else happens in this area, there will always be a place for the
analyst in the laboratory even though his/her role may change significantly in future years.

CONCLUSION

In this chapter the concepts of mechanisation and automation have been
examined. An overview has been given of the various ways in which
automation can be introduced into the analytical laboratory. Three major
aspects have been discussed: the automation of individual experiments and
analytical procedures, the automation of record keeping associated with
the bureaucracy of legislative/contractual analysis, and, the techniques
necessary to support total laboratory automation and the possible
integration of a laboratory system in such a way that it becomes a primary
component of a complex management information system.

Two basic approaches to the automation of laboratory analysis exist:
in-house development of systems and the use of turnkey packages produced
by a systems builder or instrument manufacturer. Each of these options
have been described. Most often, for laboratories not equipped with the
variety of staff expertise and special equipment required for system
building, the turnkey approach offers an attractive solution. However,
for complex sophisticated equipment, choosing between the alternatives

offered by different manufacturers can be a difficult task.   Some of the many factors that need to be considered have been outlined.

Because of the rapid developments being made in the area of automatic analysis, the final section of the chapter investigated the possible changing role of the analyst in this area.   Here it was felt that the analyst must be able to participate as a member of a multi-disciplinary team.   In this role he/she must bring to bear his/her expert knowedge of analytical chemistry for the solution of new types of problems associated with the use of highly automated machines and instrumentation.   It is only the analyst who understands the chemistry of the measuring process that the electronics expert, computer scientist and mechanical engineer attempt to automate.   Obviously, the analyst bears the ultimate responsibility of justifying the validity of the chemistry upon which the machine bases its decisions.

In this chapter nothing has been said about the use of robots within the automated laboratory.   Few laboratories have, as yet, had substantial experience with them.   Consequently, their role has still to be realistically evaluated.   There is one obvious area of application for such devices - conducting analyses and experiments that are in any way potentially dangerous or hazardous for human beings.   What other roles they have to play, other than introducing cost effectiveness and dependability into the laboratory, remains to be seen.   Many approaches to laboratory automation depend upon continuous flow techniques based upon the use of batched input.   At present, human operatives load the sample trays into the machine and unload them when they have been processed.   The manual effort involved here could easily be replaced by appropriately programmed mobile robots.   Similarly, it is not unlikely that such devices could have substantial utility in discrete methods of analysis where there is frequent movement of samples or intermediate preparations between one location in a laboratory and another.   Whether such applications are likely to materialise will, in the limit, depend upon the cost effectiveness they are able to offer and the novel approaches to analysis that they are able to produce.   The future may produce a robot analyst that is able to 'walk' around a farmer's field and poke its 'finger' into the ground at intervals of 6 feet along rows that are 3 feet apart.   On returning to the laboratory such a device might plug itself into a graph plotter and produce a graph of pH as a function of location within the field.   In addition, it might also display the results as a function of depth beneath the surface.   The potential applications of robotics to automatic chemical analysis offers many interesting and exciting possibilities.   Furthermore, the recent developments in speech comprehension and speech generation by computers is likely to produce a whole new range of automatic instruments able to speak to and understand the analyst.   The technology to support all of these advances is now available.   How and when it is used remains to be seen.

REFERENCES

Arn79   Arndt, R.W. and Werder, R.D., Automated Individual Analysis in Wet Chemistry Laboratories, Chapter 3, 73-94 in (Sto79).

Arn80   Arndt, R.W., Automation of Laboratory Automation: Where is it Going?, Journal of Automatic Chemistry, Volume 2, No. 2, 57-59, April 1980.

Aut80   Decision Criteria for the Selection of Analytical Instruments
         Used in Clinical Chemistry, Journal of Automatic Chemistry,
         Volume 2, No. 1, 22-33, January 1980.

Bag64   Bagrit, L., The Age of Automation, The BBC Reith Lectures -
         1964, Weidenfeld and Nicholson, 1965.

Bar78   Barker, P.G. and Jones, P.S., Syntactic Definition and Parsing
         of Molecular Formulae, Part 2: Graphical Synthesis of Molecular
         Formulae for Data Base Queries, The Computer Journal, Volume
         21, No. 3, 224-233, 1978.

Bar81   Barker, P.G., Data Base Interaction Using a Hand Print
         Terminal, Interactive Systems Research Group, Working Paper,
         April 1981.

Bel80   Belcher, R., The Fall and Rise of Analytical Chemistry,
         Chemistry in Britain, Volume 16, No. 12, 638-640, December
         1980.

Bet80   Betteridge, D., Analytical Chemistry - The Numbers Game,
         Chemistry in Britain, Volume 16, No. 12, 646-650, December
         1980.

But80   Büttner, H., (III) Non-monetary Criteria, 25-26, in (Aut80).

Col74   Cole, H., System/7 in a Hierarchical Laboratory Automation
         System, IBM Systems Journal, Volume 13, No. 4, 307-324, 1974.

Con81   Conklin, D. and Revere III, T.L., Reading Bar-Codes for the
         HP-41C Programmable Calculator, Hewlett-Packard Journal,
         Volume 32, No. 1, 11-14, January 1981.

Cra80   Craig, T.M., (VI) Techniques for the Economic Evaluation of
         Automatic Analysers, 31-33, in (Aut80).

Dee77   Deens, D. R., Analysis with a Purpose, (Sixth Theophilus Redwood
         Lecture), Proceedings of the Analytical Division of the
         Chemical Society, Volume 14, No.8, 199-203, August 1977.

Dob80   Dobson, B.C., Use of a Microcomputer to Control Analytical
         Instrumentation: The Autotitrator, ICI Petrochemicals Research
         and Technology Department, Wilton, Teesside, County Cleveland,
         February 1980.

Fer80   Ferranti Computer Systems, Ferranti Delphi Phoenix Computer
         Data Handling for Pathology Laboratories, Publication No.
         B00096/2, 1980.

For75   Foreman, J.K. and Stockwell, P.B., Automatic Chemical
         Analysis, Ellis Horwood, ISBN: 0-470-26619-8, 1975.

For80   Foreman, J.K., Automatic Analysis: the Laboratory Manager's
         Problems, Journal of Automatic Chemistry, Volume 2, No. 1,
         11-14, January 1980.

Haa79   de Haan, J.B., A Critique of Automated Methods of Clinical
         Analysis, Chapter 7, 208-234, in (Sto79).

Haa80   de Haaen, J.B., Automated Systems for the Clinical Laboratory:
             the User's Needs, Journal of Automatic Chemistry, Volume 2,
             No. 2, 57-59, April 1980.

Hae80   Haeckel, R., (I) Introduction, 22, in (Aut80).

Han67   Handel, S., The Electronic Revolution, Penguin Books, 1967.

Heb75   Hebditch, D.L., Data Communication - An Introductory Guide,
             Elek Science, ISBN: 0-236-31098-4, 1975.

Hen78   Henshall, A., Extended Capabilities of the SP400 Data System,
             Spectra Physics Chromatography Review, Volume 4, No. 2, 6-8,
             1978.

Hew80a   Hewlett-Packard, Series 3350 Laboratory Automation Systems,
             Publication No: 43-5953-1532, July 1980.

Hew80b   Hewlett-Packard, The 3356 B/C Laboratory Automation System,
             Publication No: 43-5953-1530, August 1980.

Hje80   Hjelm, M. and Geary, T.D., (IV) External and Internal Evaluation
             of Analytical Instruments in Clinical Laboratory Sciences,
             26-27, in (Aut80).

Hor80   Horsely, T.E.V., The Impact of Automation in the Pharmaceutical
             Industry, Analytical Proceedings of the Chemical Society,
             Volume 1, No. 5, 177-181, May 1980.

IBM69   International Business Machines, IBM Journal of Research and
             Development, Volume 13, No. 1, 2-132, January 1969.

IBM75   IBM Corporation, Laboratory Automation at Shell - KSEPL - Guide
             for Application Programming, Form: GE15-6036-0, February 1975.

IER70   Conference on Laboratory Automation, Middlesex Hospital
             Medical School, 10th-12th November 1970, IERE Conference
             Proceedings No. 20.

JEO80   JEOL (UK) Ltd., JEOL House, Grove Park, London, NW9 0JN, UK,
             FX200 - FT NMR Spectrometer, product specification, 1980.

Leu80   Leuenberger, U., Gauch, R. and Baumgartner, E., Some
             Applications of the Progammable SP4000 Data System, Spectra
             Physics Chromatography Review, Volume 6, No. 1, 9-10, May 1980.

Lev80   Levinson, S.E. and Shipley, K.L., A Conversational-Mode Airline
             Information and Reservation System Using Speech Input and
             Output, The Bell System Technical Journal, Volume 59, No. 1,
             119-137, January 1980.

Luf80   Luft, L., Automated Gravimetric Titrations, Talanta, Volume
             27, No. 2, 221-222, 1980.

Mic80a   Micropad Ltd., 10 Whittle Road, Wimbourne, Dorset, BH21 7SD, UK,
             Micropad Data Sheet, 1980.

Mic80b   Micropad Ltd., Micropad Document Design Guide, November 1980.

Mic80c   Micropad Ltd., Micropad Product Description: MPD-3000,1980.

Mit80   Mitchell, F.L., Decision Criteria for the Selection of
          Analytical Instruments Used in Clinical Chemistry. Part II -
          Definition of Problems, Types of Instruments and their
          Selection, Journal of Automatic Chemistry, Volume 2, No. 1,
          23-24, January 1980.

Per71   Perone, S.P., Computer Applications in the Chemistry Laboratory
          - A Survey, Analytical Chemistry, Volume 71, No. 10,
          1288-1299, August 1971.

Phi80   Phillips, S., Automated Determination of Ethylene Oxide,
          Spectra Physics Chromatography Review, Volume 6, No. 2, 10-11,
          Autumn 1980.

Por80   Porter, D.G., and Stockwell, P.B., In-house Design and
          Construction of Automatic Analysers for Laboratory Use, Chapter
          2, 44-72, in (Sto79).

Roy79   Roy, R.B., Application of the Technicon Auto-analyser II to the
          Analysis of Water Soluble Vitamins in Foodstuffs, Chapter 5,
          138-160, in (Sto79).

San80   Sandblad, B., (V) The Interaction of New Instrumentation with
          Laboratory Infra-structure: Modelling and Simulation for
          Planning Laboratory Functions, 28-31, in (Aut80).

Sim72   Sims, G.E., Automation of a Biochemical Laboratory,
          Butterworths, ISBN: 0-407-50500-8, 1972.

Sto79   Stockwell, P.B. and Foreman, J.K., (Eds), Topics in Automatic
          Chemical Analysis - Volume 1, John Wiley & Sons, ISBN: 0-85312-
          079-X, 1979.

Sza66   Szabadváry, F., The History of Analytical Chemistry, Pergamon
          Press, Oxford, 1966.

Sza77   Szabadváry, F., From Assaying to Analytical Chemistry: How an
          Art became a Science, Proc. of the Analytical Division of the
          Chemical Society, Volume 14, No. 7, 167-168, July 1977.

TSD80   TSD Display Products, Inc., 35 Orville St., Bohemia, New York
          11716, USA, Touch Screen Digitiser - Models: TSD 12A/15A and
          TSD 12B/15B, July 1980.

Ueb81   Uebbing, J.J., Lubin, D.L. and Weaver, E.G., Handheld Scanner
          makes Reading Bar-Codes Easy and Inexpensive, Hewlett-Packard
          Journal, Volume 32, No. 1, 3-10, January 1981.

Wad80   Wadsworth, M., The Pleasure and Problems of the Hewlett Packard
          HP3350 Series Laboratory Automation System, Paper presented at
          ANALYSIS 80 - The Installation and Management of Micro and
          Mini Computers in the Laboratory, 29th-30th September 1980.

9

Data Processing Methods

INTRODUCTION

Raw data arises as a result of activities designed to record signals that are produced by various processes in which scientists are interested. As was mentioned in Chapter 5, analytical science is essentially concerned with two broad types of data: digital and analogue. For a variety of reasons, data in digital form is often more useful than its analogue counterpart - it is easier to store in computer storage systems, it is easier to transmit via digital communication networks and, least of all, it is also usually easier to process. Digital data can be processed using a variety of hardware and software techniques. This chapter is devoted to a consideration of some of the various aspects of digital data processing using software methods. The term software was introduced in chapter 4. It was used as a generic term to describe the collections of mathematical instructions (often referred to as algorithms) that are stored within a computer and which control its data processing activity.

Software for data processing may be classified in a variety of ways. From the point of view of the present chapter the two most important categories are those of numeric and non-numeric processing. In numeric data processing (DP) the basic entities manipulated are numbers - scalars, vectors, matrices and higher order arrays - and relationships between them - curves, surfaces, equations and so on. Non-numeric processing, in contrast, deals with objects (and their inter-relationships) that are not represented in terms of numbers; typical examples of non-numeric data might be a list of people's names, chemical formulae, drug properties and so on. Later in this chapter some of the techniques that are used for processing both numeric and non-numeric data will be discussed.

Fundamental to all forms of data processing is the idea that the basic algorithms involved define processes that operate on a collection of data and transform that data into new representations of it - without necessarily destroying its original form. In order to illustrate this, consider the simple numeric data collection shown on the left of Fig. 9.1 in which a,b,c, etc represent numeric data items.

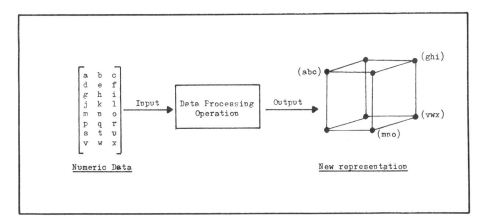

Fig. 9.1   The role of scientific data processing as a tool
for handling experimental results.

The formulation of an algorithm and subsequent production of computer
software to process the data set (shown on the left of Fig. 9.1) to
produce the image shown on the right is not difficult. Indeed, the type
of data processing involved forms the basis of many of the crystallographic programs used in crystal structure analysis from X-ray data
(Ste70, Mai78, Lun79).

Algorithmic processes operating on data collections thus represent the
fundamental nature of most scientific DP activity. A collection of data
that is capable of being shared between various users, which can be
updated and changed in various ways and whose integrity is assured is
often referred to as a data base. Exact definitions of this term will be
found in text books on this topic (Mar80, Dee77). Data bases will be
dealt with in more detail in the next chapter. However, for the moment,
it is convenient to think of data base as being a stored collection of
data capable of being processed in various ways by a computer. On this
basis the simple model shown in Fig. 9.2 may now be used to represent the
role of the data base and data processing in analytical science.

The distinction between data processing and information processing is one
of level rather than one of difference in technique. As will be described
in the next chapter, information may be thought of as data with "meaning"
added to it - that is, data plus its semantic interpretation. Thus, at
the most fundamental level a mass spectrum, NMR spectrum, chromatogram or
IR spectrum are essentially identical in that they are records of a time
varying detector response. To fit a smooth curve through a peak maximum
of any type of spectrum is not usually dependent upon its type. Consequently, this is purely a data processing activity. However, once the
time varying set of data has been assigned the semantic interpretation
'mass spectrum' it starts to become information. It takes on more meaning
as particular m/e values are assigned to particular ion fragments. The
reduction of a complete mass spectrum to its eight-peak representation is
now regarded as an information processing task that converts information
in one form to information in another. Furthermore, this particular
processing activity is one which is applicable to spectra but which would

not normally be applied to an entity such as a chromatogram, polarogram or voltammogram. In other words, data processing often involves fairly general algorithms while information processing can involve techniques that are peculiar to particular areas of activity.

A wide variety of processing techniques are available some of which are quite general purpose others being specific to particular areas of analysis. A few of the more general ones (New73) are outlined in Table 9.1. Some of these will be discussed in more detail later in the chapter.

The implementation of automated data processing techniques on digital computer systems will necessitate the availability of suitable programs. Fundamental to the production of computer software is the activity of programming - the act of designing algorithms that specify how data is to be processed and then translating the rules that have been formulated into a form that the computer can understand. Because of its importance in data processing an overview of this subject is given later in this chapter. Before this, however, it is important to look briefly at some of the different data processing tools that analysts are likely to use in order to manipulate and transform the data that they collect from their experiments.

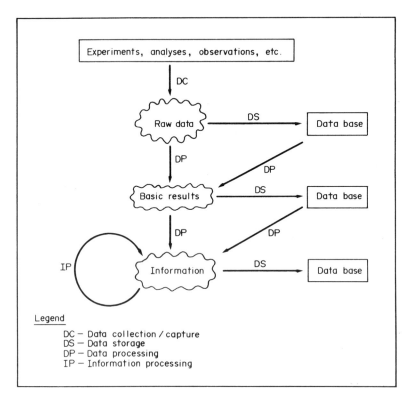

Fig. 9.2 Simple model depicting the parts played by information processing (IP), data processing (DP) and data storage (DS) in analytical science.

TABLE 9.1   A Selection of Standard Scientific Data Processing Techniques

CATEGORY	ROUTINE
Storage	Data storage Information Retrieval Search techniques Comparison
Graphics	Digital plotting CRT display Printed Plot Printed output
Correction	Axis unit modification Base-line shift Base-line drift Calibration Normalisation Scaling of data
Improvement	Point averaging Time averaging Background reduction Lagrangian interpolation Digital filtering Line sharpening Digital smoothing Summation of spectra Difference spectra
Transformation	Derivative spectrum Integral spectrum Convolution Deconvolution Fourier transform Integral transforms
Calculation	Noise level Peak location Peak integration Spectrum integration Line following Curve fitting
Analysis	Least-squares fitting Direct search minimisation Simulation Parameter improvement Normal equation solution Band-shape analysis

DATA PROCESSING TOOLS

Over the last ten years the impact of electronics on the development of calculating machines and other data processing tools has been quite substantial. Never before have analytical scientists had available such a vast range of tools to aid them with their DP tasks. These range in sophistication, power and capability from simple look-up tables - perhaps, electronic in nature - through to expensive high speed array processing computers - often used for simulation studies and spatial data processing. Some of the more important mathematical DP tools are listed in Table 9.2.

TABLE 9.2   Mathematical Data Processing Tools

1. Look-up tables and nomographs.
2. Logarithm and other Mathematical Tables.
3. Slide rule.
4. Pocket calculator.
5. Desk-top calculator.
6. Desk-top computer.
7. Microcomputer system.
8. Minicomputer.
9. Mainframe system.
10. Array processor.
11. Super-computer
12. Computer network.

The mode of operation of most of these, and their relative capabilities, were oiutlined in chapter 4. It was mentioned there that the performance of the electronic DP equipment within each of the above categories continues to increase with time while their relative costs decrease. Within each of the areas listed there will usually be a number of different vendor products available - varying in cost and capability. Many manufacturers tend to specialise in particular areas while others cover the whole range of devices listed in Table 9.2.

At the calculator end of the spectrum the trend in product advances seems to be towards increasing function, better memory capability, availability of simple peripherals and the availability of libraries of standard programs for engineering, scientific, statistical and other standard mathematical processing techniques. Increased function in a calculator or desk-top computer refers to situations wherein additional facilities are designed into the tool to enable it to be used more easily or in order to meet particular types of demand. Indeed, this is happening to such an extent that certain types of calculator are becoming orientated towards particular application areas - a calculator for an engineering application will differ from that which be used by a scientist or mathematician. Usually, the difference will be in the types of 'built-in' programs (such as 'solve', 'integrate', etc) or constants associated with the different keyboard buttons on the device. Some calculators have available user-definable keys so that once a processing method has been entered it can be assigned to one of the 'free' keys in such a way that subsequent

depression of that particular key automatically invokes the previously
stored series of operations associated with it. Improvements in memory
technology for electronic calculators are having two major impacts -
firstly, the amount of storage available within calculators is increasing
thereby enabling larger collections of processing operations to be stored
and, hence, more complex calculations undertaken; secondly, the availability of continuous memory means that data and programs entered into the
calculator will be retained after it has been switched off. Trends in the
area of add-on periperals for calculators are in the direction of
improving techniques for loading (or saving) programs from (or to) a
software library and making available facilities for printing out the
results of calculations. Increasingly, manufacturers of calculators
provide substantial libraries of DP software for use on their instruments. These are usually supplied on one or other of four different
storage media - miniature cassette tape, magnetic card, 'software books'
and ROM (read only memory). Facilities for reading cassette tape and
magnetic tape are ususally built into the calculator. Similarly, programs
stored in ROM chips are made available simply by plugging the chip into
the calculator. Software books are collections of programs written in
terms of bar-codes (see chapter 8). These can be read into the calculator
by means of a special light sensitive pen that plugs into an appropriate
surface mounted socket. The advantages of ROM and bar-code storage are
that they are not susceptible to magnetic fields and cannot easily be
destroyed by the user. However, these media are not as amenable to the
storage of user's programs as are cassette tape and magnetic card. Of
course, for larger devices such as desk-top calculators/computers flexible
disks represent a further means of storage for both data and programs. To
enable the results of calculations to be printed out there is nowadays a
wide range of both thermal and mechanical mini-printers available that
vary in both the column width (that is, number of print positions) and
character set that can be used - numeric, alphabetic or both. Some
calculators (particularly, those of the desk-top variety) provide
facilities for simple control applications. However, their capabilities
in this respect are often quite restricted. Where a DP tool for the
control of instruments is required it is often better to use a dedicated
microprocessor, a desk-top computer or a minicomputer. Depending upon the
particular application, both speed and hardware architecture may need to
be considered.

The basic architecture of microcomputers, minicomputers and mainframe
systems was described in chapter 4. Their use as DP tools has been
illustrated in the various applications that have been cited in previous
chapters. It is important to re-iterate the point that specific kinds of
application will often require tools of a particular sort and sometimes a
combination of items from a DP toolkit will be required to solve some
types of problem. Thus in situations where large amounts of data are
produced - such as in crystallography (Sas80) a mainframe computer or
array processor may be required in order to produce results with minimal
delay. In some types of application where large amounts of data are
produced - such as image processing (Dru80) or photodiode array spectroscopy (Chu78, Cod80) the power of a minicomputer or fast microprocessor
may be sufficient. Other kinds of data processing are ideally handled by
a microprocessor - an excellent illustration of this is contained in a
paper by Graneli et al who describe the use of a microcomputer system for
the processing of results produced in potentiometric stripping analysis
(Gra80). The application of a mixture of DP tools to the solution of an
analytical problem is illustrated in a paper by Kirchner et al (Kir80) who
describe the use of two microprocessors (and associated peripherals) a..d a

'phone-linked' mainframe for the analysis and presentation of data
produced in proton induced X-ray emission studies.

Computer networks enable the analyst to transmit experimental data
electronically to some (possibly unknown) remote computer. It can then be
processed by some sophisticated DP package (such as a pattern recognition
program). Afterwards the results can be automatically transmitted back to
the local computer centre from which the data originated. Networks of
this type will be discussed in more detail in chapter 12.

In addition to the hardware needed for DP there will also be a need for a
variety of software tools. Like hardware capability, software facility
usually increases as Table 9.2 is descended. Two important trends should
be noticed as this happens - better and improved programming/software
development facilities and a much wider range of pre-programmed software
for large DP applications. Each of these will be discussed in more detail
in the following section.

PROGRAMMING

Users of computers for analytical data processing may be sub-divided into
two broad categories: the parametric user, and, the program developer.
Generally, the parametric user is one who takes advantage of previously
written 'packaged' programs and uses these to produce all the necessary
results. Such a user needs to have little knowledge of programming -
his/her only requirement is to understand the limitations and capabilities
of the package being used. This is important since the user must ensure
that the act of using a particular package with a given set of data is a
valid thing to do and will lead to the production of meaningful results.
However, once this criterion has been met, it only remains for the user to
choose an appropriate parameter in order to instruct the package on how
the data is to be processed. Several different kinds of parameters may be
used although numeric (eg 1, 2, .. 10, 20, . etc) values and alphabetic
keywords (eg PLOT, SMOOTH, STORE, CURVE FIT, ... etc) tend to be the most
popular variety. Parametric approaches to data processing are often used
for the manipulation of crystallographic data (Sas80), applying statistical tests (Jor79), using simulation packages (Min80) and so on.

The second class of user mentioned above is that which represents those
who actually become involved in developing some form of computer program.
This may happen for a variety of reasons: a package does not provide the
exact facilities required so that some form of pre- or post-processing
system has to be constructed; there may not be any packaged programs
available; new equipment is being produced, designed or installed and
appropriate support software is required - perhaps to control the equipment, acquire data and then process and display it, etc. As programmable
calculators, desk-top computers and laboratory automation systems become
more commonly available, so the number of analytical scientists that
become involved in some form of program development activity is increasing. Programming is not a difficult task to undertake - indeed, it is
particularly simple if a high level language such as FORTRAN (Kat78) or
BASIC (ANS77) is used. However, in situations where assembly language or
machine code (see chapter 4) have to be used, the task becomes much more
difficult and requires a deeper detailed understanding of computer hardware and the fundamental mechanisms via which a particular system
operates.

It was suggested earlier in this chapter that a program is a dynamic
entity - it is a computational process that manipulates data and
transforms it in various ways. This process is a physical realisation of
one or more mathematical algorithms. Often, the following simple
expression is used to show the relationship between the important items
involved in the programming process.

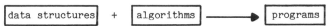

In order that it may be processed, data has to be resident within the
memory space of the computer system. A data structure may be thought of
as being a detailed specification of an area of computer memory that will
be used to store the basic data value(s) that a program is to manipulate
during its execution. The algorithm, of course, specifies the rules to be
used in processing the data. Before it can be implemented on a computer
it has to be translated into an appropriate programming language. Most
simple programs will contain essentially two sections as indicated on the
left-hand side of Fig. 9.3.

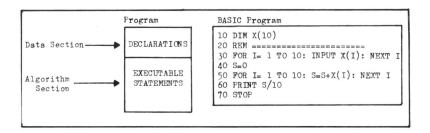

Fig. 9.3 Fundamental structure of software - data definition
and algorithm implementation.

The declaration section is used to contain definitions of all the data
structures to be used by the program while the algorithm section contains
the sequence of executable statements that specify how the data is to be
processed - that is, the implementation of the algorithm. An illustration
of an actual BASIC program is shown on the right hand side of the above
diagram - the data structure defined in line 10 (called X) is a ten
element vector; the algorithm initialises it and then computes the average
value of its elements.

There are many different types of data structure - scalars, vectors,
matrices and higher order arrays. Structures of the type listed are often
referred to as 'primitive' data structures since they are available in
most computer programming languages. Some of the more advanced languages
(for example, PL/I, PASCAL, ALGOL68, ADA, etc) provide very powerful
facilities to enable the creation of extremely sophisticated data
structures such as lists, stacks, records, trees, networks, etc (Tre76,
Aug79) via the use of special techniques involving the use of pointers
(IBM69). An important part of the programming process is thus the design
of data structures that represent the real world objects that the program
is to process. Having done this the algorithms to manipulate the data
have to be formulated, translated into an appropriate computer code and
exhaustively tested in order to ensure their correctness.

Algorithms may be designed for either real-time or post-processing of
data. It real-time situations data is processed as it is acquired, while
in post-processing it is stored away and processed after acquisition has
been completed. Writing programs for real-time applications (Pyl72,
Wex75) is often much more difficult than for post-processing situations.
Many of the difficulties associated with real-time software arise because
very often the programs concerned can involve the initiation and control
of multiple concurrent processes - called parallel or multi-processing.
These require the use of special synchronisation techniques. Most conventional programming, however, is fairly straightforward and simply involves
translating the steps specified in the algorithm into the primitive statements of the computer language being used. The majority of algorithmic
languages permit only a limited number of types of operation, namely,

1. Input/Output - that enable data values to be introduced into a
   program and the results of computations to be extracted,

2. Assignation - in which data values are moved from one location
   in the computer's memory to another; during transit between
   source and destination some intermediate data processing
   may take place,

3. Procedure Invocation - which causes particular pre-defined
   groups of operations (called procedures) to be performed
   at the point in a program at which they are referenced,

4. Looping - that causes certain groups of operations (collectively referred to as the loop body) to be repeatedly
   executed until some specified criterion (called the loop
   termination condition) is satisfied,

5. Decision Making - in which the values of particular data items
   in a program are compared and then, based upon the result
   of the comparison, some appropriate control action is taken.

Normally, flow of control through a program is sequential unless some
transfer of control statement is encountered that causes a departure from
this rule. Thus, in the example program shown in Fig. 9.3 execution
starts at line 10, proceeds to line 20, 30, ... and ultimately terminates
when the statement in line 70 is reached. Of the statement categories
listed above, the simple illustration program contains an assignation
(line 40), an input statement (as part of line 30), an output statement
(line 60) and two loops (lines 30 and 60). Further details on the
technical aspects of programming will be found in an appropriate textbook
for the language system being used for program development (Mon74, Mon78,
Cam79, Str75, Spe75, etc).

A decision to develop DP software will require making a choice of
programming language. The choice made will depend upon a number of
factors such as the type of computer system being used, the nature of the
project and the kind of limitations that are imposed - speed of program
execution, the memory space available, project costings, development
facilities and so on. In many cases there may not be any language
selection options open - if the project is simply an enhancement or
modification of existing software or if the computer system being used has
only limited software resources. However, most often, the choice to be
made is between the use of a high level language and an assembler (or
machine code). Programmer productivity using the former greatly exceeds

that achieved when the latter is used. This is a major reason for the use of high level languages. However, where maximum speed of execution and minimal storage are the prime design objectives, it is imperative to program in machine code via the use of an assembler system if one is available. For scientific programming the most popular high level languages are probably BASIC, FORTRAN and PASCAL. BASIC differs from the other two in that source language statements are interpreted (Gri71, Bro79) rather than compiled into machine code. The implications of this are that programs written in BASIC execute more slowly than those written in FORTRAN, PASCAL or machine code. BASIC is easy to learn, designed for interactive program development and, like FORTRAN, is commonly available on virtually all types of computer system - including many laboratory data acquisition systems. Transportability of programs is an important factor to consider when deciding upon a programming language to use for a project. Because of this, FORTRAN and BASIC have most often been the prime choice for the development of scientific application programs. However, the increasing availability of PASCAL and the better program development facilities that it offers are factors which merit its further consideration.

Examples of the use of each of the above language options will be found in the scientific literature. Clark et al (Cla80a, Cla80b, Cla80c) have presented a series of articles describing the use of PASCAL - running on a microcomputer - for the manipulation and graphic display of molecular structures. The use of FORTRAN is illustrated in work that has been described in papers by Weber (Web80), Pottle et al (Pot80) and Kirchner et al (Kir80) while an example of the use of BASIC is contained in a paper by Randic et al (Ran80). Some interesting examples of assembly language programming of a microcomputer and of machine code programming of a programmable calculator will be found in articles by Graneli (Gra80) and Rechberger (Rec80), respectively. Often, large machines are used to prepare software for small computer systems via the use of special language translators called cross-assemblers (see chapter 4). Cheng et al (Che80) describe the use of this approach to develop assembler language programs for an INTEL 8085 microprocessor system. The mainframe was a remote Honeywell Level 66 accessed via a telephone link. This approach to program development is becoming increasingly popular for small laboratories.

NUMERICAL METHODS

Numerical data processing methods are probably most commonly used for the treatment of experimental analytical data. A numerical DP technique may be considered to be a set of one or more mathematical transformations that are applied to a collection of numeric data. There are usually two basic reasons for the application of such transforms. Either it is regured to modify the data in some way - smooth it, apply corrections for known perturbation effects, eliminate unlikely values, and so on; alternatively, there may be a need to analyse it in some way in order to deduce the mathematical relationship between the experimental variables that the data points connect. A typical illustration of the application of numerical methods to the processing of experimental data is illustrated in the sketches presented in Fig. 9.4. These illustrate the use of derivative techniques for peak analysis during computer based chromatographic data acquisition - peak shape, type and area allocation are determined by taking account of the appearance of minima, maxima changes in sign in the first differential and maxima in the second differentials. Digital

smoothing of data (Sav64, Ste72, Pro80), differentiation and integration
are each examples of commonly used numerical DP techniques.

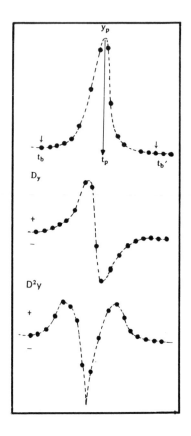

Fig. 9.4 Numerical processing of experimental data.

The variety of problems within analytical chemistry to which numerical
techniques can be applied is probably unlimited. Sometimes the methods
used may be quite empirical. More often, however, the method employed
will be based upon the application of rigorous theoretical considerations.
The type of technique employed may be an elementary least squares
regression analysis (Pol77, Moo80) implemented on a programmable
calculator or the application of a more sophisticated technique such as a
Fourier (Str73, Sci78, Bin80) or Hadamard (See80) transform via the use of
a minicomputer system. Obviously, numerical methods will fall into two
major types - those which are of a generally applicable nature and those
that are peculiar to a particular area of analysis - electrochemistry,
thermal analysis, kinetic methods, RIA and so on. Many of the more well
established 'general purpose' numerical DP techniques were listed in Table
9.1. Descriptions of these will be found in appropriate textbooks devoted
to numerical methods (Joh80, Pol77, Baj75, Bev69). Details of methods of
processing applicable to specific areas of analysis will be found in

specialist texts (for example, Chapman (Cha78) - mass spectometry, Leiva (Lei78) - RIA, etc) and journal articles (such as Reese (Ree80) - chromatography, Gans (Gan80) and Leggett (Leg77) - spectra, Bobst (Bob80) - ESR, and so on). A review of some of the newer numerical methods will be found in Ana80.

An important area in which the analytical scientist often becomes involved is that of mathematical modelling, that is, the representation of a system and all its subsystems (components) in mathematical form. The type of model that is derived to represent any particular physical system will depend upon the questions that are asked and the limitations that are imposed. However, once a basic qualitative model has been formulated the mathematical equations for the model can be derived from fundamental physical principles or from measurements conducted on the components of the system. In general, the mathematical equations that describe a system may assume many forms: they may be linear or non-linear equations, they may involve ordinary or partial differential equations, integral equations and difference equations and others. If information is to be derived from a model then equations of the form previously mentioned have to be solved. Many of these have no analytical (in the mathematical sense) solution. Consequently, this is an area where computer based numerical DP methods are able to play a substantial role. Typical examples of the techniques involved have been described by Peterson (Pet73), Schaumberg (Sch80), Zlatev (Zla80) and Vanderslice et al (Van81). The latter work by Vanderslice discusses numerical techniques for the solution of the diffusion-convection equation that describes dispersion in a cylinderical tube - this is of much importance in analytical methods based upon the currently popular Flow Injection Analysis technique.

Because many procedures within analytical chemistry involve quantities and measurements about which there is some element of uncertainty, another class of numerical tool is required which is capable of handling this type of data - statistical methods. Broadly, statistics may be thought of as the science of collecting, classifying, presenting and interpreting numerical data. The field can be roughly sub-divided into two areas: descriptive statistics and inferential statistics. The former deals with the collection, presentation and description of data while the latter is more concerned with the techniques of interpreting the data and making decisions based on the results. One of the reasons for the need to apply statistical processing to analytical measurements stems from the fact that they are usually subject to a wide variety of errors - random, systematic, gross, etc. The presence of such errors need to be detected and their contribution to and influences upon a result determined. Appropriate statistical models (based upon the Binomial, Poisson, Gaussian or non-parametric approaches) and metrics derived from these are required in order to express the reliability of an estimate and the confidence that the scientist(s) puts in the measurement - good examples of the basic approach will be found in Jordan's treatment of inter-laboratory result comparison (Jor79) and Schwartz's approach to instument calibration curves with non-standard variance (Sch79). A wide variety of statistical techniques are available (see Cal80, Baj78) such as the mean, standard deviation, chi-square, t-test, F-test and a host of others. Particularly useful for control applications are the use of statistical control charts (Ghe80) and cumulative sum techniques (CUSUM - see Cha80, Spr77). When applying these tests for the analysis of data it is important to bear in mind the assumptions and limitations inherent in their derivation. In view of this it is always worthwhile applying several of these tests whenever such an approach is feasible (Kel79, Ghe80). Many statistical processing and testing procedures are available as software packages on computer systems (see below) and, in some cases, as function buttons on pocket/programmable calculators. Again, it is important to stress that their ease of application should be treated with due respect and the results produced accepted with appropriate caution. Pollard (Pol77) recommends that when using packaged statistical (or numerical) procedures for the first time the user should always check them with a data set for which the results are know. Such an approach should reflect any major flaws in the procedure being tested.

Statistical and numerical methods often form the basis of a wide variety of other processing techniques - particularly pattern recognition (see later) and optimisation. The principles of the latter will be familiar to most experimental scientists involved in the development of both conventional and instrumental analytical methods. Thus, in the laboratory, many situations arise where it is necessary to perform a few preliminary experiments in order to determine which variables or factors are important in a procedure. Then, after the important factors have been determined, optimisation is effected by varying factors one at a time while keeping the others fixed until an optimum is reached. Examples of this procedure are well know in many analytical situations - in spectroscopy (Bro80), in elution chromatography (Wat79), in atomic absorption analysis (Dow79), X-ray analysis (Jab79) and so on. Several different approaches to locating optimal situations exist (such as the Simplex method and derivatives thereof) and are adequately described in text books devoted to this topic (Wil64, Kow68, Adb74). Two interesting examples that are representative of methods commonly used will be found in papers by Ernest (Ern68) and Rubin (Rub79). The former describes the use of optimisation methods for seeking optimal gradient and curvature settings to control magnetic field homogeneity during computer based high resolution nuclear magnetic resonance experiments. The latter paper outlines the use of statistically designed and evaluated factorial experiments in order to locate the optimum operating conditions of a gas chromatographic detector. With the increasing availability of powerful microcomputers/minicomputers in the laboratory, automatic optimisation is gaining considerable popularity. Descriptions of some of the recent developments in this area have been given by a variety of reporters (Rya80, Hol80, Swa80). Of particular importance is the development of a high speed algorithm (Bri79) that is particularly well suited to pattern matching operations. This type of application will be described in greater detail in a later section.

Previous mention has been made, both in this section and in chapter 4, of the wide range of pre-packaged computer software available for performing both statistical and numerical data processing. Whenever software of this type can be employed it can offer substantial savings in development time and programming effort - provided it is well documented and easy to use. Sometimes occasions arise when such software needs to be modified and tailored to meet particular individual requirements. The extent to which this happens often depends upon the origin of the software and the function it performs - random number generators sometimes cause problems. Two broad approaches to the use of pre-packaged DP software currently exist:

(A)   the use of subroutine libraries, and,
(B)   the use of program packages.

As was indicated earlier, a subroutine (or procedure) is a pre-defined set of progam language statements that perform some DP function. The collection of statements (usually written in machine code for speed and efficiency) can be incorporated into a user's program and then invoked each time the subroutine's name is used within a program. Subroutines exist for most of the commonly used statistical and numerical techniques. A collection of such routines provided by a computer manufacturer (IBM67, IBM68) or other source is often referred to as a scientific subroutine package. The type of routines likely to be included in a typical small machine library are listed in Table 9.3

TABLE 9.3   Contents of a Scientific Subroutine Library

Among the statistical functions provided are:
  Data screening
  Moments for grouped data
  T-statistics
  Correlation
  Multiple linear regression
  Polynomial regression
  Canonical correlation
  Analysis of variance
  Discriminant analysis
  Factor analysis
  Time series
  Nonparametric statistics
  Random number generation

Mathematical functions include:
  Eigenvalue and eigenvector determination
  Matrix addition, subtraction, products
  Transposition of matrices
  Operations on rows and columns of matrices
  Integration and differentiation
  Fourier analysis
  Bessel functions
  Elliptical integrals
  Fresnel integrals
  Exponential integrals
  Gamma function
  Legendre polynomials
  Linear equations
  Nonlinear equations
  Roots of polynomials
  Operations on polynomials

When using a routine from such a library all the user has to know is (a) the name of the routine to be used, (b) its function, (c) its calling convention, and (d) how it returns its result(s). The calling convention refers to the manner in which the subroutine expects to receive the numerical values it is to process. Thus, if a routine is going to perform numerical integration of a function defined in terms of pairs of (x,y) values, then the calling convention may require that the set of x-values be stored in one vector and the y-values in another separate one. Alternatively, the routine may require a single data structure to be used - perhaps a one dimensional array in which x and y values alternate or a

two dimensional array in which x-values occupy odd numbered rows and
y-values occupy even numbered ones. The exact subroutine calling
requirements will be described in the written documentation associated
with the software library system. Two types of scientific subroutine
package are likely to be available on large and medium sized computers:
those provided by the computer manufacturer (described above) and those
supplied by independent software organisations. One commonly used set of
routines that fall into the latter category are the NAG routines
(Numerical Algorithms Group - see NAG78). They are used in exactly the
same way as manufacturer supplied routines but have the advantage that
they are both manufacturer (and hence) machine independent. As an example
of their use, suppose it is required to use a Runge-Kutta technique as a
means of integrating a set of first order differential equations,

$$Y_i' = F_i(T, Y_1, Y_2, Y_3, \ldots Y_N) \qquad \text{for i} = 1,2,3 \ldots N$$

over a particular range from T = X to XEND to some tolerance TOL. By
including a statement of the form,

CALL D02BAF(X,XEND,N,Y,TOL,FCN,W,IFAIL)

in a program, the user can gain access to code capable of performing the
appropriate numerical processing - provided values are assigned to all the
entities specified in the calling convention (see NAG78). Further illustrations and a more detailed discussion of routines of this type will be
found in a paper by Zlatev et al (Zla80a).

The use of pre-packaged subroutines for data processing requires some form
of programming ability on the part of their user. To remove this
necessity, the second of the previously listed approaches to software is
often favoured - the use of pre-packaged program suites. In principle,
the use of such a collection of programs enables the user to simply enter
data values into a computer system along with a list of parameters that
specify which DP options are required and how the final results are to be
presented. A wide variety of packages exist for various types of
application (for example, mathematical programming, linear programming,
statistical analysis and so on). Typical packages that are often used for
statistical processing include:

MIDAS   - Michigan Interactive Data Analysis System (Fox76),
SPSS   - Statistical Package for Social Science (Nie70),
GENSTAT - General Statistical Processing (Alv77), and
BMDP   - Bio-Medical Data Processing (Dix79, Jor79).

In addition to program suites for general statistical analysis and
numerical processing there is a considerable amount of packaged software
for processing data within particular subject areas. Examples of such
systems include:

GAUSSIAN70 - electronic structure calculations (Mez80),
MINDO   - calculation of molecular properties (Bai69),
MMI   - molecular mechanics (Jef80), and
CHEMICS   - organic compound structure elucidation (Yam77).

Very large program packages such as those listed above are often enhanced,
maintained and distributed from a central source - called a Program
Exchange Depository. A typical example of such an establishment is the
QCPE centre (Quantum Chemistry Program Exchange - see Qua81) that operates

at the University of Indiana, USA. Usually, such centres will distribute
copies of software to any non-profit making organisation interested in
using them.

NON-NUMERIC DATA PROCESSING

There are many classes of computational problem that do not involve the
manipulation and processing of large sets of numerical data in the ways
that were described in the last section. Thus, given a randomly ordered
list of instrument manufacturer's names and addresses it may be required
to construct a computer program that will sort these into alphabetical
order by name; sort them into decreasing order of product code (0010
denotes a GC, 0020 denotes an IR spectrometer, 0035 denotes an NMR
machine, etc); or, rearrange them into geographical location of supplier
(Switzerland, USA, Germany, UK, etc) and then sort them alphabetically by
supplier within the previously arranged groupings. Similarly, it may be
required to process a collection of data in such a way that it is possible
to select all those items which possess a particular set of pre-defined
attribute values. For example, search a list of instruments and find all
those that are non-dispersive IR spectrometers, weigh less than 10Kg and
which are suitable for outdoor operation; analyse a list of patients and
report all those who attended a clinic between 4th June and 8th July and
who had a low blood pH, weighed less than 200 lbs and were prescribed a
course of drug X; list all the research papers on flow injection analysis
that were not written by Ruzicka and that were published in Talanta or
Analytical Chemistry during 1981. Each of the examples outlined above are
illustrations of applications of non-numeric data processing. Many other
examples could be cited, and the reader is referred to appropriate texts
(Win77, Wai73, Hal75) for further details of these.

Sorting and searching (Knu73, Lor75) are probably two of the most
frequently used non-numeric DP operations - along with record keeping and
report generation. A sorting operation is used to assemble stored items
of data into order (ascending or descending) so that some subsequent
processing step is facilitated - perhaps, for efficient updating of a
stored collection of data or for fast access to it by means of some form
of retrieval process. As a simple illustration of this, consider a simple
data retrieval system implemented on a microcomputer and which contains
physical constants of laboratory chemicals. An analyst requests the
computer to retrieve the solubility at $t^{\circ}C$ of a compound X. Suppose
that the data base contains appropriate details for a total of N
compounds; assume that the items are not ordered in any way. Using a
sequential search strategy in which items can only be examined in the
order in which they appear in the data base, an average of $(N+1)/2$ items
will need to be examined each time an information retrieval request is
made. If, however, the data items are sorted into order based upon some
unique attribute value (such as compound name), then it is possible to
apply a more efficient search technique based upon a binary search
(Tre76); this would require an average of only $FLOOR(\log_2 N)-1$ items to
be examined. It would be feasible to reduce the number of items searched
even further if a suitable key to address transformation algorithm could
be formulated which would enable the data to be retrieved directly (Knu73,
Tre76, Aug79).

Searching operations are important in many 'library search' systems for
retrieving a variety of different types of spectra (Ven77, Del79, Mar79,

Sur77). The principle is straightforward: the spectrum of the unknown compound is compared with those contained in a standard library until a match is found within a given tolerance limit. Alternatively, the search may enable the spectra of several 'candidate' molecules to be retrieved. Unfortunately, this approach cannot always be used since for very large library collections the search time involved would be too great. Thus, Biller et al (Bil77) have reported using GC-MS techniques for the analysis of peptide mixtures. Because a library of all possible di- to penta-peptides would contain in excess of 3.4 million spectra they have not used search techniques. Instead, they have devised a set of interpretive algorithms for automatically interpreting the observed spectra - along similar lines to the DENDRAL program used in mass spectrometry (Win77). This, of course, is another important area of non-numeric data processing - the formulation of models to facilitate the analysis and generation of spectra. In this context, an important area of study is that of simulation; this will be described in more detail in the next section.

A large number of non-numeric DP applications - such as graphic displays, data base management implementation, knowledge based systems, etc - often require the use of data structures that are more complex than those used for storing straightforward numeric data. The most useful of these is probably the list structure. In an array the individual elements usually occupy contiguous memory locations. Thus, knowing the address of the first element that of any other can be calculated. Furthermore, once the address of an element is known its contents can be directly accessed. Sometimes an array is referred to as a linear list. In a more generalised list structure, the only known location is that of the first number of the list (called the 'head'). Subsequent elements of the list are obtained by following a chain of pointers; each element of a list always contains one or more pointers (or addresses) that specify the address of its successor (forward chaining) and predecessor (backward chaining). Figure 9.5 illustrates some examples of list structures.

In these diagrams, the rectangles represent the elements (or nodes) of the list structures. Each element may contain two basic types of information: a data value and an address that specifies the location of the next element in the list, these are the pointers referred to above. In the diagrams, pointers are represented by arrows; a null pointer is used to indicate the end of a chain of addresses. Null pointers are indicated by "-" symbols in the lists shown. Given a list of N items, there are nine basic operations that can be performed on it:

(a) access the kth element ($1 \leq k \leq N$),
(b) insert a new element before or after the kth,
(c) delete the kth element,
(d) concatenate (that is, link together) two or more simple
     lists,
(e) split a list into two or more lists,
(f) make a copy of a list in terms of its contents and
     structure,
(g) count the number of elements in a list,
(h) sort the elements according to some criterion,
(i) search for an element that has a particular value.

Performing operations of this type using conventional programming languages such as FORTRAN and BASIC is not easy because of the difficulties that are encountered when an attempt is made to obtain the actual machine addresses at which the list elements are to be stored.

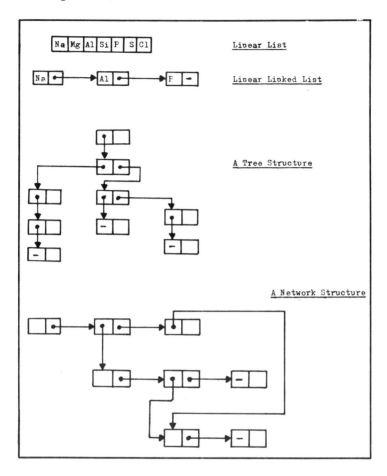

Fig. 9.5 Examples of non-numeric data structures.

However, it is possible to implement list structures and operations upon them using these languages. Further details on the application of FORTRAN to non-numeric processing of this type will be found in the book by Day (Day72). Usually, however, the more advanced type of non-numeric DP application are programmed in languages such as PL/I (IBM69) or PASCAL (Wir78) because these languages make available a primitive pointer data type which makes the construction of list structures an extremely simple process.

Applications of list processing and artificial intelligence programming techniques arise in many areas of data handling within analytical chemistry. Woodruff and Smith (Woo80) have illustrated the application of these methods in the design of a program for the interpretation of infrared spectra. They describe the construction of a system in which the chemist inputs basic physical details of a substance, its IR spectrum and the set of rules that describe how to interpret the spectrum. The inter-

pretation rules are represented in terms of a language called CONCISE
(Computer Orientated Notation Concerning Infrared Spectral Evaluation).
The program produces a list of probabilities of the liklihood of
particular functionalities being present in a compound/mixture. The
advantages of the system are claimed to lie in the fact that the user does
not need to have any knowledge of programming and, as in the case of
Biller's system (described above), no library search is needed.

Many other examples of this type of processing exist. The basic techniques are applicable to any area involving the use of a data base system,
the graphic representation (both static and animated) of information or
the design of systems requiring the application of decision making rules.
Some further examples will be found in Ran80, Cla80a, Cla80b, Cla80c and
Wai73.

SPECIAL TECHNIQUES

There are several types of 'special' data processing technique that, to
the analytical chemist, are either fairly well established tools, or, are
currently generating substantial interest. Some of these will be outlined
in this section.

Simulation and Modelling
Some indication of the priciples behind modelling were given earlier in
this chapter and so the technique of simulation will be discussed here.
Many types of simulation experiment exist. The present discussion,
however, is concerned with only those which are computer based. In
principle, the computer is an ideal tool for simulation since it enables a
model of virtually any physical process to be created. Once defined, the
properties of the model can be investigated by varying appropriate design
parameters that were built into the model at the time it was constructed.
The importance of simulation lies in the fact that if the parameters are
selected in such a way that the model upon which it is based is an
accurate mirror of reality then it is possible that:

(a)   predictions about reality can be made on the basis of the
      model, and,
(b)   it may be possible to obtain estimates of quantities (such
      as rate constants, concentrations, etc) that cannot be
      measured in any practical way.

Suppose a model is proposed which purports to describe the observed colour
(and its intensity) of a solution of copper sulphate as a function of
temperature, concentration, pH, and so on. If the model is an accurate
one then it should enable the variation of colour intensity with dilution
to be predicted. Furthermore, it should permit estimates to be made of
various parameters relevant to the processes involved in the physical
system - the equilibrium constants for ionic equilibria, absorption
coefficients for light interaction and so on.

In most approaches to simulation there are usually six steps involved:

(1)   develop a mathematical model based on data and experience,
(2)   represent the model in terms of suitable analogue, digital
      or hybrid computers,
(3)   run a computer simulation for a number of situations where
      the real system behaviour is known,

(4) adjust the model parameters and structures until it acts
     like the real system,
(5) execute the simulation to predict the system behaviour
     in new steps,
(6) from the output results develop a closed feed-back loop
     and let the new output be input until the desired results
     are obtained.

This methodology is summarised in Fig. 9.6.

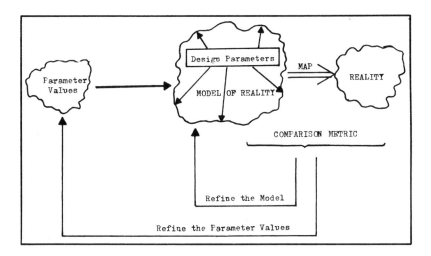

Fig. 9.6 Model representing the basic steps involved in
the application of simulation techniques.

The type of techniques involved in the simulation will depend critically
upon the nature of the process(es) involved. If they are deterministic
then conventional mathematical analysis can be applied. On the other
hand, if they are stochastic in nature then probability and statistical
analysis must be employed. Two types of simulation are used in practice.
That which is used will depend upon whether the system is controlled by
the influence of discrete events or the continuous variation of some
system variable. In discrete event simulation the occurrence of a
particular event is used to cause a change in the value of some attribute
of the system. It may start or stop an activity or create/destroy an
entity of the system (such as an excited state or radioactive particle).
In contrast, models in which variables show continuous variation require
the use of techniques for handling differential or difference equations.
These are used to describe the rate of change of given processes and,
during the simulation, need to be solved in order to effect a simulation
of the system under study.

Although analogue computers are still used for simulation studies, the
majority of simulations are nowadays conducted on digital machines. It is
possible to formulate models and solve simulation problems using
conventional programming languages. However, the task is made much
simpler by the wide range of simulation packages and special purpose

simulation languages that are now available on most large computers. Some
of these are listed in Table 9.4, further details will be found in the
references cited.

TABLE 9.4 Some Popular Simulation Packages

GPSS   - General Purpose Simulation System   (Sch74)
               (Package for Discrete Simulation)

SIMSCRIPT   - special simulation language   (Mar63)

SIMULA   - programming language with enhanced
               facilities for simulation   (SIM73)

HOCUS   - discrete event simulation package   (Poo77)

DYNAMO   - continuous, closed loop information-
               feedback systems   (Pug76)

CSMP   - Continuous System Modelling Program   (IBM72)

The use of these special purpose simulation languages/packages is to be
recommended if they are within easy access. If not, a conventional
programming language such as FORTRAN or PASCAL would need to be used.
This latter approach is illustrated in a paper by Lindsay et al (Lin80)
who have written a FORTRAN program to simulate in vivo chronoamperometry.
The motivation for the construction of this program arose from the
researchers' need to be able to better understand the relationship between
observed responses and experimental variables in experiments involving
electrical and chemical (via drugs) stimulation of the brain of a rat.
Many other examples of simulation programs written in FORTRAN will be
found in the literature. SIMULA is an example of a language that is more
orientated towards the writing of programs designed for the solution of
simulation problems; its use is illustrated in a paper by Sandblad
(San80). Illustrations of the application of simulation packages will be
found in articles by Miner (Min80) and Shah (Sha77).

A useful introduction to simulation techniques will be found in the book
by Tocher (Toc63). Interesting applications of the various methods
available will be found throughout the literature of analytical
chemistry. Monte Carlo techniques are particularly popular and well known
- applications in the area of chemical kinetics have been described by
Manock (Man73). The many advantages of simulation studies have been
indicated in papers that originate from most of the authors that use the
technique. Thus, de Jong et al (Jon80) have illustrated the value of
computer simulation in their studies of the factors that influence the
optimum conditions in preparative scale high performance liquid chromatography - here elution functions were obtained by computer simulation of
the chromatographic transport as column plate number was varied.
Similarly, Dillard et al (Dil76, Dil77, Dil79) have extensively studied
the application of simulation to differential pulse polarography; the
simulations enable the prediction of polarographic peak shape and position
as a function of controlled experimental variables such as pulse height,
pulse time and drop time. In this application the use of simulation
offers the analyst a means of selecting and optimising the experimental
conditions without having to resort to lengthy empirical studies.

When using simulation techniques it is important to remember that the results produced are only as good as the model will permit. The ability to simulate observed results is of little value if the basic model upon which the calculations are based is incorrect as a result of making invalid suppositions or approximations. Accuracy in model building is thus of prime significance.

Array Processing

A brief introduction to the topic of array processing was given in chapter 4 in the section on super-computers. There, it was suggested that computers capable of supporting this technique were ideally suited to data processing applications in which high degrees of parallelism could be exploited. The advantage of using such a facility lies mainly in the substantial reductions in computation time that can be achieved. Some typical results, reported by Pottle et al (Pot80) are shown in Table 9.5 below - the values on the left are for an array processor and those on the right for a conventional mainframe.

TABLE 9.5 Performance Comparison of Mainframe Computer
and Array Processor

	AP-120B/Prime 350	IBM 370/168
Energy evaluation	1.12 sec (AP)	12.7 sec
Gradient evaluation	190 sec (AP)	636.5 sec
Total CPU time	55 sec (Prime)	10 min, 52 sec
Total elapsed time	5 min	33 min
Cost of job	N/A	$118.64

These results were produced in benchmark studies in which energy levels and gradients for protein structures were calculated on each of the machines. Considerable time reduction can thus be achieved. Invariably, these reductions lead to a decrease in the cost of solving those types of problems to which the method can be applied. Furthermore, problems whose solution was previously not possible, because of the substantial computation times involved, become feasible with the advent of array processors. Two major areas where these techniques are likely to have substantial impact are pattern recognition (applied to robot vision, signal analysis and data processing of large data collections) and modelling/simulation.

Applications of Fuzzy Mathematics

Accuracy and precision are fundamental objectives within all branches of science. Despite this, many situations arise in which there are considerable degrees of uncertainty associated with events, situations and reasoning. Thus, humans are susceptible to being imprecise and often inaccurate in the way they reason, talk and make decisions. In order to handle 'real-life' situations of this sort there is growing interest in a branch of mathematics that relates to fuzzy (or 'ill-defined') concepts and which deals with topics such as fuzzy logic, fuzzy set theory, fuzzy algorithms and so on. There are many areas of science to which this branch of mathematics is applicable, particularly in pattern recognition and information retrieval, wherein precise specification or definitions of concepts are difficult to formulate. Fuzzy mathematics provides appropriate tools for dealing with these types of situation.

While classical mathematics is well suited to mechanistic systems that do not involve people, the great complexity associated with humanistic systems requires approaches that are significantly different; fuzzy mathematics is believed to be of significant value for the treatment of such systems. Because a single individual and his/her thought processes may be regarded as a humanistic system, the scientist and scientific activity are open to analysis by these methods. Similarly, studies of scientific communities and the impact of science on society each represent examples of humanistic systems that could be modelled using the methods embodied in this branch of mathematics. There is a vast literature associated with studies in this area. The papers by Zadeh (Zad76) and Gaines (Gai75) offer reasonable introductions to the subject.

Pattern Recognition

In 1973, Isenhour and Jurs (Ise73) wrote:

'The interpretation of experimental data, and the corresponding establishment of cause and effect relations, is an essential aspect of experimental chemistry. Perhaps the interpretation of experimental data is the most pressing current problem in analytical chemistry. The advent of the high speed digital computer along with instrumentation that produces massive volumes of data are making information theory and related topics more and more essential to modern chemistry.'

To these ends, automatic pattern recognition techniques may be of substantial value to the analytical chemist. In view of their importance, some of the approaches used will be briefly outlined in the remaining part of this chapter.

When experimental data and observations are analysed and interpreted, two basic methods are commonly used: theoretical and empirical. Application of the theoretical approach involves attempting to explain the observed results in terms of explicit causal relations derived from previous experiments or from logically constructed models. In contrast, empirical methods do not depend upon established theory and seek to look for relationships between data values, from which theories may be deduced. Most pattern recognition techniques fall into the latter category.

Some of the commonest approaches to pattern recognition are embodied in a set of analytical procedures known collectively as Cluster Analysis - sometimes referred to an 'unsupervised pattern recognition'. The goals of cluster analysis are simply to separate a set of data items into groups or clusters (Dub80). Everitt (Eve74) formulates the problem as follows: given a sample of N objects, each of which is measured on p variables, devise a classification scheme for grouping the objects into g classes, the number of classes and the characteristics of the classes to be determined. Situations of this type often arise in analytical science. Analysts are frequently confronted with the problem of having to analyse large amounts of data - produced, perhaps, by some highly automated chemical analysis. Because of the large number of measurements involved or their complexity then, unless the data can be classified into more manageable groups, and the groups treated as units, the problem can become intractable. However, by reducing the information on the whole set of N observations to information about g groups (where g N) it may be possible to simplify the problem considerably and thereby give a more concise account of the results under consideration. There are many applications of cluster analysis - data reduction, model building, hypothesis testing, and so on. The books by Everitt (Eve74) and Tyron (Tyr70) offer useful introductions to the subject. Various computer program packages are available which implement a variety of clustering algorithms. The most

well known of these is probably the CLUSTAN package (Wis78). This system
was originally developed in the 1960s for the collective study of
different cluster analysis methods. Since then it has become established
in a large number of research centres as the main tool for helping to
solve classification problems.

Another approach to automatic pattern recognition is the 'learning
machine' approach. Chemical applications of this method have been
described by Isenhour and Jurs (Ise73, Jur75). The learning machine
method offers an empirical approach to data interpretation in which the
decision making process used to classify the data improves as a result of
experience. This is achieved through the application of negative feedback
which causes the decision process to be modified so as to discriminate
against wrong answers - thereby improving its performance with time. In
most automatic pattern recognition processes there are essentially four
logical steps involved:

     (1)   measurement

     (2)   preprocessing

     (3)   feature selection

                                   Feedback

     (4)   classification

The first phase, the measurement step, is essentially a data collection
operation in which relevant observations are collected together into an
ensemble of scalar measurements comprising an n-tuple called the pattern
vector:

$$X = x_1, x_2, x_3, x_4, \ldots\ldots x_n$$

in which each element in the vector represents a physically measurable
quantity. In the preprocessing step the measurements are manipulated in
various ways by the application of mathematical transforms that are
designed to minimise the irrelevant information in the raw data while
preserving sufficient information to allow discrimination among the
pattern classes. Often the transformations may be applied in ways that
enhance those features which are likely to be most useful in the classification of the unknowns. Sometimes, the transforms produce new features,
for example, by multiplying each element in the pattern by a weighting
factor or by taking linear combinations of the original measurements.
Alternatively, the pattern vectors may be subject to principle component
analysis (Karhunen-Loëve - see Men70) in order to reduce the dimensionality of the data, Fourier transformation or the Hardamard transform.
Once the preprocessing step has been performed the next stage involves
selecting those features that are likely to be most useful for classification. The cost of classifying can be reduced if the minimum number of
features are used, hence it is important to eliminate as many features as
possible without adversely affecting the classification performance. The
transformed patterns are classified in the final stage of the pattern
recognition process. Here, the classifier is used to assign data to
classes based upon the application of some decision rule. The classifications are usually always made by considering the location of the
patterns in a hyperspace formed by using each of the features as an axis
(Bye80). Two of the most popular classifiers for chemical applications
are the linear discriminant rule (Jur75) and the k-nearest neighbour
classification (Jur75, Bye80).

The pattern recognition process is usually carried out on a set of known samples called the training set. The true identity of each pattern is then compared to the identity assigned in the classsification step and the results fed back into one or more of the previous steps for error correction. This may take place totally automatically or it may also allow for input from the analyst. In contrast to the cluster analysis methods outlined earlier, this technique of classification is called 'supervised learning'. The performance of a learning machine is reflected in its ability to correctly classify members of the training set. Four properties are often used to assess performance: (1) recognition - the ability of a discriminant function to correctly classify those patterns with which it was developed, (2) convergence rate - the speed with which the training algorithm converges towards 100% recognition, (3) reliability - the ability of the classifier to classify correctly members of the training set which have been distorted, and, (4) prediction - which refers to the ability of the classifier to correctly classify patterns which were not members of the training set.

The amount of literature available on pattern recognition is growing continually. In addition to the sources cited previously, further details on the general techniques and approaches to the subject will be found in books by Mendel (Men70) and Batchelor (Bat74, Bat78). Specific applications within chemistry are dealt with by Varmuza (Var80) who presents a comprehensive description and bibliography. Some of the more popular areas of interest include gas chromatographic analysis (Cla79, Zie79), spectrometric analysis (Sim77), electrochemistry (Per79a, Per79b) and acoustic emission studies (Bet80). Kryger has produced a concise overview of the principles involved in pattern recognition (Kry80) and a survey of techniques used in analytical chemistry (Kry81). These serve as useful introductions to the subject.

CONCLUSION

Data processing in its many varied forms is of paramount importance in the treatment of analytical measurements. The types of processing used and the extent to which it is employed will depend upon the nature of the analytical investigation and the kind of results that are produced.

Within the confines of this chapter it would have been impossible to give detailed descriptions of each of the different approaches that are involved. Consequently, a significant number of citations are presented in the reference section so that the details of particular methods can be found either by consulting an appropriate research paper or by reference to one of the authoritative text books dealing with the topic concerned.

Undoubtedly, as computing machines become much cheaper and the function per unit cost of various electronic calculating devices falls, so there is likely to be a substantial increase in gadgets and bench-top tools designed for use in conjuntion with particular types of analysis. Similarly, as microprocessors and substantial amounts of electronic memory are added to instruments, two significant trends are likely to take place.

Firstly, in many instruments the DP function is likely to become 'built-in'. This will be achieved by means of a microprocessor that has access to all the necessary processing programs contained within its memory - the programs being stored within a data base that is an integral part of the machine. Selection of a processing method will be via touch button or via light pen interaction.

Secondly, in laboratories where there is a central laboratory minicomputer and suitable connections to individual computerised instruments, then it is feasible to store computational/control methods in a centralised data base held on the minicomputer. Appropriate segments of the data base can then be transmitted to particular instruments in order to process data for specific experiments and analyses. Once processing is complete the final results may be transmitted from the instrument back to the central minicomputer for subsequent result reporting and archival. As will be discussed in chapter 12 the technology to achieve this type of processing now exists and is rapidly starting to appear within many laboratories.

REFERENCES

Adb74   Adby, P.R. and Dempster, M.A.H., Introduction to Optimisation Methods, Chapman and Hall, London, ISBN: 0-412-11040-7, 1974.

Alv77   Alvey, N.G., et al., GENSTAT - A General Statistical Program, Release 4.01, October, 1977. Available from: The Program Secretary, Statistics Department, Rothampstead Experimental Station, Harpenden, Hertfordshire, UK.

Ana80   Analytical Proceedings of the Chemical Society, A series of articles on "New Numerical Methods, Optimisation and Pattern Recognition", Volume 17, No. 4, 120-138, 1980.

ANS77   American National Standards Insitute, Proposed American National Standard Programming Language Minimal BASIC - ANSI BSR X3.60, American National Standards Institue, New York, 1977.

Aug79   Augenstein, M. and Tenenbaum, A., Data Structures and PL/I Programming Prentice-Hall Inc., ISBN: 0-13-197731-8, 1979.

Bai69   Baird, N.C. and Dewar, M.J.S., Ground States of Bonded Molecules. VI Benzene and its Isomers. Inductive Effects in Benzene, Journal of the Americaal Chemical Society, Volume 91, 352-355, January 15th, 1969.

Baj75   Bajpai, A.C., Calus, I.M. and Fairley, J.A., Numerical Methods for Engineers and Scientists: A Student's Coursebook, Taylor and Francis, ISBN: 0-471-99542-8, 1975.

Baj78   Bajpai, A.C., Calus, I.M. and Fairley, J.A., Statistical Methods for Engineers and Scientists, John Wiley, Chichester, ISBN: 0-471-996-440, 1978.

Bat74   Batchelor, B.G., Practical Approach to Pattern Classification, Plenum Press, ISBN: 0-306-30796-0, London, 1974.

Bat78   Batchelor, B.G., (Ed.), Pattern Recognition Ideas in Practice, Plenum Press, ISBN: 0-306-65274-2, 1978.

Bet80   Betteridge, D., Lilley, T. and Cudby, M.E.A., Acoustic Emissions
             from Polymers. Part II. Use of Pattern Recognition Methods,
             Analytical Proceedings of the Chemical Society, Volume 17, No.
             10, 434-436, October 1980.

Bev69   Bevington, P.R., Data Reduction and Error Analysis for the
             Physical Sciences, McGraw-Hill, New York, 1969.

Bil77   Biller, J.E., Herlihy, W.C. and Biemann, K., Identification of
             the Components of Complex Mixtures by GC-MS, Chapter 2, 18-25,
             in Computer Assisted Structure Elucidation, (Ed. Smith, D.H.),
             ACS Symposium Series No. 54, ISBN: 0-8412-0384-9, 1977.

Bin80   Binkley, D.P. and Dessey, R.E., Linear Parameter Estimation of
             Fused Peak Systems in the Spatial Frequency Domain, Analytical
             Chemistry, Volume 52, No. 8, 1335-1344, July 1980.

Bob80   Bobst, A.M., Sinka, T.K. and Langemeier, P.W., Digital Analysis
             of ESR Spectra of Spin Labelled Nucleic Acid Systems, Computers
             and Chemistry, Volume 4, No. 1, 45-50, 1980.

Bri79   Brissey, G.F., Spencer, R.B. and Wilkins, C.L., High Speed
             Algorithm for Simplex Optimisation Calculations, Analytical
             Chemistry, Volume 51, No. 13, 2295-2297, November 1979.

Bro79   Brown, P.J., Writing Interative Compilers and Interpreters,
             John Wiley, ISBN: 0-471-27609-X, 1979.

Bro80   Brown, J.R., Saba, C.S., Rhine, W.E. and Eisentraut, K.J.,
             Particle Size Independent Spectrometric Determination of Wear
             Metals in Aircraft Lubricating Oils, Analytical Chemistry,
             Volume 52, No. 14, 2365-2370, December 1980.

Bye80   Byers, W.D. and Perone, S.P., k-Nearest Neighbour Rule in
             Weighting Measurements for Pattern Recognition, Analytical
             Chemisty, Volume 52, No. 13, 2173-2177, November 1980.

Cal80   Calus, I.M., Review of Statistical Methodology Applied to the
             Results of Chemical Analysis, Analytical Proceedings of the
             Chemical Society, Volume 17, No. 4, 120-123, April 1980.

Cam79   Camp, R.C., Smay, T.A. and Triska, C.J., Microprocessor Systems
             Engineering, Matrix Publishers, Inc., Portland, Oregon, ISBN:
             0-916460-26-6, 1979.

Cha78   Chapman, J.R., Computers in Mass Spectrometry, Academic Press,
             ISBN: 0-12-168750-3, 1978.

Cha80   Chamberlin, J.D., Cumulative Sum Techniques, Analytical
             Proceedings of the Chemical Society, Volume 17, No. 5, 172-176,
             May 1980.

Che80   Cheng, H.Y., White, W. and Adams, R.N., Microprocessor
             Controlled Apparatus for In Vivo Electrochemical Measurement,
             Analytical Chemistry, Volume 52, No. 14, 2445-2448, December
             1980.

Chu78   Chuang, F.S., Natusch, D.F.S. and O'Keefe, K.R., Evaluation of a
           Self-scanned Photodiode Array Spectrometer for Flame Absorption
           Measurements, Analytical Chemistry, Volume 50, 525-530, 1978.

Cla79   Clark, H.A. and Jurs, P.C., Classification of Crude Oil Gas
           Chromatograms by Pattern Recognition Techniques, Analytical
           Chemistry, Volume 51, No. 6, 616-623, May 1979.

Cla80a   Clark, D.D. and Schuster, S.M., Microcomputer Manipulation and
           Graphic Display of Molecular Structures - I. Introduction,
           Computers and Chemistry, Volume 4, No. 2, 75-78, 1980.

Cla80b   Clark, D.D. and Schuster, S.M., Microcomputer Manipulation and
           Graphic Display of Molecular Structures - II. File Structure and
           Maintenance, Computers and Chemistry, Volume 4, No. 2, 79-82,
           1980.

Cla80c   Clark, D.D. and Schuster, S.M., Microcomputer Manuiplation and
           Graphic Display of Molecular Structures - III. Display,
           Computers and Chemistry, Volume 4, No. 2, 83-85, 1980.

Cod80   Codding, E.G., Ingle, J.D. and Stratton, A.J., Atomic Absorption
           Spectrometry with a Photodiode Array Spectrometer, Analytical
           Chemistry, Volume 52, No. 13, 2133-2140, November 1980.

Day72   Day, A.C., Fortran Techniques With Special Reference to Non-
           numerical Applications, Cambridge University Press, ISBN:
           0-521-08549-7, 1972.

Dee77   Deen, S.M., Fundamentals of Data Base Systems, Macmillan
           Press, ISBN: 0-333-19739-9, 1977.

Del79   Delaney, M.F. and Uden, P.C., Statistical Prediction of File
           Searching Results for Vapour Phase Infrared Spectrometric
           Identification of Gas Chromatographic Peaks, Analytical
           Chemistry, Volume 51, No. 8, 1242-1249, July 1979.

Dil76   Dillard, J.W. and Hanck, K.W., Digital Simulation of
           Differential Pulse Polarography, Analytical Chemistry, Volume
           48, No. 1, 218-222, January 1976.

Dil77   Dillard, J.W., Turner, J.A. and Osteryoung, R.A., Digital
           Simulation of Differential Pulse Polarography with Incremental
           Time Change, Analytical Chemistry, Volume 49, No. 8,
           1246-1250, July 1977.

Dil79   Dillard, J.W., O'Dea, J.J. and Osteryoung, R.A., Analytical
           Implications of Differential Pulse Polarography of Irreversible
           Reactions from Digital Simulation, Analytical Chemistry,
           Volume 51, No. 1, 115-119, January 1979.

Dix79   Dixon, W.J. and Brown, M.B., BMDP - Biomedical Computer
           Programs, University of California Press, Berkeley, California,
           ISBN: 520-03569-0, 1979.

Dow79   Dowd, G.F. and Hilborn, J.C., Improved Accuracy in Atomic
           Absorption Analysis by Optimisation of Cell Parameters,
           Analytical Chemistry, Volume 51, No. 9, 1578-1580, August
           1979.

Dru80   Drummer, D.M. and Morrison, G.H., Digital Image Processing for
           Image Quantification in Ion Microscope Analysis, Analytical
           Chemistry, Volume 52, 2147-2153, 1980.

Dub80   Dubes, R. and Jain, A.K., Clustering Methodologies in
           Exploratory Data Analysis, 113-228, in Volume 19 of Advances in
           Computers, Ed: Yovitts, M.C., Academic Press Inc., ISBN:
           0-12-012119-0, 1980.

Eve74   Everitt, B., Cluster Analysis, Heinemann, ISBN: 0-435-822-977,
           1977.

Ern68   Ernst, R.R., Measurement and Control of Magnetic Field
           Homogeneity, Review of Scientific Instruments, Volume 39,
           998-1012, 1968.

Fox76   Fox, D.J. and Guire, K.E., Documentation for MIDAS,
           Statistical Research Laboratory, The University of Michigan, Ann
           Arbor, USA, 3rd Edition, September 1976.

Gai75   Gaines, B.R., Multivalued Logics and Fuzzy Reasoning, 26-38, in
           Proc. of the BCS - AISB Summer School, Cambridge, UK, 1975.

Gan80   Gans, P., Improved Methods for Numerical Differentiation of
           Spectroscopic Curves, Analytical Proceedings of the Chemical
           Society, Volume 17, No. 4, 113-135, April 1980.

Ghe80   Ghersini, G., Ghini, M. and Salghetti, F., Systematic Approach
           to the Control of Accuracy and Precision in Routine Analysis,
           Analytical Proceedings of the Chemical Society, Volue 17, No.
           5, 197-200, May 1980.

Gra80   Granéli, A., Jagner, D. and Josefson, M., Microcomputer System
           for Potentiometric Stripping Analysis, Analytical Chemistry,
           Volume 52, No. 13, 2220-2223, November 1980.

Gri71   Gries, D., Compiler Construction for Digital Computers, John
           Wiley, ISBN: 0-471-32771-9, 1971.

Hal75   Hall, P.A.V., Computational Structures: An Introduction to Non-
           numerical Computing, Elsevier North-Holland Inc., New York,
           ISBN: 0-444-19522-X, 1975.

Hol80   Holt, M.J.J., Norris, A.C., Pope, M.I. and Selwood, M.,
           Numerical Optimisation of Kinetic Parameters by Unidirectional
           Search, Analytical Proceedings of the Chemical Society, Volume
           17, No. 4, 127-131, April 1980.

IBM67   IBM Corporation, System/360 Scientific Subroutine Package,
           (360A-CM-03X), Version III, Programmer's Manual, Form:
           GH20-205, 1967.

IBM68   IBM Corporation, System/360 Scientific Subroutine Package,
           (360A-CM-07X), Program Description and Operations Manual, Form:
           GH20-0586-0, 1968.

IBM69   IBM Corporation, Introduction to the List Processing Facilities
           of PL/I, Form No: GF20-0015-0, 1969.

IBM72   IBM Corporation, System/360 Continuous System Modelling Progam
          III (CSMP III): Program Reference Manual, Form: SH19-7001,
          1972.

Ise73   Isenhour, T.L. and Jurs, P.C., Learning Machines, Chapter 8,
          285-330, in Volume 1 of Computers in Chemistry and
          Instrumentation (Computer Fundamentals for Chemists), (Eds,
          Mattson, J.E., Mark, H.B. and MacDonald, H.C.), Marcel Dekker,
          New York, ISBN: 0-8247-1432-6, 1973.

Jab79   Jablonski, B.B., Wegscheider, W. and Leyden, D.E., Evaluation of
          Computer Directed Optimisation for Energy Dispersive X-ray
          Spectrometry, Analytical Chemistry, Volume 51, No. 14,
          2359-2364, December 1979.

Jef80   Jeffery, G.A. and Taylor, R., The Application of Molecular
          Mechanics to the Structure of Carbohydrates, Journal of
          Computational Chemistry, Volume 1, No. 1 99-109, Spring 1980.

Joh80   Johnson, K.J., Numerical Methods in Chemistry, Marcel Dekker,
          Inc., 270 Maddison Avenue, New York, NY 10016, 1980.

Jon80   de Jong, A.W.T., Smit, J.C., Poppe, H. and Kraak, J.C., Optimum
          Conditions for High Performance Liquid Chromatography on the
          Preparative Scale, Analytical Proceedings of the Chemical
          Society, Volume 17, No. 12, 508-512, December 1980.

Jor79   Jordan, D.C. and de Alvare, L.R., Maximum Likelihood Estimation
          Evaluation of a Material with Unequal Number of Replicates,
          Analytical Chemistry, Volume 51, No. 7, 1079-1080, June 1979.

Jur75   Jurs, P.C. and Isenhour, T.L., Chemical Applications of Pattern
          Recognition, John Wiley, ISBN: 0-471-45330-7, 1975.

Kat78   Katzan, H., FORTRAN 77, Van Nostrand Reinhold Company, New
          York, ISBN: 0-8144-5462-3, 1978.

Kel79   Kelter, P.B. and Carr, J.D., Limitations of the Normal Variate
          as a Statistical Evaluation of Kinetic Data, Analytical
          Chemistry, Volume 51, No. 11, 1857, September 1979.

Kir80   Kirchner, S.J., Oona, H., Perron, S.J., Fernando, Q., Lee, J.J.
          and Zeitlin, H., Proton Induced X-ray Emission Analysis of Deep
          Sea Ferromanganese Nodules, Analytical Chemistry, Volume 52,
          No. 13, 2195-2201, November 1980.

Kow68   Kowalik, J. and Osborne, M.R., Methods for Unconstrained
          Optimisation Problems, Elsevier, New York, 1968.

Knu73   Knuth, D.E., The Art of Computer Programming Volume 3: Sorting
          and Searching, Addison-Wesley Publishing Company, ISBN: 0-201-
          03803-X, 1973.

Kry80   Kryger, L., Computational Practice in Pattern Recognition,
          Analytical Proceedings of the Chemical Society, Volume 17,
          No.4, 135-138, April 1980.

Kry81   Kryger, L., Interpretation of Analytical Chemical Information by
         Pattern Recognition Methods, A Survey, Talanta, in press, 1981

Leg77   Leggett, D.J., Numerical Analysis of Multicomponent Spectra,
         Analytical Chemistry, Volume 49, No. 2, 276-281, February
         1977.

Lei78   Leiva, W., Quint, J., Selesky, M. and Ulrich, W., Data
         Reduction and Quality Control in the RIA Laboratory, 53 slides
         (49 min) + 31 page printed text, SAVANT, PO Box 3670, Fullerton,
         Calif. 92634, 1978.

Lin80   Lindsay, W.S., Justice, J.B. and Salamone, J., Simulation
         Studies of In Vivo Electrochemistry, Computers and Chemistry,
         Volume 4, No. 1, 19-26, 1980.

Lor75   Lorin, H., Sorting and Sort Systems, Addison-Wesley Publishing
         Company, ISBN: 0-201-14453-0, 1975.

Lun79   Lundgren, J.O., Crystallographic Computer Programs, Report
         UUIC-B13-4-04, Institute of Chemistry, University of Uppsala,
         Sweden, 1979,

Mai78   Main, P., Hull, S.E., Lessinger, L., Germain, G., Declercq, J.P.
         and Woolfson, N.M., MULTAN78 - A System of Computer Programs
         for the Automatic Solution of Crystal Structures from X-ray
         Diffraction Data, University of York, England and Louvain,
         Belgium, 1978.

Man73   Manock, J.J., Application of Monte Carlo Techniques to Chemical
         Kinetics, Chapter 5, 267-291, in Volume 3 of Computers In
         Chemistry and Instrumentation (Spectroscopy and Kinetics),
         (Eds. Mattson, J.S., Mark, H.B. and MacDonald, H.C.), Marcel
         Dekker, Inc., ISBN: 0-8247-6058-1, 1973.

Mar63   Markowitz, H., Hausner, B. and Kan, H., SIMSCRIPT: A Simulation
         Programming Language, Prentice-Hall, Englewood Cliffs, 1963.

Mar79   van Marlen, G., Dijkstra, A. and van't Klooster, H.A., Influence
         of Errors and Matching Criteria Upon the Retrieval of Binary
         Coded Low Resolution Mass Spectra, Analytical Chemistry,
         Volume 51, No. 3, 420-423, March 1979.

Mar80   Martin, D., Data Base Design and Implementation on Maxi and
         Mini Computers, Van Nostrand Reinhold Company, ISBN:
         0-442-30430-7, 1980.

Men70   Mendel, J.M. and Fu, K.S., Adaptive Learning and Pattern
         Recognition Systems, Academic Press, New York, 1970.

Mez80   Mezey, P.G., Strausz, O.P. and Gasavi, R.K., A Note on Density
         Matrix Extrapolation and Multiple Solutions of the Unrestricted
         Hartree-Fock Equations, Journal of Computational Chemistry,
         Volume 1, No. 2, 178-180, Summer 1980.

Min80   Miner, R.J., Wortman, D.B. and Cascio, D., Improving the
         Throughput of a Chemical Plant, SIMULATION, Volume 35, No. 4,
         125-132, October 1980.

Mon74   Monro, D.M., Interactive Computing with BASIC - A First
         Course, Edward Arnold, ISBN: 0-7131-2488-1, 1974.

Mon78   Monro, D.M., Basic BASIC - An Introduction to Programming,
         Edward Arnold, ISBN: 0-85012-233-3, National Computing Centre
         Publications, United Kingdom, 1978.

Moo80   Moore, B.A., Correlation and Regression Analysis: Applications
         to the Analysis of Chemical Data, Analytical Proceedings of the
         Chemical Society, Volume 17, No. 4, 124-127, April 1980.

NAG78   Numerical Algorithms Group, NAG Fortran Library Manual, Mark
         7, (FLM7), Volume 1, 1978.

Nie70   Nie, N., Bent, D.H. and Hull, C.H., Statistical Package for the
         Social Sciences, McGraw-Hill Book Company, ISBN: 07-046530-4,
         1970.

New73   Newman, L., The Use of the Large Computer in Chemical Instrumen-
         tation - Application to Magnetic Resonance Spectroscopy, Chapter
         1, 3-41, in Computers in Chemistry and Instrumentation, Volume
         3, Spectroscopy and Kinetics (Eds. Mattson, J.S., Mark, H.B. and
         MacDonald, H.C.), Marcel Dekker Inc., ISBN: 0-8247-6058-1, 1973.

Per79a   DePalma, R.A. and Perone, S.P., On-line Pattern Recognition of
         Voltammetric Data: Peak Multiplicity Classificiation,
         Analytical Chemistry, Volume 51, No. 7, 825-828, June 1979.

Per79b   DePalma, R.A. and Perone, S.P., Characterisation of Heterogene-
         ous Kinetic Parameters from Voltammetric Data by Computerised
         Pattern Recognition, Analytical Chemistry, Volume 51, No. 7,
         829-832, June 1979.

Pet73   Peterson, N.C. and Butcher, H.J., Integration of Complex Rate
         Equations using Infinite Series, Chapter 6, 293-313, in Volume 3
         of Computers in Chemistry and Instrumentation (Spectroscopy and
         Kinetics), Eds. Mattson, J.S., Mark, H.B. and MacDonald, H.C.,
         Marcel Dekker, Inc., ISBN: 0-8247-6058-1, 1973.

Pol77   Pollard, J.H., A Handbook of Numerical and Statistical Tech-
         niques, Cambridge University Press, ISBN: 0-521-21440-8, 1977.

Poo77   Poole, T. and Szymankiewicz, J., Using Simulation to Solve
         Problems, McGraw-Hill Book Company, ISBN: 0-07-084472-0, 1977.

Pot80   Pottle, C., Pottle, M.S., Tuttle, R.W., Kinch, R.J. and
         Scheraga, H.A., Confirmational Analysis of Proteins: Algorithms
         and Data Structures for Array processing, Journal of Computa-
         tional Chemistry, Volume 1, No. 1, 46-58, Spring 1980.

Pro80   Proctor, A. and Sherwood, P.M.A., Smoothing of Digital X-ray
         Photoelectron Spectra by an Extended Sliding Least-squares
         Approach, Analytical Chemistry, Volume 52, No. 14, 2315-2321,
         December 1980.

Pug76   Pugh, A.L., DYNAMO User's Manual, (5th Edition), MIT Press,
         Cambridge, Massachusetts, ISBN: 0-262-66029-6, 1976.

Pyl72   Pyle, I.C., Hooton, I.N. and MacDonald, R.J., (Eds), Computing
          With Real Time Systems - Volume 1, Proceedings of the First
          European Seminar, AERE, Harwell, Transcripta Books, ISBN:
          0-903012-02-2, 1972.

Qua81   Quantum Chemistry Program Exchange, Room 204, Chemistry Dept.,
          University of Indiana, Bloomington, Indiana 47401, USA.

Ran80   Randic, M., Brissey, G.M., Spencer, R.S. and Wilkins, C.L., Use
          of Self-avoiding Paths for Characterision of Molecular Graphs
          With Multiple Bonds, Computers and Chemistry, Volume 4, No. 1,
          27-43, and Volume 4, No. 2, 101-102, 1980.

Rec80   Rechberger, P. and Linert, W., Analysing Electrolytic
          Conductivity Data with a Programmable Calculator, Computers and
          Chemistry, Volume 4, No. 2, 61-68, 1980.

Ree80   Reese, C.E., Chromatographic Data Acquisition and Processing,
          Part I - Data Acquistion, Journal of Chromatographic Science,
          Volume 18, 201-206, May 1980 - (Part II - to be published).

Ric79   Richards, J.A. and Griffiths, A.G., On Confidence in the Results
          of Learning Machines Trained on Mass Spectra, Analytical
          Chemistry, Volume 51, No. 9, 1358-1361, August 1979.

Rub79   Rubin, I.B. and Bayne, C.K., Statistical Designs for the
          Optimisation of the Nitrogen - Phosphorus Gas Chromatographic
          Detector Response, Analytical Chemistry, Volume 51, No. 4,
          541-546, April 1979.

Rya80   Ryan, P.B., Barr, R.L. and Todd, H.D., Simplex Techniques for
          Non-linear Optimisation, Analytical Chemistry, Volume 52, No.
          9, 1460-1467, August 1980.

San80   Sandblad, B., Decision Criteria for the Selection of Analytical
          Instruments: V. The Interaction of New Instrumentation with
          Laboratory Infra-structure: Modelling and Simulation for
          Planning of Laboratory Functions, Journal of Automatic
          Chemistry, Volume 2, No. 1, 28-31, January 1980.

Sas80   Sasaki, S., Fujino, K., Takeuchi, Y. and Sadanaga, R., On the
          Estimation of Atomic Charges by the X-ray Method for Some Oxides
          and Silicates, Acta Crystallographica, Volume A36, Part 6,
          904-915, November 1980.

Sav64   Savitsky, A. and Golay, M.J.E., Smoothing and Differentiation of
          Data by Simplified Least Squares Procedures, Analytical
          Chemistry, Volume 36, No. 8, 1627-1634, July 1964.

Sch74   Schriber, T.J., Simulation Using GPSS, John Wiley, ISBN:
          0-471-76310-1, 1974.

Sch79   Schwartz, L.M., Calibration Curves With Non-Uniform Variance,
          Analytical Chemistry, Volume 51, No. 6, 723-727, May 1979.

Sch80   Schaumberg, K., Wasniewski, J. and Zlatev, Z., The Use of Sparse Matrix Techniques in the Numerical Integration of Stiff Systems of Linear Ordinary Differential Equations, Computers and Chemistry, Volume 4, No. 1, 1-12, 1980.

Sci78   FFT - A New Rapid Method for Fourier Transforms, SCIENCE, Volume 199, No. 4335, 1326-1327, 24th March 1978.

See80   Seelig, P.F. and Levie, R., Double Layer Capacitance Measurements with Digital Synchronous Detection at a Dropping Mercury Electrode, Analytical Chemistry, Volume 52, No. 9, 1506-1511, August 1980.

Sha77   Shah, S.N., A CSMP Example for Chemical Engineering, SIMULATION, Volume 29, No. 3, 88-92, September 1977.

SIM73   SIMULA - User's Guide - IBM System/360, Norwegian Computer Centre, 1973

Sim77   Simon, P.J., Griessen, B.C. and Copeland, T.R., Categorisation of Papers by Trace Metal Content Using Atomic Absorption Spectrometric and Pattern Recognition Techniques, Analytical Chemistry, Volume 49, No. 14, 2285-2288, December 1977.

Spe75   Spencer, D.D., A Guide to BASIC Programming, Addison-Wesley, ISBN: 0-201-07106-1, 1975.

Spr77   Sprent, P., Statistics in Action, Penguin Books, ISBN: 0-1402-1955-2, 1977.

Ste70   Stewart, J.M., Kundall, F.A. and Baldwin, J.C., The X-Ray 70 System, Computer Science Centre,, University of Maryland, College Park, Maryland, USA, 1970.

Ste72   Steiner, J., Termonia, Y. and Deltour, J., Comments on Smoothing and Differentiation of Data by Simplified Least Squares Procedure, Analytical Chemistry, Volume 44, No. 11, 1906-1909, September 1972.

Str73   Strobel, H.A., Chemical Instrumentation: A Systematic Approach to Instrumental Analysis, Addison-Wesley, ISBN: 0-201-7301-3, 1973.

Str75   Struble, G.W., Assembler Language Programming: the IBM System/360 and 370, Addison-Wesley, ISBN: 0-201-07322-6, 1975.

Sur77   Surprenant, H.L. and Reilley, C.N., Computerised Structural Predictions from ^{13}C NMR Spectra, Chapter 6, 77-91, in Computer Assisted Structure Elucidation, (Ed. Smith, D.H.), ACS Symposium Series No. 54, ISBN: 0-8412-0384-9, 1977.

Swa80   Swann, W.H., Numerical Methods for Non-linear Optimisation, Analytical Proceedings of the Chemical Society, Volume 17, No. 4, 127-131, April 1980.

Toc63   Tocher, K.D., The Art of Simulation, English University Press, ISBN: 0-340-11452-5, 1963.

Tre76   Tremblay, J.P. and Sorenson, P.G., _An Introduction to Data
           Structures with Applications_, McGraw-Hill, ISBN: 0-07-065150-7,
           1976.

Tyr70   Tyron, R.C. and Bailey, D.E., _Cluster Analysis_, McGraw-Hill
           Book Company, 1970.

Van81   Vanderslice, J.T., Stewart, K.K., Rosenfeld, A.G. and Higgs,
           D.J., Laminar Dispersion in Flow Injection Analysis, _Talanta_,
           Volume 28, 11-18, 1981.

Var80   Varmuza, K., _Pattern Recognition in Chemistry_, Lecture Notes
           in Chemistry, Volume 21, Springer-Verlag, ISBN: 3-540-10273-6,
           Berlin-Heidelberg-New York, 1980.

Ven77   Venkataraghavan, R., Dayringer, H.E., Pesyna, G.M., Atwater,
           B.L., Mun, I.K., Cone, M.M. and McLafferty, F.W., Computer
           Assisted Structure Identification of Unknown Mass Spectra,
           Chapter 1, 1-17, in _Computer Assisted Structure Illucidation_,
           (Ed. Smith, D.H.), ACS Symposium Series No. 54, ISBN:
           0-8412-0384-9, 1977.

Wai73   Waite, W.M., _Implementing Software for Non-numeric Applica-
           tions_, Prentice-Hall Series in Automatic Computation, New York,
           ISBN: 0-134-518-985, 1973.

Wat79   Watson, M.W. and Carr, P.W., Simplex Algorithm for the
           Optimisation of Gradient Elution High Performance Liquid Chroma-
           tography, _Analytical Chemistry_, Volume 51, No. 9, 1835-1842,
           September 1979.

Web80   Weber, J., Lacroix, R. and Wanner, G., The Eigenvalue Problem in
           Configuration Interaction Calculations: A Computer Program Based
           on a New Derivation of the Algorithm of Davidson, _Computers and
           Chemistry_, Volume 4, No. 2, 55-60, 1980.

Wex75   Wexelblat, R.L., Programmed Control of Asynchronous Program
           Interrupts, 1-41, Volume 13, _Advances in Chemistry_, (Eds.
           Rubinoff, M. and Yovits, M.C.), Academic Press, ISBN: 0-12-
           012113-1, 1975.

Wil64   Wilde, D.J., _Optimum Seeking Methods_, Prentice-Hall,
           Englewood-Cliffs, New Jersey, 1964.

Win77   Winston, P.H., _Artificial Intelligence_, Addison-Wesley, ISBN:
           0-201-08454-6, 1977.

Wir78   Wirth, N. and Jensen, K., _PASCAL - User Report and Manual_,
           (2nd Edition), Springer-Verlag, New York, ISBN: 0-387-90144-2,
           1978.

Wis78   Wishart, D., _CLUSTAN User Manual_, (3rd Edition), Program
           Library Unit, Edinburgh University, January 1978.

Woo80   Woodruff, H.B. and Smith, G.M., Computer Program for the
           Analysis of Infrared Spectra, _Analytical Chemistry_, Volume 52,
           No. 14, 2321-2327, December 1980.

Yam77   Yamasaki, T., Abe, H., Kudo, Y. and Sasaki, F., CHEMICS: A
         Computer Program System for Structure Elucidation of Organic
         Compounds, Chapter 8, 108-125, in Computer Assisted Structure
         Elucidation, (Ed. Smith, D.H.), ACS Symposium Series No. 54,
         ISBN: 0-8412-0384-9, 1977.

Zad76   Zadeh, L.A., A Fuzzy Algorithmic Approach to the Definition of
         Complex or Imprecise Concepts, International Journal of Man-
         Machine Studies, Volume 8, 249-291, 1976.

Zie79   Ziemer, J.N., Perone, S.P., Caprioli, R.M. and Seifert, W.E.,
         Computerised Pattern Recognition Applied to Gas Chromatography/
         Mass Spectrometry Identification of Pentafluoropropionyl Dipep-
         tide Methyl Esters, Analytical Laboratory, Volume 51, No. 11,
         1732-1738, September 1979.

Zla80   Zlatev, Z., Wasniewski, J. and Schaumberg, K., Classification of
         the Systems of Ordinary Differential Equations and Practical
         Aspects in the Numerical Integration of Large Systems,
         Computers and Chemistry, Volume 4, No. 1, 13-18, 1980.

Zla80a   Zlatev, Z., Schaumberg, K. and Wasniewski, J., Implementation of
         an Iterative Refinement Option in a Code for Large Sparse
         Systems, Computers and Chemistry, Volume 4, No. 2, 87-99,
         1980.

10

Data Bases and Data Centres

INTRODUCTION

Data and Information

In most areas of modern science, business and technology there are two
words that are probably used more frequently than any others: <u>data</u> and
<u>information</u>. As a consequence of their widespread use the meaning asso-
ciated with each of these has tended to become somewhat blurred. Indeed,
in many cases, people use the words virtually interchangeably. A similar
criticism applies to the term <u>data base</u>. This is often applied, quite
liberally, as a means of describing a computer based collection of stored
material. It seems to matter little whether the basic commodity being
stored is data, information or knowledge. Occasionally, the term
<u>information base</u> appears in the literature but its usage is far less
frequent than data base. A related term <u>knowledge base</u> is sometimes
used in the context of Expert Systems (see chapter 2) but it is not as
often encountered. One of the objectives of this introduction is to
remove some of the ambiguity that surrounds terms like data, information
and data base.

Data is a basic commodity that is produced as a result of observations
made on some process or entity. The most common type of data is probably
numeric - sequences of numbers selected from an appropriate number system.
However, any symbol set could be used as a means of representing data;
indeed, non-numeric symbolisms are also quite commonplace. The advantages
of using numbers as a means of expressing data values lies in their
ability to express magnitude relationships between the basic entities they
describe. Chapter 9 emphasised the point that information is produced as
a result of applying mathematical transformations to raw data. Information is thus a more valuable commodity than the precursor from which it is
derived. Its value stems from its usefulness in decision making. Indeed,
a simple, commonly accepted definition for the term is: that which
enables decisions to be made. The difference between data and information
is now easy to see - it is summarised in the following equation:

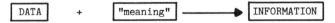

$$\boxed{\text{DATA}} \quad + \quad \boxed{\text{"meaning"}} \quad \longrightarrow \quad \boxed{\text{INFORMATION}}$$

As an illustration of this, consider a sequence of numbers such as (23, 18, 23). These could represent many things: a point in space, female statistics, a triangle, the dimensions of a box .... There is an infinity of possibilities. However, if some "meaning" is added to the data values by providing an appropriate context (say: triangles) then it is immediately possible to deduce that the numbers represent a triangle. More specifically, an equilateral triangle. Similarly, a mass spectrum (as depicted in Fig. 3.5) is essentially just a tabulation or graphic representation of data (as are all histograms) until the context 'mass spectrum' is added. Then, it starts to become information since it enables certain decisions to be made about the nature of the compound(s) that it represents. The relationship between data, information and decision making is summarised schematically in Fig. 10.1.

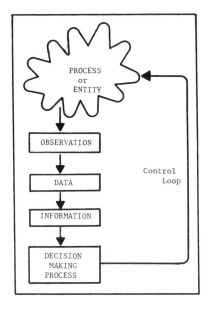

Fig. 10.1 Relationship between data and information.

Information Storage and Retrieval

Data and information are costly commodities since a considerable amount of research effort has to go into their generation. Because of the importance of information in decision making it becomes essential to be able to store it in such a way that it can be easily retrieved when it is required for use. Storage and retrieval operations are probably two of the most important types of information handling procedures in all branches of human activity. A variety of media may be used to store information: printed materials, microfilm, optical disk or any of the computer based storage facilities that were described in chapter 5 (data collection). The advantage of having data/information stored on the latter kinds of device lies in the fact that the computer may be used to automatically retrieve the information when it is required for use. It also adds a great deal of flexibility to the information in that it is very easy to convert it from one form to another. Furthermore, being in an electronic form it is easy to transmit between different locations, and so, is more easily shared across geographically remote centres. The electronic movement of information will be described in more detail in chapter 12 when computer networks are discussed. Another attractive feature of electronic storage is the compact way in which it permits information to be held and the speed with which it may be processed and retrieved. Once retrieved, various types of display devices are available to enable the information to be presented in attractive and appealing ways (as was outlined in chapter 2). The availability of appropriate document printing and word processing equipment can easily be used to convert the electronic information into its 'paper form' whenever the need arises.

There are two broad types of information retrieval with which most scientists are concerned: bibliographic and scientific. The former refers to information contained in books, journals, reports and other types of printed sources. Bibliographic information services will be described in the next chapter. This one will concentrate on the storage and retrieval of scientific information that relates to the physical and chemical properties of chemical substances, for example, their melting points, boiling points, infrared spectra, mass spectra and so on. Quite often the term 'Data Centre' is used to refer to those installations and laboratories that specialise in handling this type of information (for example, the Cambridge Crystallographic Data Centre, the Aldermaston Mass Spectral Data Centre, etc). Some of these will be discussed in more detail later.

Information Retrieval Systems

Information retrieval systems are primarily designed to help analyse and describe the items contained in an information store, to organise them and search among them, and finally to retrieve them in response to a user's query. In general, designing and using a retrieval system involves four major activities: information analysis, information organisation and search, query formulation and retrieval/dissemination of information. An introduction to the basic principles of information storage and retrieval systems has been given by Senko (Sen69). An early example of a simple chemical information retrieval system has been described by Lefkovitz and Powers (Lef66). This system, based upon the use of list-structures, permitted four basic types of query: molecular formula, structural formula, structural fragments and various descriptive keys. Since this early work a wide variety of 'chemical' information systems have been described in the literature. An attempt to broadly classify these is shown in Fig. 10.2.

In general, information retrieval systems may be manual or computer based. Deductive systems that employ artificial intelligence methods necessarily imply the use of some sort of computer based technique. These are closely related to the Expert Systems mentioned in chapter 2. Document retrieval tends to refer to printed bibliographic material such as books, reports, journals, and so on. Retrospective systems enable the user to retrieve documents that may have been produced at any stage during the history of a subject. Current awareness facilities attempt to alert the users to the availability of new material. When such a service is orientated towards particular individuals, groups or departments it is often referred to as Selective Dissemination of Information. Bibliographic information retrieval will probably be the most well known since the princples involved

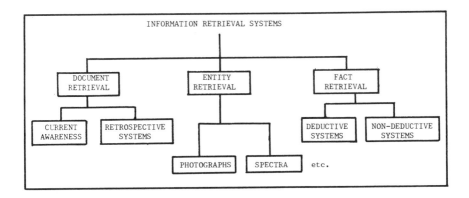

Fig. 10.2   Classification of information retrieval systems.

form the basis upon which many libraries operate. Entity retrieval has
much in common with conventional retrieval but applies to information that
may be stored on media different to that used for documents. Furthermore,
the information being stored may not necessarily be expressed in terms of
the printed word. Such systems include those methods used for retrieving
information stored on slides, filmstrips, photographs, sound recordings,
spectrometer traces, and so on. Often these can form the basis of many
specialised retrieval units. As was mentioned earlier, information
retrieval processes may be imple- mented via manual techniques or may
involve the use of computers. Undoubtedly, most of the modern systems
depend upon computing equipment in order to perform many of their
functions. An introduction to the use of computers in information
retrieval will be found in the books by Paice (Pai77) and Heaps (Hea78).
A discussion of recent advances in automatic information retrieval,
consequent upon current developments in specialised hardware systems and
new algorithms for text searching, has been given by Salton (Sal80).

Information retrieval systems were the fore-runner of the modern data
base. Indeed, even today, a large number of the commercially available
data bases give the appearance of being sophisticated information
retrieval systems to the majority of their users. However, as will be
discussed in more detail in the next section there is a significant
difference between the two.

Data Bases and Information Systems

A data base may be thought of as being composed of several components
which taken together provide:

   (a)   a time varying collection of data/information stored in
         such a way as to minimise any duplication,
   (b)   a means of controlling and administering access to the
         stored data, and,
   (c)   the provision of facilities to ensure the security,
         privacy and integrity of the stored data.

It is the nature of the additional access operations (such as updating, modification, deletion, sharing, etc) that provide the essential differences between a data base and an information retrieval (IR) set-up. Of course, by inhibiting or 'hiding' these additional access modes the data base can be made to look like an information retrieval system. A data base, however, is probably a more fundamental concept since it can often be used as a building block for the latter.

Modern IR technology provides the means of storing all types of information until its retrieval is required. It then provides techniques for retrieving the information. However, once the information has been stored, an IR system does not provide any facility to enable the user to update or modify it. A library exhibits many of the basic properties of a classical information retrieval system in that books are deposited in it and users then attempt to retrieve the books that they are interested in by means of author/subject indexes or by browsing. Users of the library are not at liberty to change the contents of the books in any way.

To refer to the library as just an information retrieval system is a vast over-simplification since it is very much more than this. Because of the wide variety of services it provides and the complex nature of the interactions in which it is involved, a library is more accurately described as an 'information system'. This more complex type of entity will be described in the next section. For the moment, however, it will be defined as being 'a complex network of interacting parts the function of which is to provide its users with the information they require in order to achieve their short term and long term goals'. Information systems vary widely in their sophistication and capabilities depending upon the nature of the function that they have to perform. Those that involve the use of computer systems invariably utilise data bases as fundamental building blocks.

Consider the example of the library again. Libraries that are based on the use of a computer usually use the available facilities to augment or replace many of the manual library operations. For example, the computer may be used to provide a computerised author index or subject index. As new books are added to the library, become lost, stolen or replaced, so the contents of these indexes need to be modified in order to reflect these time varying changes. The ability to cater for this type of situation is an ideal application for the data base. Similarly, there is no reason why all the details of the users of a library should not be held on yet another data base in such a way that records of the borrowing transactions are maintained. This example has been chosen to illustrate the fact that a complex information system, such as a library, may depend upon several quite independent data bases. Each of these may perform quite different information handling functions relevant to the overall management of the library system.

The Importance of Data and Scientific Information Collections

This section is concerned only with scientific data and information as produced by the results of experimentation and research. Bibliographic collections of information are covered in more detail in the next chapter. Chemical data collections (and scientific data banks in general) are of vital importance in a wide variety of application areas such as the food and pharmaceutical industries, clinical and medicinal chemistry, instrument manufacture and many others. There are a number of reasons why so much importance is attached to the accumulation and storage of scientific data. Some of these are outlined below:

(1) Avoid Duplication of Effort
It is important to be able to store physical and chemical data relating to chemical compounds and other phenomena such as reactions and processes that are of industrial or academic significance. This argument is particularly relevant when the effort involved in making measurements is costly and time consuming. Provided suitable and acceptable values can be stored they may then be made available to those others wishing to use them.

(2) Provision of Acceptable Standards
On both a national and international basis it is imperative to be able to make available acceptable (perhaps legally enforceable) standards with respect to the composition of substances such as foodstuffs, drugs (Ben79) and other materials that are important to world health and safety. Similary, as a consequence of this, it becomes equally important to store a wide variety of data on the performance and pitfalls of the various analytical methods and standards that are used in performing analyses of this type. The significance of this lies in situations where a wide variety of instruments is used by different organisations to conduct particular types of analysis.

(3) Provide Historical Data Trends and Projected Forecasts
The storage of experimental data over a period of time is useful if some type of trend analysis is to be performed or the likely future effects of changing conditions is to be predicted. This type of data collection is very valuable in the areas of environmental pollution and other studies of time-varying changes in the environment (Pac80). Thus, a researcher may be interested in looking at the concentration of various chemical substances in the air or the sea over a wide geographic area as a function of time over a period of several months or years. Once the data has been collected and stored the researcher, and any others who are interested, can analyse it in various ways to look for trends and forecast future situations.

(4) Legislative Purposes
Certain situations can arise in which chemical analysis may be required in order to provide enforcement of some legislative procedure. Consequently, in many areas of scientific analysis the organisations responsible for conducting the analysis are required by law or written agreement to keep copies of the results of all analyses performed in their laboratories. The time span involved might be quite considerable - perhaps several years. Furthermore, it may be that these results must be available for inspection in order to ensure that contractual or legal standards are being maintained.

(5) Records of Special Events
Often, during the evolution of the human species events take place that in all probablility may never happen again. Obviously, the recording and storage of the scientific data associated with these events is of sub-stantial value if later generations of scientists are to analyse the results and formulate theories based upon them.

These five examples provide just a few illustrations of the need for and importance of scientific data collections. The data centres responsible for the maintenance of the data will vary in complexity and size depending

upon their exact function. Some of these will be discussed in more detail
in a later section of this chapter.

RECORDS, FILES AND DATA BASES

Physical, chemical and analytical data relating to chemical compounds may
exist in a wide variety of types and formats, for example,

(a) numeric information such as melting point, boiling point,
     solubility data, heat of reaction, crystal structure coordinates,
     etc,

(b) textual information in the form of coded descriptions of physical
     characteristics, properties or different types of nomenclature used
     to describe the substance such as its molecular formula, Wiswesser
     notation (Smi68), standard name, common names and commercial
     synonyms,

(c) graphical and sonic information in the form of special types of
     photographs of various sorts (such as X-rays, electron micrographs,
     oscilloscope traces, etc), instrument traces of spectra,
     chromatograms, thermograms, polarograms, sound recordings, and so on.

Of these different types of data, numeric and textual have been the
easiest forms to computerise. However, modern technology (through the
advent of powerful digitising methods) now makes it possible to
computerise most other forms of information - such as those listed in
category (c).

Individual pieces of data such as a melting point or a solubility are
often referred to as elementary data items since they are not able to be
resolved into anything simpler. Various elementary items may be assembled
to form a more complex unit referred to as a record. This will usually
have a structure that indicates the way in which individual items of
data are organised and united to form the whole. Thus, within a record
there are three important factors to consider for each elementary item:
its type, size and position within the record. In order to illustrate how
records are constructed, suppose it is required to design one suitable for
holding information relevant to the chemical elements. The first step
would involve listing all the elementary items that are to be included in
the record. Then, the type and size of each of these would be listed.
Finally, the order in which they are to appear in the record would be
specified. Usually a record will have a name that reflects the type of
data or information that it contains.

A simple example of a record structure called ELEMENT-DETAILS is illustrated in diagram A of Fig. 10.3. Diagram B shows two instances of records
having this structure. Both records use a special code (SP436 and AA612)
to represent the analytical method contained in the last field. The
meaning of this code would be found from a suitably designed analytical
methods dictionary. A collection of records, brought together for some
particular purpose, is referred to as a file of data or information.
Thus, the two records contained in diagram B may be combined with three
others in order to make up a file containing five records. This is
illustrated in diagram C.

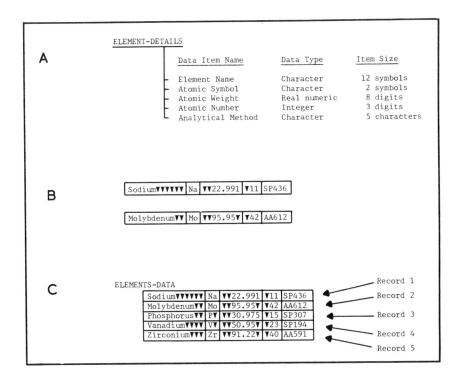

Fig. 10.3 Files and records; A: a record structure;
B: some instances of records having this logical structure;
C: a simple file of records.

The file illustrated in Fig. 10.3 was called ELEMENTS-DATA and it contained five instances of a record called ELEMENT-DETAILS. The relationship
between the file structure (arrangement of records within the file) and
record structure (arrangement of data items within the record) may be
represented using a notation similar to that shown in Fig. 10.4.

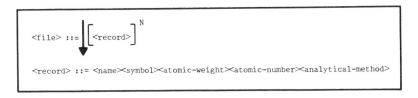

Fig. 10.4 Notation for representing files; the file contains
N records and the field descriptors of the record (shown within
angular brackets) denote the position of the fields within the
record.

In addition to files containing a single record type it is also possible to construct files having several different types of record within it. Some examples of file structures containing different types of record are illustrated in Fig. 10.5.

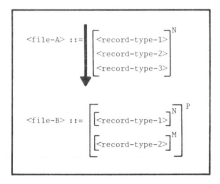

Fig. 10.5 Examples of complex file structures expressed in terms of the notation introduced in Fig. 10.4.

The first example shows a file containing three different types of record and a total of 3xN records. In the second example there are two basic record types and a total of (N+M)xP records in the overall file. Usually the various records in a file will bear some form of relationship to each other. One of the most common types of relationship is that in which a parent record 'owns' one or more filial records thereby imposing a hierarchical structure on the file. This is illustrated in Fig. 10.6. Here the file contains three basic types of record: an author record, a record giving particular details about the author and then a group of records (called a repeating group) corresponding to the different analytical methods that the person concerned has authored. Hierarchically organised files such as this form the basis of hierarchical data base management systems (these will be described later) that attempt to express relationsips between data items in terms of tree-like structures.

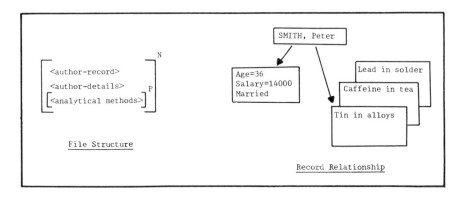

Fig. 10.6 Hierarchically organised file; both file structure and record relationship are indicated.

Previously (see Fig. 10.3), the file ELEMENTS-DATA was introduced as an example of a simple file. All of the records in this file had the same number of data items and a given item was the same length in every record occurrence. Files having this property are said to contain fixed length records. Situations can arise, however, in which a particular field varies in length from one record to another. Any file that exhibits this type of record structure is described as containing variable length records. Referring to the previous example (ELEMENTS-DATA), suppose that for each atomic element in the file there was more than one method of analysis that could be used depending upon the exact requirements of the application. This would create a situation that would necessitate the use of variable length records. This is illustrated in Fig. 10.7.

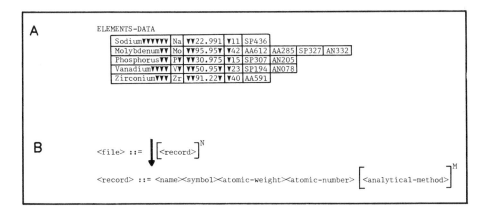

Fig. 10.7 Variable length records; A: instance of a file containing records of this type; B: logical representation of this file using the notation of Fig. 10.4.

The notation used in Fig. 10.7 (B) indicates that the file contains only one type of record but within this record there is a field that may be absent or may be repeated up to M times corresponding to as many analytical methods that are available for the analysis of the substance concerned.

Once data and information have been stored in a file it is important to be able to gain access to it or retrieve it. Computer based file storage devices usually permit access in either one or both of two possible access modes:

(a) sequential, and,
(b) random or direct.

Magnetic tape (see chapter 5), for example, permits only sequential access to records of a file whereas magnetic disks and drums (the so-called 'random access devices') allow both sequential and random retrieval of the stored data. Corresponding to each of these access methods there is an equivalent file processing mode. When sequential file processing is performed each record is processed in turn in the order in which it

appears in the file. However, in random file processing, records may be
processed in any order depending upon the requirements of the application.
If only a fraction of the records in a file are to be processed the random
processing technique is usually more efficient. In contrast, if every
record in the file has to be processed the sequential technique is
probably better. So that these methods of file processing can be
explained further, consider the file depicted in Fig. 10.8 - which
contains a list of analytical methods for various compounds.

Record	ANALYTICAL-METHODS	
001	Compound-1	Analytical-methods-list-1
002	Compound-2	Analytical-methods-list-2
003	Compound-3	Analytical-methods-list-3
004	Compound-4	Analytical-methods-list-4
005	Compound-5	Analytical-methods-list-5

Fig. 10.8 An analytical methods file.

On the extreme left-hand side of the diagram there is a vertical column of
figures; these correspond to the 'logical record numbers' associated with
the records in the file. During sequential processing the records would
be accessed and processed in the following order:

        001, 002, 003, 004, 005

whereas, in random processing the records could be retrieved and processed
in any of over 120 different ways of which the following represent just a
few examples:

        005, 004, 003, 002, 001
        004, 002, 003, 001, 005
        002, 005, 001, 004, 003

Sequential processing is important when a file has to be accessed in order
to answer a search specification such as

        Find all compounds whose analysis
        involves the use of reagent XYZ

In situations where a special index does not exist, each record in the
file would need to be examined in turn to test if it satisfied the search
criterion.   This is a classical sequential search application and would
be very time consuming if a large number of records had to be examined.
In contrast to the above type of sequential search, a strategy to answer a
query of the form:

        Find methods of analysis for compound X

could ideally be handled by examining just one record. Usually, to answer
a question of this type some form of process (either an index or

algorithm) will exist to enable one or more record attribute values to be converted into the address (or record number) of the record within the file - thereby enabling it to be accessed directly.

Files are important because they are the fundamental building block from which data base systems are constructed. A data base is essentially a collection of data files organised in such a way as to minimise any data duplication (or redundancy). Furthermore, it defines and controls the ways in which the stored data may be accessed and who is allowed to have access to it. Figure 10.9 presents a simple schematic illustration of the relationship between files and the parent data base management system to which they belong.

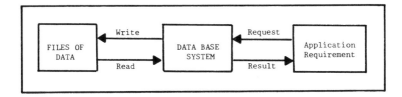

Fig. 10.9   A simple data base management system.

Only the simplest situation is depicted in this diagram. It shows how the data base services some application requirement for data - perhaps an analytical method is sought or a new one is to be added to the stock available. All data maintenance requests are channelled through the data base system. Appropriate programs within the data base then manipulate the files in various ways so as to satisfy the requirements of the application. This could involve data retrieval or storage.

When designing and building data base systems two basic approaches may be used,

(a)   Hand Coding, and,
(b)   Use of a Data Base Package.

The first of these, (a), is probably the most difficult and time consuming since this usually involves writing a considerable number of programs to perform all the necessary operations that will be required. Using a data base management package is the easiest even though it may not necessarily be the best. Most computer vendors of large, medium and small computer systems usually make available some form of data base management package. This can be used to generate a tailor made data base system for any particular application area. Figure 10.10 illustrates the steps involved.

The user designs the data base and then encodes this design in terms of a special type of high level language (see chapter 4) called a Schema Definition Language or SDL. The term schema is the technical name given to the detailed data base specification. The schema is processed by a special translator that converts the user's definition of the data base into a series of programs and tables. These form the basic skeleton of the required data base. This skeleton is then combined with the initial data to be contained in the system to form the operational data base. As time progresses the data in the operational data base may change as a

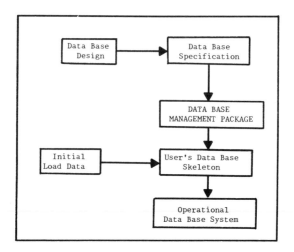

Fig. 10.10   Generation of a data base by means of a
data base management package.

result of the various responses made by the system to meet user's requirements.

Commercial data base management packages of the type just outlined may be
broadly classified into three major types:

    (a)   Hierarchical,
    (b)   Network, and,
    (c)   Relational.

Details of each of these approaches will be found in books on data base
management such as those by Tsichritzis (Tsi77), Kroenke (Kro77) and
Wiederhold (Wie77), the latter being the most comprehensive of those
listed.   An in depth description of the evolution of data bases and a
detailed comparison of the three different types of data base will be
found in a special issue of the journal Computing Surveys (ACM76).
Thalman and Thalman (Tha79) have also given a comparison of the three
different approaches within the context of chemical data.

Earlier in this chapter it was suggested that the hierarchical approach to
data base design attempts to model relationships using hierarchical (tree-
like) data structures.   This implies that there will be a difference in
importance between the different items of data.   This can be seen from the
date base illustrated in Fig. 10.11.

In the hierarchical data base model each record has just one owner
except the topmost level (the record Chemical Element) which is called the
root node and has no explicit owner.   The multiplicity of the parent-
sibling relationship is indicated by the number of arrow-heads in the link
between a parent and its sibling.   In the above example, each chemical
element has only one atomic number but could have more than one isotope.

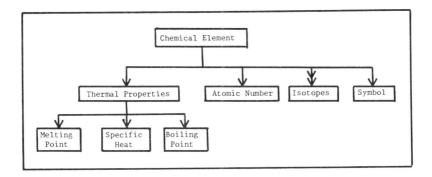

Fig. 10.11   An example of a hierarchical bata base.

Not all data relationships can be modelled by tree-like structures.
Consequently, the network approach tends to model relationships by means
of more complex graph structures. Using this type of model, records can
have more than one owner and there is no hierarchy implied between
records. This approach is the most popular amongst computer vendors and
is the one upon which the CODASYL Data Base Task Group Report (Tsi77) is
based. Figure 10.12 shows an example of a data base involving a network
structure.

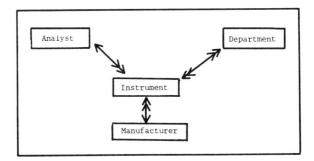

Fig. 10.12   A data base containing a network structure.

This simple data base depicts the relationship between four types of
entity: analysts, departments (within an organisation), analytical
instruments and the manufacturers of those instruments. The diagram
relects how each analyst may be responsible for one or more instruments
but each instrument is the responsibility of only one analyst. Similarly,
each instrument has only one manufacturer but each manufacturer may
produce more than one instrument. Finally, instruments may belong to one
or more departments (that is, shared between them) and each department may
own one or more instruments.

The relational data base model is based upon the use of mathematical
entities called <u>relations</u> upon which are defined various transformation
operations. A relation is essentially just a two dimensional table of
single valued data. Each column of data has a unique name and the order
of columns is immaterial. Columns of a relation are referred to as
<u>attributes</u>. Each attribute has a <u>domain</u>, which is the set of values
that can appear in the attribute. Horizontal cross-sections across all
columns in a relation are referred to as <u>tuples</u> (or 'records'). The
number of tuples in a relation is called its degree. An example of a
relation of degree 4 is given Fig. 10.13.

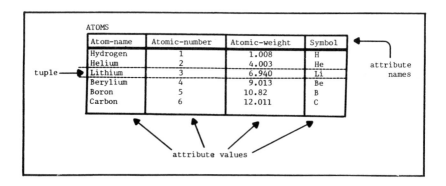

Fig. 10.13   An example of a relation of degree four.

In this example the relation is called ATOMS and the four column headings:
Atom-name, Atomic-number, Atomic-weight and Symbol are referred to as
<u>attribute names</u>. An attribute (or combination of attributes) that
uniquely identifies a tuple is referred to as a <u>candidate key</u>. One of
the candiate keys is selected to be used as a unique identifier and is
called the primary key of the relation. The relation shown in Fig. 10.13
may be represented by the following expression:

ATOMS(Atom-name,<u>Atomic-number</u>,Atomic-weight,Symbol)

in which the Atomic-number attribute (shown underlined) is used as the
primary key. Various types of operations (such as JOIN, PROJECT,
DIVISION, DIFFERENCE, etc) are defined on the relations. These enable new
relations to be created from existing ones and existing relations to be
reorganised and processed in various ways. Further details on relational
data bases will be found in any of the reference texts cited at the
beginning of this section. The relational model is very appealing both
because it can encompass all the properties of hierarchical and network
models and because of its mathematical rigour.

Whatever underlying data base system is used to store data and information, it must provide certain basic facilities, namely,

(1)   the ability for the user to update, modify and destroy
        stored items of data or information,
(2)   the ability to store data with minimal redundancy,

(3)   the ability to create new relationships between the
      stored data items,
(4)   the ability to share data in various ways with other
      users, and,
(5)   security and privacy of shared data.

These aspects are summarised in Fig. 10.14.

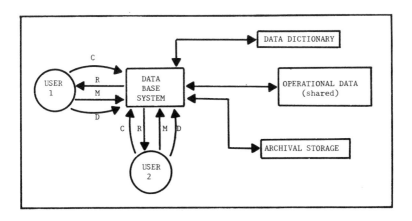

Fig. 10.14   Basic functional requirements of a data base system;
C: creation of new relationships or data; R: retrieval of stored
data; M: modification of stored data; D: deletion of stored
items or relations.

To the data base user these facilities will appear in the form of a
command language or query facility. These will usually be embedded in a
conventional progamming language (such as FORTRAN, COBOL or PL/I - see
chapter 4) or may be made available through a special purpose interactive
query language. Languages of this type provide very sophisticated
facilities and enable data manipulation requests to be posed in almost
'natural language' terms. Some typical examples of user commands to a
relational data base query handler are illustrated in Table 10.1.

Further details on this type of query facility will be found in (Tsi77),
Kro77, Wie77, ACM77 and Tha79). Similar types of human-machine dialogue
are used in the on-line retrieval of bibliographic and other types of
scientific information; these will be described in more detail in the next
chapter.

TABLE 10.1   Examples of Human-Machine Dialogue involved
in Data Base Interaction

(A)
```
SELECT RELATION ATOMS;
PRINT ATOMIC-NAME
     WHERE ATOMIC-WEIGHT<100
     AND ATOMIC-WEIGHT>80
```

(B)
```
SELECT RELATION BLOOD-SAMPLES
DELETE ANALYSIS
     WHERE DATE = 23.8.80
     AND NAME = SMITH
```

(C)
```
SELECT RELATION INSTRUMENTS
UPDATE LAST-SERVICE = 24.9.80
     WHERE INSTRUMENT-NAME = MASS SPECTROMETER
     AND SERIAL-NUMBER = 17364A
```

(A) - Retrieval;   (B) - Deletion;   (C) - Update

INFORMATION SYSTEMS

An information system is a more complex entity than a data base. Indeed,
as was mentioned in the introduction, data bases may be thought of as
being the fundamental building blocks from which information systems are
constructed. The relationship between the two is illustrated in the
simple model depicted in Fig. 10.15. The diagram shows two categories of
user - A and B. User A is regarded as being typical of those who interact
with the system by means of simple information manipulation requests.
Depending upon the purpose of the system and the way in which it is
designed, this type of user may also wish to add information to the
system. The second category of user, User-B, plays more of a management
role and is responsible for maintaining the information system. This
involves monitoring its performance and modifying it - if and when
necessary. This activity requires the collection of statistics relating
to appropriate performance metrics - response times for queries, the type
of requests that are made, the nature of any unsatisfied requests for
information and so on.

The internal architecture of the information system reveals a variety of
components: a command handler, various query handlers, data bases, files
of different types and a communication facility. The command handler is
responsible for implementing the user's overall requirements. The
majority of users will require only simple data/information manipulation
facilities. Provided these can be serviced by the system's own data bases
most of the user's requests will be processed by the query handlers associated with these. Other handlers will be responsible for allowing the

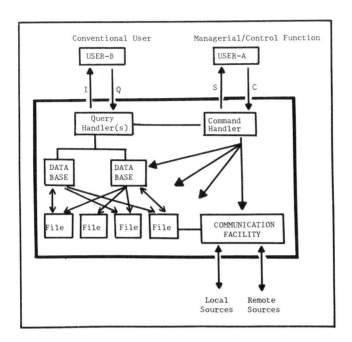

Fig. 10.15   Architecture of an information system; I: retrieved
information; Q: retrieval query; S: statistical details;
C: control commands.

user to select a data base and will then permit a dialogue with it to be
sustained in the way described in Table 10.1.   If need be, switching to
other internal data bases may be easily facilitated.   The role of data
bases in the handling of large collections of information (Chemical
Abstracts Service) has been described by Huffenberger (Huf75) while an
example of a sophisticated information system that depends upon the type
of architecture outlined above has been reported by Uchida (Uch79).

The command handler provides facilities which differ from those of the
query processors.   They permit the construction of new data bases that are
tailored to the individual requirements of a particular user or user
group.   By means of the communication facilities, the command handler is
able to provide access to other local and remote information systems.
These may involve on-line instruments each having their own local storage
and built-in data bases.   Such a facility (shown in Fig. 10.16) might be
extremely important in situations where expensive exquipment has to be
shared in order to satisfy a regional or national need.

It is the communication facilities that enable individual information
systems to be interconnected in order to construct world-wide networks of
shared information and data.   Techniques to achieve this will be discussed
in greater detail in chapter 12.

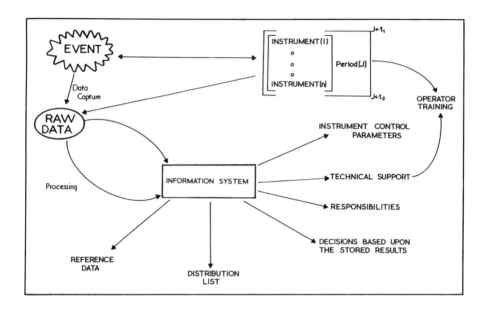

Fig. 10.16 Information system used to coordinate data collection and instrument control.

DATA CENTRES

A data centre may be defined as an establishment that is responsible for handling scientific data collections relevant to particular branches of science. They act as references centres that are responsible for maintaining and documenting a given data collection. In addition, they have the responsibility of ensuring the quality of the data, (that is, its accuracy and correctness), its integrity and security. Another of the important tasks of the data centre is that of making the data available (either freely or for an appropriate fee) to other potential users. There is a wide variety of different types of data centre throughout the world. In this section, some of those of relevance to analytical chemistry will be briefly outlined.

As far as scientific data collections are concerned there are various levels of involvement depending upon the function that the data centre has to perform. There are five basic levels:

(A) The Local Level
for example, a local public analyst's collection of infrared or mass spectral data files that are used only by the analyst and his/her colleagues. These may be available in machine readable form and processed with the aid of a small desk-top computer attached to the instrument to which they relate.

(B) The Corporate Level
in this case a collection of data may be shared between the various
divisions of a large corporation. The data files associated with the ICI
CROSSBOW system provide an example of this type of involvement. This type
of system will be briefly described in the next chapter in the section on
indexing.

(C) The Regional Level
in which scientific data relevant to a particular region might be held
centrally in order to service requests relevant to regional research and
industrial processes taking place in particular geographical locations.

(D) National Data Centres
each country will have its own appointed centres to maintain collections
of data in various subject areas. In addition to making the data
available throughout the parent country they will also be responsible for
fostering collaboration and data exchange with corresponding centres in
other countries. Table 10.2 lists some of the data centres that operate
within the United Kingdom (Rob76). The Cambridge Crystallographic Data
Centre at the University of Cambridge (Ken75) is typical of those listed
in the table. This was established in 1965 and specialises in data
produced by X-ray and neutron diffraction studies of crystal and molecular
structures. In addition to those listed in the table, other typical
examples of data collections include X-ray powder diffraction files and
ASTM/DOW Infrared Spectral Files (Ric79).

(E) International/World-wide Centres
these operate across national boundaries and provide facilities for the
compilation, evaluation and dissemination of data on a world-wide basis.
The construction and maintenance of such centres often necessitates a high
degree of cooperation and much collaborative effort (Row79). As computer
based telecommunications facilities become more easily available and more
reliable so there is an increasing demand for the services that this type
of data centre is able to provide. Access and retrieval of data is particularly easy when it is transmitted in electronic form - perhaps via
satellite.

The data centre operated by the National Institute of Health and the
Environmental Protection Agency (NIH/EPA) in the United States of America
is probably one of the most well known international data centres (Hel77,
Fel77, Mil78). The NIH/EPA Chemical Information System (CIS) has been
developed by various agencies of the United States Government in
cooperation with other governments and organisations. It consists of a
collection of scientific data bases which are made available internationally through a communication network similar to that depicted
schematically in Fig. 10.17. The system provides a diverse range of
numeric (as opposed to bibliographic) data on over 140,000 chemical
substances.

Figure 10.18 summarises some of the data files that are available for
searching and the ways in which these may be searched. CIS has a unique
linking system the heart of which is the Structure and Nomenclature Search
System (SANSS). This allows the user, in a single operation, to search 41
different files. Some of the different types of search available include:

         - specific complete structure or substructure,
         - complete or partial name (Chemical Abstracts Service
            Index name, Trade names, Common names and Synonyms),

TABLE 10.2 Scientific Data Centres located within the United Kingdom

1. London University,
    Department of Statistics,
    London School of Economics and Political Science,
    Houghton Street,
    Aldwych,
    London, WC2A 2AE.   (Data Banks and their systems)

2. Department of Chemistry,
    Birmingham University,
    P.O. Box 363,
    Birmingham, B15 2TT.   (Kinetic Data)

3. Cambridge University Crystallographic Data Centre,
    Chemical Laboratory,
    University of Cambridge,
    Lensfield Road,
    Cambridge.   (Crystallographic Data)

4. Leeds University,
    Department of Fuel and Combustion Science,
    Leeds, LS2 9JT.   (High Temperature Processes)

5. Leeds University,
    Department of Physical Chemistry,
    Leeds, LS2 9JT.   (Reaction Rate Data)

6. Bristol University,
    Senate House,
    Bristol, BS8 1TH.   (Mass Spectral Data Files)

7. IUPAC Thermodynamic Tables Project Centre,
    Imperial College of Science and Technology,
    Department of Chemical Engineering and Chemical Technology,
    Prince Consort Road,
    London, SW7.   (Thermodynamics)

8. School of Molecular Sciences,
    Sussex University,
    Falmer,
    Brighton, BN1 9QH.   (CATCH Tables - Thermodynamics)

9. Biodeterioration Information Centre,
    Aston University,
    Department of Biological Sciences,
    Gosta Green,
    Birmingham, B4 7ET.   (Biodeterioration)

10. QUODAMP,
    Belfast, Queens University,
    Department of Computer Science,
    Belfast, BT7 1NN.   (Atomic and Molecular Physics)

11. Imperial Chemical Industries Limited,
    Petrochemicals Division,
    Billingham,
    Teesside.   (Chromatographic Information)

TABLE 10.2 (Continued)   Scientific Data Centres located within
the United Kingdom

12. Pfizer Limited,
    Ramsgate Road,
    Sandwich,
    Kent   (Biological/Biochemical)

13. Mass Spectrometry Data Centre,
    Building A8.1A,
    Ministry of Defence,
    Aldermaston,
    Reading, Berks.   (Mass Spectral Data)

14. University College, Cardiff,
    P.O. Box 78,
    Cardiff, CF1 1XL.   (Pyrrole Data)

15. National Physical Laboratory,
    Teddington,
    Middlesex.   (Thermodynamics)

16. British Trust for Ornithology,
    Beech Grove,
    Tring,
    Herts, HP23 5NR   (UK Avifauna & Relationships with Man)

17. Commonwealth Institute of Entomology,
    56, Queens Gate,
    London, SW7 5JT.   (Applied Entomology)

18. Biological Records Centre,
    Monks Wood,
    Abbots Ripton,
    Huntingdon.   (Flora and Fauna Distribution)

19. The Meteorological Office Library,
    London Road,
    Bracknell,
    Berkshire, RG12 2SZ.   (Environmental Data Records, etc.)

20. Commonwealth Bureau of Soils,
    Harpendon, AL5 2JQ.   (Soil Science)

21. Water Data Unit,
    Reading Bridge House,
    Reading, RG1 8PS.   (Flow Data, etc.)

22. Geotitles Weekly,
    P.O. Box 1024,
    Westminster,
    London, SW1P 2JL.   (Geosciences)

23. Scientific Documentation Centre,
    Halbeath House,
    Dunfermline, KY12 0TZ.   (Spectral and other data)

24. INSTAB,
    Water Research Centre,
    Stevenage Laboratory,
    Elder Way,
    Stevenage, Herts, SG1 1TH. (Toxicity and Biodegradability)

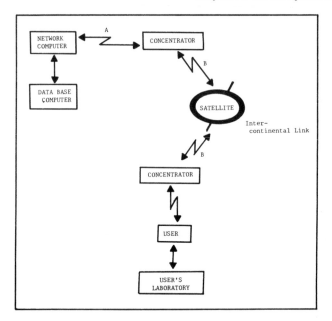

Fig. 10.17 Communication facilities used by the NIH/EPA Chemical
Information System (CIS); A: local data links (leased line or
microwave); B: inter-continental links via satellite.

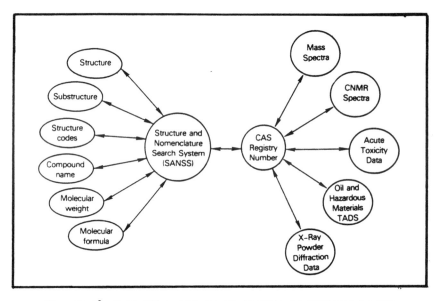

Fig. 10.18 Data files and search techniques employed by the
NIH/EPA Chemical Information System.

- specific atom-centred fragments,
- specific rings or ring structures,
- specific molecular weights,
- specific complete or partial molecular formulae,
- compounds having a specific number of rings,
- compounds having a specific number of atoms, and,
- Chemical Abstracts Registry Numbers.

The CIS system is undergoing continual developent and modification in order to extend and improve the facilities that it is able to offer. Improvements in search algorithms, for example, can be used to reduce the amount of time spent searching thereby minimising the cost of a search. Similarly, improvements in graphic terminal display devices should enable much easier input and output of structural diagrams of molecules, etc.

Many other national and international data centres exist both in the private and public sectors. Some of these have been described by Magrill (Mag80).

CONCLUSION

The basic nature of data and information has been described and the need to be able to store and retrieve each of these commodities has been emphasised. Data for storage may originate from many different sources. These may be sub-divided into two broad categories: theoretical and experimental. Theoretical values may be produced as a result of long, complex and involved calculations that are too time consuming to be repeated each time the values are required. Consequently, it is easier to store them. Experimental results are produced by observations made using a variety of conventional, instrumental and automated analytical techniques. These results may need to be stored for reference or for control applications at local, national and international levels.

The manner in which data is organised into records, files data bases and, ultimately, information systems has been briefly outlined. Current models used in the design of data bases have been presented. The descriptions given should provide the background necessary to understand their use as tools to aid the sharing of scientific data and information. There are two important areas of application for these tools: the handling of bibliographic and numeric data. The latter forms the basis of many scientific data collections and the data centres that support them. Some examples of the different types of data centre have been outlined briefly in this chapter. The next chapter will describe some of the bibliographic information services that the computer is used to provide while chapter 12 is devoted to a discussion of computer based communication networks that enable the implementation of world-wide information systems.

REFERENCES

ACM76   Association for Computing Machinery, COMPUTING SURVEYS, Volume
             8, No. 1, March 1976.

Ben79   Benson, W.R. and Wright, W.M., Chemical Data: An Essential Tool
             in the Regulation of Drugs, Journal of Chemical Information and
             Computer Science, Volume 19, No. 1, 3-8, 1979.

Fel77   Feldmann, R.J., Milne, G.W.A., Heller, S.R., Fein, A., Miller,
             J.A. and Koch, B., An Interactive Substructure Search System,
             Journal of Chemical Information and Computer Sciences, Volume
             17, No. 3, 157-163, 1977.

Hea78   Heaps, H.S., Information Retrieval - Computational and
             Theoretical Aspects, Academic Press, ISBN: 0-12-335750-0, 1978.

Hel77   Heller, S.R., Milne, G.W.A. and Feldmann, R.J., A Computer Based
             Chemical Information System, SCIENCE, Volume 195, No. 4275,
             253-259, 21st January 1977.

Huf75   Huffenberger, M.A. and Wigington, R.L., Chemical Abstracts
             Service Approach to Management of Large Data Bases, Journal of
             Chemical Information and Computer Sciences, Volume 15, No. 1,
             43-47, 1975.

Ken75   Kennard, O., Watson, D., Allen, F., Motherwell, W., Town, W. and
             Rodgers, J., Cambridge Crystallographic Data Centre, Chemistry
             in Britian, Volume 11, No. 6, 213-216, June 1975.

Kro77   Kroenke, D., Database Processing - Fundamentals, Modelling,
             Applications, Science Research Associates, ISBN: 0-574-21100-4,
             1977.

Lef66   Lefkovitz, D. and Powers, R.V., A List-based Chemical Informa-
             tion Retrieval System, 109-129 in Proc. of the 3rd Annual
             Colloquium on Information Retrieval, May 12-13th, 1966,
             Philadelphia, Pennsylvania, Thompson Book Company, Washington,
             D.C., (Ed: Schector, G.).

Mag80   Magrill, D., A Name for Every Chemical, New Scientist, Volume
             87, No. 1208, 16-18, July 1980.

Mil78   Milne, G.W.A., Heller, S.R., Fein, A.E., Frees, E.F., Marquart,
             R.G., McGill, J.A., Miller, J.A. and Spiers, D.S., The NIH-EPA
             Structure and Nomenclature Search System, Journal of Chemical
             Information and Computer Sciences, Volume 18, No. 3, 181-186,
             1978.

Pac80   Pack, D.H., Precipitation Chemistry - A Two Network Dataset,
             SCIENCE, Volume 208, No. 4488, 1143-1145, 6th June 1980.

Pai77   Paice, C.D., Information Retrieval and the Computer, McDonald
             and Jane's Computer Monographs, ISBN: 0-354-04095-2, 1977.

Ric79   Richman, J.A., Chemical Information Sources: Aids in the Review
             of Drug Applications, Journal of Chemical Information and
             Computer Sciences, Volume 19, No. 1, 1-3, 1979.

Rob76   Robson, A., A Preliminary Study of Data Handling Techniques in
            the United Kingdom, British Library Research and Development
            Report, No. 5296, April 1976.

Row79   Rowlett, R.J., International Sharing of the Production and
            Distribution of Chemical Information Services, Journal of
            Chemical Information and Computer Sciences, Volume 19, No. 4,
            193-195, 1979.

Sal80   Salton, G., Automatic Information Retrieval, IEEE COMPUTER,
            Volume 13, No. 9, 41-56, 1980.

Sen69   Senko, M.E., Information Storage and Retrieval Systems, Chapter
            4, 229-281, in Volume 2 of Advances in Information Systems
            Science, (Ed: Tou, J.T.), Plenum Press, New York, 1969.

Smi68   Smith, E.G., The Wiswesser Line Formula Chemical Notation, New
            York, McGraw-Hill, 1968.

Tha79   Thalmann, N. and Thalmann, D., A Problem Orientated Analysis of
            Data base Models, Journal of Chemical Information and Computer
            Sciences, 86-89, Volume 19, No. 2, 1979.

Tsi77   Tsichritzis, D.C. and Lochovsky, F.H., Data Base Management
            Systems, Academic Press, ISBN: 0-12-701740-2, 1977.

Uch79   Uchida, H., Information System, Data Bases and On-line Services
            of the Japan Information Centre of Science and Technology
            (JICST), Journal of Chemical Information and Computer Sciences
            Volume 19, No. 4, 199-201, 1979.

Wie77   Wiederhold, G., Database Design, McGraw-Hill, ISBN: 0-07-
            070130-X, 1977.

11

Information Services

INTRODUCTION

The fundamental motivation for a substantial amount of human activity stems from the basic, intrinsic biological need for the species to generate information. This need arises for a variety of reasons. The most important of these relates to the fact that information is a vital prerequisite for logical scientific decision making. The most common source of information is research activity. Research produces a wide variety of knowledge in the form of human experience and published works such as reports, articles, and books. Recorded knowlege is usually stored in some form of information centre such as a library where it can be made available to those others who are interested in using it. A simple model that depicts the processes of information production, storage (or archival) and dissemination is illustrated in Fig. 11.1.

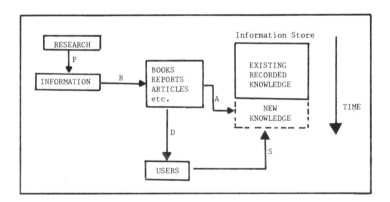

Fig. 11.1 Information generation, storage and dissemination,
P: production; R: reading; A: archival; D: dissemination,
S: searching.

In this diagram P denotes the production of information, R its recording, A its archival and D its dissemination to users - a type of current awareness situation. The conventional 'literature search' is denoted by S. This will operate on both the recent (or new) knowledge and the accumulated wealth of retrospective information. The simple model illustrates how the continual generation of knowledge necessitates an increase in the volume of the information store as time progresses - unless some means of compaction is used such as storage on microfilm or the utilisation of computer stores. Most people are aware of the substantial increases in the volume of printed material that appears in libraries - particularly in the areas of science and technology. A most convincing indication of the growth rate of the chemical literature can be obtained by comparing the physical sizes of the cumulative indexes of a publication such as Chemical Abstracts. If a graph of the relative physical size versus time is plotted a curve similar to that shown in Fig. 11.2 is obtained (Bar76). This curve reflects an exponentially increasing variation with time - often referred to as the 'information explosion'.

Both Murdock et al (Mur67) and King (Kin78) have presented comprehensive models for scientific and technical information generation and transfer. Each of these is based upon the classic 'sender-channel-receiver' concept.

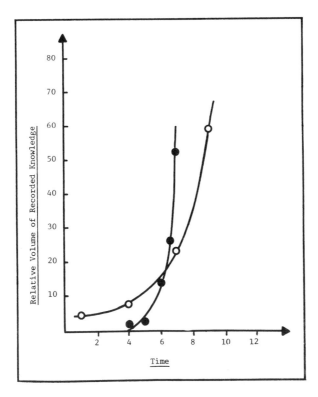

Fig. 11.2 Growth of scientific and technical knowledge; ●:computer journals during the period 1940 - 1969: o: chemical information during the period 1907 - 1976.

They both encompass the idea that the entire information transfer process
is cyclic - as described by the generalised model of information systems
described by Yovits and Ernst (Mur67). Figure 11.3 depicts King's cyclic
model.

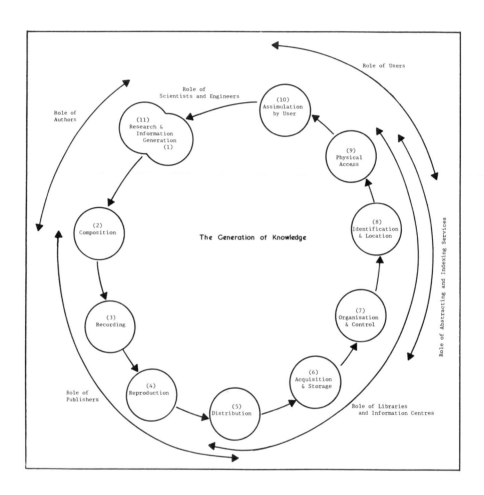

Fig. 11.3 King's model for scientific and technical information transfer.

The overall information transfer mechanism involved in this model depends,
essentially, upon eleven basic steps. These are numbered 1 through 11 in
the diagram. Most of the steps are self-explanatory; further details,
however, are given in the original paper (Kin78). It is imperative to
stress that the retrieval of information (steps 5-8) is as important as
its generation (steps 1 and 11) and storage (steps 2 through 4). In the
context of retrieval operations, Fig. 11.3 introduces two important tools:
abstracting and indexing services. A wide variety of aids of this type
are available to assist the researcher in handling the ever increasing

volume of information that is produced annually (see Fig. 11.2). In many
cases the indexes that are used are manually maintained and are often
based upon the use of card files. However, because of the substantial
volume of material currently being generated a large number of information
centres now use computer generated indexes. Some of these will be
discussed later in this chapter.

Information is produced by people for use by other people. Figure 11.3
gives an indication of the roles played by some of the various entities
involved in the information transfer process. However, it gives no
indication of the characteristics of either the people who use the
information or those who generate it. An understanding of each of these
species is important from the point of view of implementing computer based
information transfer methods. There are two areas of particular significance. Firstly, that of Expert Systems (as was discussed in chapter 2):
and, secondly, that of on-line retrieval (which will be described later in
this chapter). As a means of understanding the nature of those who
generate information a study of publishing habits is a useful way to
proceed. An interesting investigation of eminent chemists has been
undertaken by Baglow and Bottle (Bag79). They have constructed a profile
of the rate of publication of papers as a function of age and background.
There were no conclusive results other than an indication that eminence
was related to the number and rate of publication and that this was
achieved more often in an academic environment than anywhere else. As far
as the author is aware no corresponding profile of the typical
'information user' has been formulated.

There are many guides and conventional sources available to help the
chemist handle some of the problems associated with searching the
literature of chemistry and related disciplines (Cra57, Mel65, Dav74,
Woo74, Mai79, Bot79, Ant79). Readers will find many of these sources
useful introductory guides to the chemical literature. In this chapter,
however, emphasis will be given to some of the information services
provided by the computer.

COMPUTER PREPARED INDEXES

It was shown in the previous section that the annual growth rate of
technical literature is quite substantial. The problem which this produces
is further complicated by the publication of new periodicals and irregu-
larly issued publications such as books, reports, monographs, Proceedings
of Conferences and Symposia, etc. Obviously, if engineers and scientists
are to keep abreast of this rising tide of information, appropriate
services need to be made available. Indexes and abstracts of various
types are the most frequently used answer. However, the consistent 8-9%
annual growth rate in the number of primary papers abstracted in its
publications like Chemical Abstracts has caused a tremendous growth in its
volume (see Fig. 11.2). It became evident to many organisations that the
traditional manual system for processing, publishing, and using the
secondary literature was too slow, too expensive, too rigid and too
wasteful. Consequently, some years ago many information agents such as
Chemical Abstracts Services (CAS) began moving towards the use of computer
based processing systems. In 1961 CAS began producing Chemical Titles
on magnetic tapes in addition to the traditional hardcopy. Since then the
utilisation of the computer has increased substantially. The ultimate aim

of this organisation is to produce a system in which all information will be input to a unified store, and from which a range of general or more specialised services will be supplied. In this section two typical computer based indexes will be described.

The KWIC Index

One of the earliest of the machine made indexes was devised by Luhn (Luh59) called a KWIC index. KWIC is an acronym for Key Word to Context. Since the early index produced by Luhn, many versions of the basic system have appeared. KWIC indexing is most often applied to the <u>titles</u> of documents. In an index of this sort one line is assigned per title and a keyword in the title is printed in context, in a vertically aligned position. Usually the keyword is printed in the middle of the line and sometimes emphasised by over-printing or by the use of bold typeface (see Fig. 11.4).

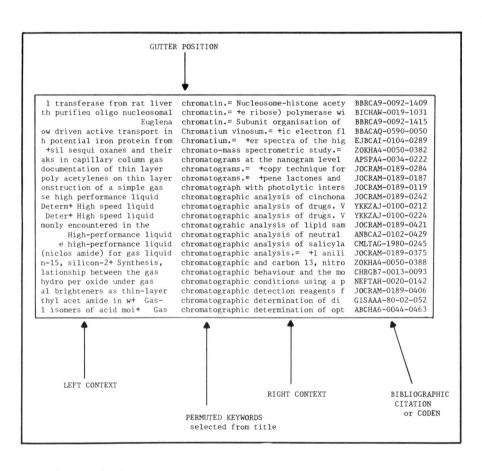

Fig. 11.4 Example of a KWIC index; section of CHEMICAL TITLES, page 37, no. 7, April 7th, 1980.

In the index the keywords are sorted alphabetically. Keywords are those
words in the title which convey information concerning the subject content
of the paper. Words which add no meaning (such as 'and', 'the', 'if',
etc) to the title (or are non-informative) are called 'noise-words',
'forbidden words' or, more commonly, 'stopwords' (see Fig. 11.5).

extent	individual	many	perfect
extraordinary	individuality	material	perfection
extreme	individuals	materials	performance
exremely	induced	matter	performed
f	industrial	may	period
fabrication	industry	mean	permanent
factors	ineffectiveness	meaning	permanently
failure	inexpensive	means	permissible
fall	influence	measure	permitting
false	influenced	measurable	persistance
familiar	influences	measured	persistence
far	influencing	measurement	phenomena
fast	ingredient	measurements	phenomenon
favor	ingredients	measures	place
feasibility	inherently	measuring	placed
feature	initial	media	played
featured	initially	medium	plus
features	initiated	member	point
fed	initiation	membered	points
few	inner	members	policies
fifth	inside	method	poor
fight	instead	methods	populated
figures	interaction	MeV	population
fill	interactions	middle	portion
filled	interest	minimum	position
filling	intermediate	minus	positions
fills	interpretation	minute	positive
final	interrelation	miscellaneous	possessing
finding	interrelations	missing	possibilities
findings	interrelationships	mixed	possibility
finds	into	mixture	possible
fine	introduced	mixtures	practicability
finely	introducing	mm	practical
firmly	introduction	mmu	practice

Fig. 11.5 Section of a stopword list; stopwords are words that are
prevented from appearing in the gutter position of the KWIC index.

Each of the keywords in a title will give rise to an entry in the index.
The general format of an entry is as follows:

left context	KEYWORD	right context	CITATION

To the right of each permuted title is printed a reference code which
provides a highly condensed bibliographic citation for the source of the
article or document. The most effective way to use the index is to scan
vertically the alphabetical keyword list, pausing at words of interest to
examine the horizontal context. If the context confirms relevancy, then
the bibliographic citation can be used to obtain further details of the
article.

Allowing only one line for the title means that its beginning or end (or both) may be truncated. This is not a serious disadvantage since experienced users of such a system are usually able to determine the relevance of a title based upon just one or two keywords and their immediate leading or trailing context. The advantage of printing the keyword in context is that the words preceding and following the keywords can be regarded as sub-headings or modifiers which quickly indicate to the user how pertinent the title is to his/her area of interest. In addition to the KWIC Index proper most KWIC systems contain other supporting indexes such as a Bibliography and an Author Index. The format of each of these will depend upon the particular KWIC index to which they apply.

One of the most well known examples of a KWIC index is that produced by the Chemical Abstracts Service called CHEMICAL TITLES (Fre63), a sample extract from which was presented in Fig. 11.4. Chemical Titles is essentially a concordance to chemical research papers selected from over 700 journals of pure and applied chemistry and chemical engineering. It is published (via the aid of computer typesetting) fortnightly and distributed on a world-wide basis. As was mentioned above, the publication consists of three parts. The KWIC Index has already been described in some detail and so only the Bibliography and Author Index now need to be discussed.

The Bibliography is that part of the publication in which the titles indexed in the KWIC section are listed in the form of tables of contents of the journals covered in the issue. The complete journal citation (in heavy type) precedes each table of contents. The bibliography is arranged alphabetically by the CODEN for each journal. A CODEN is a unique, unambiguous five character identifier assigned to a journal. This is extended to six characters by the addition of a special 'check character' to enable computer detection of errors. The CODEN assigned to a particular journal never changes. The bibliography is used to obtain the complete title, all authors names, and the inclusiive pagination for all papers selected for Chemical Titles from that issue of the journal. In addition, the Bibliography may be used as an alternative entry point into the chemical literature since it enables users to access chemical information by journal of interest (for example, Analytical Chemistry or Talanta) rather than via keyword.

The third part of Chemical Titles is the Author Index. This is a listing of all the authors of the papers indexed in the issue arranged alphabetically by the author's last name. This index is useful for searching for works by particular authors (for example, West or Deans). Details of the coverage of Chemical Titles and the CODENs associated with individual journals are contained inside the cover of each issue. Each edition also provides explicit instructions on how to use the indexes.

In addition to the use of KWIC systems by large organisations (such as CAS) for handling 'world-wide' literature, other approaches have also been used either on an institutional basis (run on an in-house minicomputer) or on an individual basis - the personal KWIC. Software packages exist to cater for a wide variety of demands (IBM62, Bar79a, All77, Has65, Pet71, Sko70). The PERSONAL KWIC is summarised in the sketch presented in Fig. 11.6. In this system, information of interest to the analyst is coded in the form of records consisting of two parts, a descriptive title and a locator code.

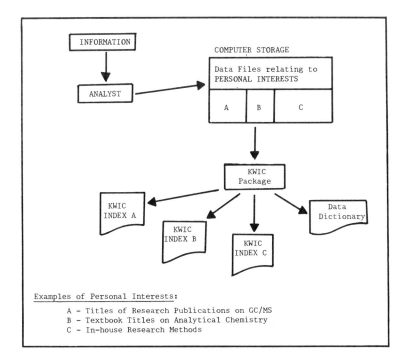

Fig. 11.6   Organisation of a personal KWIC index.

A typical record might appear as follows:

THE ANALYSIS OF METALS BY ATOMIC ABSORPTION SPECTROMETRY	AA371

In the above example the descriptor part of the record is shown on the
left and the locator part (AA371) is contained on the right. The keywords
contained in the descriptive title specify the nature of the object being
classified. These are used by the KWIC package to form a KWIC index which
is essentially an alphabetical listing of permuted keywords for each
descriptor record in the file. The type of output produced from such a
package is illustrated in Fig. 11.7. The diagram illustrates the entries
that would appear in the KWIC index for the single descriptor/locator
record described earlier in this section.

The locator code is used to specify where to locate the document (or other
entity) to which the locator refers. The meaning of the locator code
(which is arbitrarily assigned by the user) may be found from appropriate
entries made in the data dictionary. As can be seen from Fig. 11.6, the
user of the system uses various files to store the information relevant to
the different facets of his/her interests. A, B and C each represent
three different files that may be independently processed by the package
to produce separate KWIC indexes, one for domain A, another for domain B

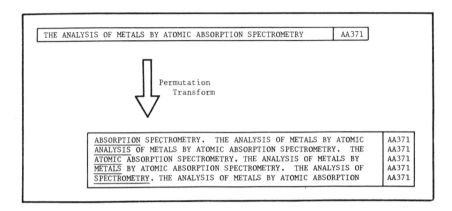

Fig. 11.7   Output produced from a personal KWIC index.

and a third for domain C.   A system such as this can often be used quite
successfully for setting up small project data bases.   This is easily
achieved since each facet of interest can have its own KWIC index, for
example, people, supplies, equipment, etc.   Further details of the
PERSONAL KWIC will be found in Bar79a.

The Science Citation Index

The Science Citation Index (SCI) is another example of an index that is
produced via the aid of a computer (ISI81, Gar64).   It is published by the
Institute for Scientific Information (ISI) and indexes every significant
article and editorial item from every issue of over 2,700 of the world's
most important scientific journals and covers over 100 different subject
areas.   In total, more than 400,000 articles are processed each year.
These are taken from the following general areas of science:

> Agricultural, Biological and Environmental Sciences,
> Engineering and Technology,
> Medical and Life Science,
> Physical and Life Science,
> Social and Behavioural Sciences.

The SCI is orientated towards pure sciences and Western languages.   Publication commenced in 1961 and ever since its introduction it has been a
valuable tool for scientific literature searching.   Its unique feature is
that searches starting from a paper of interest lead forward to more
recent papers on the subject as well as backwards to previous papers.   The
basic principle of citation indexes has been described by Weinstock
(Wei71) while their particular application to chemistry has been outlined
by Cawkell (Caw70).

The mechanism upon which the SCI is based depends upon the proven fact
that almost every scientific paper cites previous research work.   Most
reports or papers contain a list of references (or citations) at the end
which refer to the works of other authors.   These may be used to trace

backwards or forwards for related work. The citation index thus works in the opposite direction to a conventional index. From one key paper (which might have been published many years ago) the Citation Index identifies later published research. The only assumption is that there is a subject connection between the citing article and the paper being cited. In summary,

A citation index traces from the author, title, (or whatever), of any given paper "P1", all the later papers which have cited paper P1. It is therefore possible to locate paper P1 (published in, say 1963) in the Citation Index and to trace all the later papers (1965, 1968, 1974 ..... 1980) which have cited it. If necessary, paper P2 (which cited paper P1) can also be used as an entry.

The three principal sections of the SCI are:

(1) The Source Index which is arranged alphabetically by the names of the authors (that is, the later papers which cite the earlier ones). For each article indexed, it provides a complete bibliographic description that includes the full title, the names of all authors, journal name, volume, page number and year of publication.

(2) The Citation Index arranged alphabetically by the names of the authors being cited (that is, those receiving the citations). Following each cited author is a list of the current articles that are giving the citations.

(3) The Permuterm Subject Index in which every significant word from the title of every article covered during the indexing period is paired with every other word in that title. These permuted pairs are alphabetically listed as two-level indexing entries and linked to the names of the authors who used them in the titles of their articles.

The format of the major sections of the Science Citation Index is illustrated in Fig. 11.8. In addition to the three main parts there are other supplementary indexes (for example, the Corporate Index, Patent Citations Index, Anonymous Citation Index and so on). Each index is published quarterly, the final volume being an annual cumulation. Each quarterly or annual set forms an independent unit and must be used as such. Thus, the 1979 Permuterm subject Index is of no use with the 1978 Source Index.

When using the SCI to search the literature, four basic methods of searching are most frequently used:

(1)   By Citation
(2)   By Author
(3)   By Subject
(4)   By Organisation

It is worthwhile outlining the utility and general procedure adopted with each of these techniques.

(1) Searching By Citation
This method is most applicable when an earlier key paper, book or other published material relevant to the search subject is known. The steps involved are as follows:

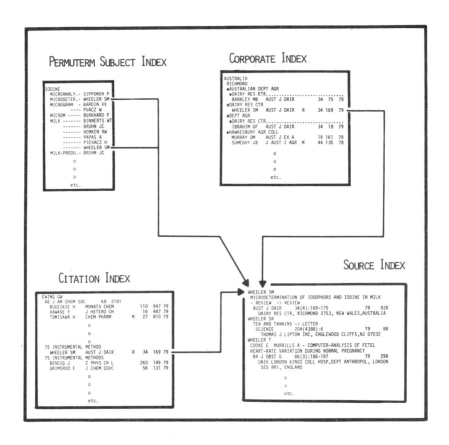

Fig. 11.8 Principal indexes of the Science Citation Index
(January/Febuary 1980 edition).

1.1   Select a paper (if possible, a key paper) which is central to
      the subject and which is as specific as possible.
      If a paper cannot be found then do a subject search in order to
      locate one.

1.2   Look up the author in the Citation Index. Following his/her
      name will appear a list of papers. Locate the key paper
      selected in step 1.1 and observe the names of the authors who
      have cited it.

1.3   Check each citing author in the Source Index. Under each author
      there will be an entry giving full bibliographic details of the
      citing articles. The titles of the articles should enable their
      relevance to be assessed.

1.4   Locate relevant articles in the local library or obtain them via
      an inter-library exchange scheme.

(2) Searching By Author
This technique is best used when the name of an author in the field of
interest is known (for example, Ewing, Strobel, Deans, Perone etc). The
search method simply involves looking up the name of the specified author
in the Source Index section of the SCI. This strategy enables the
researcher to check if an author has published anything during the period
covered by the index. This is particularly effective if the author has a
reputation for publishing important material in a particular area.

(3) Searching By Subject
This approach is used to start a search when no earlier relevant papers on
a subject are known. Typical situations would be background searches for
a researcher entering a new field or a graduate student starting a new
project. The basic procedure is as follows:

3.1 Select terms which are descriptive of the subject. Where
     possible, include synonyms and near synonyms.
3.2 For each of the terms selected in 3.1, look-up its entry in
     the Permuterm Index and note the names of the authors who
     have used these terms linked to co-terms which further refine
     the area of interest.
3.3 Look up each author in the Source Index and determine whether
     the items are of interest. If necessary, the search can be
     extended by using the Citation Index.

(4) Searching By Organisation
This method of searching is used when the researcher knows of one or more
organisations that are involved in the type of work that is the subject of
the search (for example, the Food Research Association, Lawrence Livermore
Lab, University of Houston - Chemistry Department etc). This index
enables the researcher to identify all the authors from a given organisation who have had something published during the period indexed. Once
the names of the authors who have published from the organisation have
been obtained, full information on the items they have published can be
obtained from the Source Index.

As a communication system the network of journals plays an important role
in the exchange of scientific and technical information. This is
important if the duplication of expensive research effort is to be
avoided. Taken together the three SCI indexes described above provide a
powerful means of performing various types of literature search for both
ongoing and new research projects. It will be interesting to see what
influences the electronic journal (based upon computer communication
networks - see chapter 12) will have upon the classical means of
information dissemination currently being maintained by the printing
press.

Through continual progress in research and development new types of index
have become available. One recently introduced (ISI78) is the index to
scientific and technical conference proceedings. These enable both
current awareness and retrospective searching of the conference literature
using a variety of indexes: category index, Permuterm subject index,
author (or editor), sponsor, corporate or meeting location indexes. Other
indexes include subjects such as research projects, scientific instruments, equipment and so on.

Other Types of Index and Abstracts

Abstracts (like 'titles' - see Figs. 11.4 and 11.7) represent a very
concise way of summarising the information content of a research report,
book or other 'storage device'. They are usually more informative than
titles because of the extra information that they contain. Within science
and technology a wide variety of abstracts services exist. CHEMICAL
ABSTRACTS produced by the American Chemical Society is probably familiar
to all chemists. For many years computers have played a fundamental role
in the preparation of the printed versions of these. Of course, the
computer is indispensible for the processing of the electronic versions
held on magnetic tape. Like any other form of information, if abstracts
are to be useful they need to contain pointers to the original source
documents and they also need to be indexed in an effective way if they are
to be used as an aid to retrospective information retrieval. A simple
scheme similar to that shown below is often quite an effective tool.

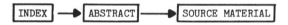

The index can be based upon any of the attributes of the source material
(author, title, date, affiliation, nationality, location, cost etc) and
may be manual or computer based. The abstracts contain a summary of the
contents of the source material. These too (and in some cases the source
material) may be available in computerised form.

Indexing is one of the most powerful tools available to those involved in
information storage and retrieval activities. Unfortunately, the design
of indexes relating to chemical substances was thwarted by the difficulties associated with the ambiguities in chemical nomenclature systems.
To overcome these difficulties various unambiguous notations have been
derived (Bon63). The most well known of these is the Wiswesser chemical
line notation (Wis54, Smi68, Pal70). Using this system, three dimensional
molecular structures can be represented by unique, unambiguous one-
dimensional notations. As there is only one notation for a given compound
there is thus only one place to look for it in an index. A chemical line-
notation is inherently better suited than a nomenclature system for the
preparation of indexes relating to chemical reactions and compounds -
particularly, for organic materials. Wiswesser notation is employed in a
wide variety of contexts relevant to the construction of chemical information systems. Some of these are described in Wip74, Lyn74, Ash75, Ken76
and Oko77.

An important aspect of chemical information systems is the retrieval of
structural information. There are two main approaches to the identification and searching of chemical structures by computer. One method (used
by the Chemical Abstract Service) is based upon the use of a computer
based compound registry system that allocates a registry number to each
unique chemical structure. The other approach is based on the use of
Wiswesser notations.

One of the easiest ways of using Wiswesser line notation (WLN) for the
retrieval of data about compounds is probably through a permuted index of
notations - this is similar to a KWIC index (Pal70). Using this technique
generic sub-structure searching (that is, looking for references to
compounds containing given types of structure) can easily be performed. A
typical query might be "Find all substances containing the XYZ group", or,

"Find all substances that contain the group PQR and the group RST".
Searches such as these can be performed manually on permuted indexes that
have been produced by computer. Alternatively, computer based searches
may be performed on conventional unpermuted files of WLN strings. Many
computer based chemical information systems utilise WLN notation. One of
the most well know of these is the CROSSBOW system operated by Imperial
Chemical Industries (Eak74, Eak75).

The CROSSBOW project began in 1966 and since its inception has accumulated
a large amount of information on a wide range of organic molecules. At
present there are about 130,000 compounds represented in the data base.
Access to the stored data may be via WLN, molecular formula, and company/
divisional reference numbers. Because of the difficulty of providing a
comprehensive search service using only WLN text searching facilities, the
CROSSBOW system uses a multi-level search technique. The system enables
various types of retrieval operations to be performed at three basic
levels. These are shown in Fig 11.9.

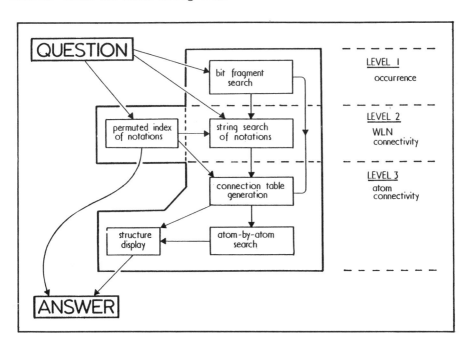

Fig. 11.9 Levels of searching in the CROSSBOW chemical
information system.

The three basic methods of searching the chemical files are as follows:
(a) a search of program generated fragments that are permanently stored in
a bit file, (b) string searching of WLN and/or other parameters such as
molecular formulae, and (c) an atom-by-atom search of a connection table
generated at the time the search is initiated. The principle underlying
this philosophy is that each level can act as a screen so that only
relevant compounds are searched by the separate techniques. In order to
perform these types of search the CROSSBOW system needs to hold the

following data on each compound:

(1) WLN,
(2) a 152 attribute fragment code stored as a bit screen, and
(3) molelcular formulae, varying from 3-18 characters.

The fragment code is generated only once - when the compound is registered
into the collection. Algorithms exist to generate the fragment screen
from the WLN. It shows occurrence without connectivity. In all 148
chemical fragments are coded - there being four spare bit positions for
other uses. Searches are carried out with the basic AND, OR and NOT logic
between fragments or groups of fragments, for example,

AND, BITS "137 50".   will search for all compounds having a benzene
                              ring (137) and a substituent halogen (50).

AND, BITS "97 114 50".   searches for all halogeno-substituted
                              naphthalenes containing only one ring system.

The output from the fragment search is usually examined at the second
search level. This entails examination of the WLN for a string or a
number of strings of notation symbols. There may also be a search of
molecular formula. As in the case of fragment searching, Boolean logic
can be applied. Some typical searches might be:

AND, MF "P0101".   locate all phosphorus compounds containing
                              only one phosphorus atom,

AND, MF "N0499".   search for nitrogen compounds with more than
                              four nitrogen atoms,

NOT, MF "P", "N0599".   search for all compounds that do not contain
                              phosphorus and which contain five or less
                              nitrogen atoms.

Some examples of WLN string searches might be:

AND, WLN "R DR".

OR, WLN "L66J" "R", "R"-"R", "L66J"-"L66J"

while mixed searches might appear as follows:

AND, BITS "137". AND, WLN "R".END

AND, BITS "137". OR, WLN "R".END

String searching is slower than fragment searching but can often produce a
conclusive answer. The third level of searching is atom-by-atom. This is
a time consuming process since it involves generating the connection table
(Bur75, Oko77) for a molecule and then performing a detailed network
search. The atom-by-atom search is specified as a set of nodes connected
in a particular pattern. This type of search is used when an enquiry
requires retrieval of compounds containing particular types of atom
connected in particular ways. Because of the time which such searches
usually take, it is quite common to adopt various screening techniques (of
which the previous two search methods might constitute two examples) to
minimise the number of candidate structures to be processed. Further

details of these techniques as used in the CROSSBOW system are described
in War75, Bir75 and Bur75.

Throughout the world there are many chemical information centres (Row80) -
such as UKCIS in the UK (Bat70), NIH/EPA in the USA (Mar80, Ber79) and so
on - that provide a wide variety of different types of chemical information and data services. Some examples of data centres were given in the
last chapter. Their operation depends upon the use of good, well designed
indexing systems. One of the most important of these is the inverted file
index since this forms the basis of many on-line retrieval facilities.
These will be discussed in more detail in the next section.

ON-LINE SEARCHING

The Basic Principles

Historically, the most commonly available on-line retrieval systems have
been used for retrieving bibliographic information such as abstracts or
full text reports stored in machine readable form. Furthermore, a large
number of on-line systems were developed as a natural progression of
initial involvement with batch orientated current awareness and retrospective retrieval systems. The UKCIS system mentioned in the previous
section is typical of systems that developed in this way. The basic
principles of on-line searching have been described by a number of authors
including Barker (Bar76), Doszkocs (Dos80), Magrill (Mag78), Williams
(Wil80) and others (Kre78, Rog78). Unfortunately, as a result of the
widespread applicability of on-line searching techniques a considerable
problem of terminology has arisen - each particular system having its own
collection of words and phrases to refer to much of what is essentially
common to all systems. An attempt to present some standard terminology
and definitions has been made in a paper by Hawkins and Brown (Haw80).
Most of the discussion that follows is in accord with the recommendations
of these authors.

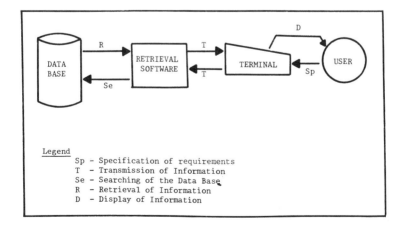

Fig. 11.10   Mode of operation of an on-line data base system.

On-line searching refers to situations in which a computer terminal (such as a visual display unit or a graphics device) is used to specify to a computer program the details of information that is to be retrieved from a data base containing stored (chemical) information. The information to be retrieved may be numeric, textual or of a graphic nature. Visual display units are suitable for numeric and textual data but more sophisticated devices having a greater screen resolution are usually needed for manipulating graphic data such as molecular structures, spectra or three dimensional models. The arrangement of components in a typical on-line (or interactive) retrieval system are depicted in Fig. 11.10.

In this system the user specifies (Sp) to the retrieval software the nature of the information to be retrieved from the data base. That is, a 'search query' is formulated. The computer program responsible for retrieval operations interprets this query and then performs an appropriate search (Se) through the information contained in the data base. Items of information that satisfy the original search specification posed by the user are retrieved (R) and transmitted (T) back to the terminal where they are appropriately displayed (D).

Figure 11.11 illustrates the nature of the human-machine dialogue involved in performing a retrieval operation. It shows the appearance of the VDU screen at different stages during the user's interaction with the retrieval programs. Instructions or information that originate from the computer program appear in upper case letters while information entered by the user is shown in lower case symbols - usually preceded by a special symbol called a 'prompt' character which in this instance is a colon. The object of the dialogue shown is the retrieval of some information about computers from a database containing research abstracts. In order to achieve this goal the user of the system must construct a 'search profile' which will specify the nature of the information that is to be retrieved.

The first entry in the dialogue (1) assigns a name to the search profile; the one used here is 'pgbsearch'. The computer then requests the user to state the concept of interest (entry 2). The reply made is '*computers'. The asterisk included in this expression is used to indicate term truncation. At stage 3 the computer prints out all the indexing terms that contain the stem '.....computers'. The output produced consists of three columns that have been labelled A, B and C. Column C is an arbitrarily assigned descriptor, column B is the list of indexing terms and column A gives the number of documents/abstracts that have been indexed under each of the terms - thus, there are 533 reports indexed by the keyword 'COMPUTERS' and 137 indexed by the term 'DIGITAL COMPUTERS'. At stage 4 in the dialogue the machine requests instructions on what it has to do next. The user now tells the computer to select some of the terms (k004, k009, k010 and k015) for future use.

The second frame shows how the user requests a summary of the contents of the search profile (line 1) and then goes on to specify the Boolean relationships that must be satisfied by the indexing terms before a report abstract is selected for printing. For example, line 3 is used to state that a document is to be printed only if it has been indexed under the terms 'TIME SHARED COMPUTERS' and also under either 'ON LINE COMPUTERS' or 'ON-LINE COMPUTERS'. In frame 3 the user makes a final check of the search profile (line 1) and then requests the computer to initiate a search on the abstracts data base (line 2). The result of the search is displayed (see line 3) and then, in frame 4, the user specifies how the search results are to be displayed - in this case, on a hard copy printer

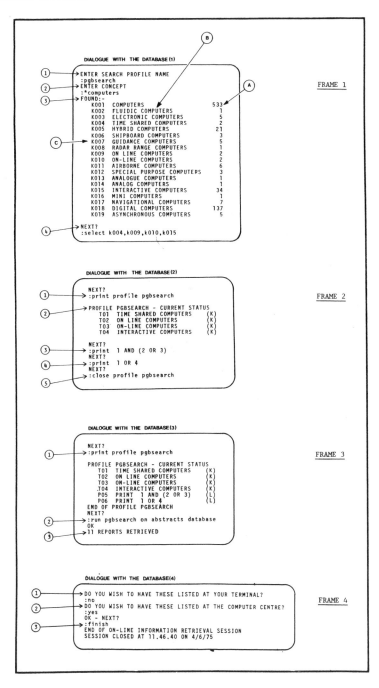

Fig. 11.11 Automatic generation of data base search programs
by means of appropriate human-computer dialogues.

at the local computer centre.

The dialogue described above is characteristic of the way in which on-line
searches are made. The exact mode of usage and the facilities offered by
the retrieval software will vary from one system to another (see Con80) as
will the details of the dialogue protocol. However, as further progress
is made in the area of on-line searching techniques some considerable
standardisation should result. Thus, there is substantial effort
currently being devoted within the European countries to produce a set of
standard on-line retrieval commands.

On-line data bases may be of three basic types: commercial (or public),
in-house (corporate) and personal. They may be based upon a variety of
indexing and/or searching methods - keywords in titles, abstracts or full-
text, manually or automatically assigned indexing terms or any other
useful retrieval attribute. Systems may be orientated towards the
retrieval of bibliographic material, scientific information relevant to
chemical substances or particular concepts such as scientific instrumentation (Rud77).

The use of public/commercial data base vendors has become quite popular
and commonplace mainly because the organisations responsible for their
creation and distribution are sufficiently large to enable reliably
comprehensive coverage of material. Typical examples of organisations
involved in this type of work include Chemical Abstracts Services - (CAS)
- (Huf75), Lockheed (Sum75), Systems Development Corporation (Cua75) and
UKCIS (Bat70). There is a slight difference between the first of these
and the others in that CAS is a data base producer while the other three
are data base service vendors. The most common type of service offered by
these is bibliographic retrieval. In exchange for the facilities that
they produce the organisations concerned usually levy appropriate service
charges. These are based upon both the 'connect time' (that is, the
period of time for which the user is 'conversing' with the system) and the
volume of information retrieved. Connection with the computer system that
holds the data base is usually made via a telecommunication network.
Computer networks will be discussed in more detail in the next chapter and
a specific example (the System Development Corporation) of one of the
above data base vendors will be described. The number of data bases
available for on-line searching grows continually. Comprehensive lists of
these can be obtained from data base vendors or from current issues of
publications dealing with on-line searching, for example, the ONLINE or
DATABASE journals.

Non-bibliographic information is usually available from various organisations that have created appropriate data bases of chemical information.
The most well know of these is probably the NIH/EPA Chemical Information
System which was described in the last chapter. The overall system
consists of a collection of scientific data bases available through an
interactive computer program. It provides a wide variety of information
on over 140,000 chemical substances. The data bases can be searched using
a two dimensional structure representation generated by the user, CAS
index name, Trade names, common names and synonyms - these can be searched
in complete or truncated form. In addition, the files can be searched by
CAS registry number which also gives access to Chemical Abstracts. The
Chemical Information System has been developed by agencies of the United
States government (National Institute of Health and the Environmental
Protection Agency) in cooperation with other governments and
organisations. The facilites are made available globally by means of a

telecommunication network.

As the cost of computer facilities decreases and the amount of information that an individual/organisation has to manipulate increases, so the idea of an 'in-house' (in the case of an organisation) or personal (in the case of an individual) data base becomes more appealing. The concepts underlying the personal data base system have been described by Barker (Bar79b) and van Ree (Ree76). The latter system is essentially a computerised reference retrieval system for personal use. It is based upon the use of a programmable calculator and is used to store up to 900 literature references. These are stored on a tape cassette and each reference is indexed by up to five abbreviated keywords. The system described by Barker is designed to run on a large mainframe computer system and may be accessed via a telecommunication network using a teletype or visual display unit. It is a disk storage based system that uses magnetic tape for archival and back-up purposes. The system evolved as a natural extension of the KWIC index which has been previously described. Based on the use of titles or descriptors, it uses an inverted file of keywords to enable titles to be retrieved along with appropriate locator information to enable the documents concerned to be located. Boolean search logic (AND, OR, NOT) is available to enable the scope of the retrieved results to be modified in various ways. New information can be introduced and old information deleted or modified while the system is on-line. Although the system operates on a mainframe system there are plans to make it available on a microcomputer system.

INFORMATION DISSEMINATION SYSTEMS

As its name suggests, the basic purpose of an information dissemination system is to transmit information or data from a source to various recipients located at different geographical destinations. These may be individual departments within an organisation, different organisations within a country (or distributed world-wide) or the various members of a research team. The basic principles involved in the information dissemin-ation process are embodied in the simple model depicted in Fig. 11.12. In this diagram the destinations (four are shown) are connected to the information source(s) by means of appropriate communication channels (C1, C2, ...Cn). These enable data/information to be transmitted from the source to the many possible destinations. The type of material transmitted might be analytical data or bibliographic material. Regarding the way in which the information is disseminated, two basic possibilities exist:

      (a)   all destinations receive the same information from the source, or,

      (b)   there is selective transmission from the source to the destination so that each only receives that which is necessary to perform its function.

In case (b), the selectivity of information transfer is achieved by imposing an information 'filter' between the transmitter and receiver of information. This is shown in channel 4 in which F4 represents the information filter. Its specification and design will determine the type of information that is passed along the channel from source to destination. Responsibility for its specification may lie with the

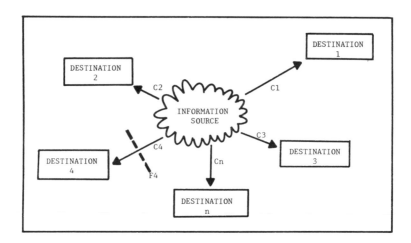

Fig. 11.12 Principle of information dissemination; Ci denote
communication channels and Fi represent information filters.

destination (that is, the recipient specifies the type of information
required for a particular task) or with the source (that is the transmitter decides the type of information that it will send to the destination).
The model presented in Fig. 11.12 represents a gross simplification of
reality and so a more realistic model is presented in Fig. 11.13.

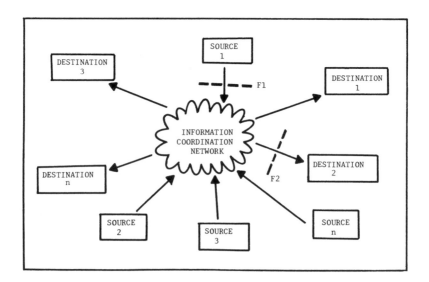

Fig. 11.13 Sources and destinations of information in a
coordinated network; F1 and F2 are information filters
associated with source 1 and desination 2, respectively.

In this model both sources and destinations are represented. These are 'connected' by an information coordination network the function of which is to ensure that information from the various sources reaches the destinations for which it is intended. This model contains two types of information filter. One of these is placed between the source and the network (as illustrated by F1) and the other between the network and the recipient (as in the case of F2). The positioning of the filters will depend upon the exact nature of the function they perform. A model such as that depicted above may be used to explain both the conventional printed 'journals' approach to information dissemination and many of the computer based SDI (Selective Dissemination of Information) systems that have sprung up to support the journal system. Furthermore, it may also be used to explain the various mechanisms by which 'computer conferencing'

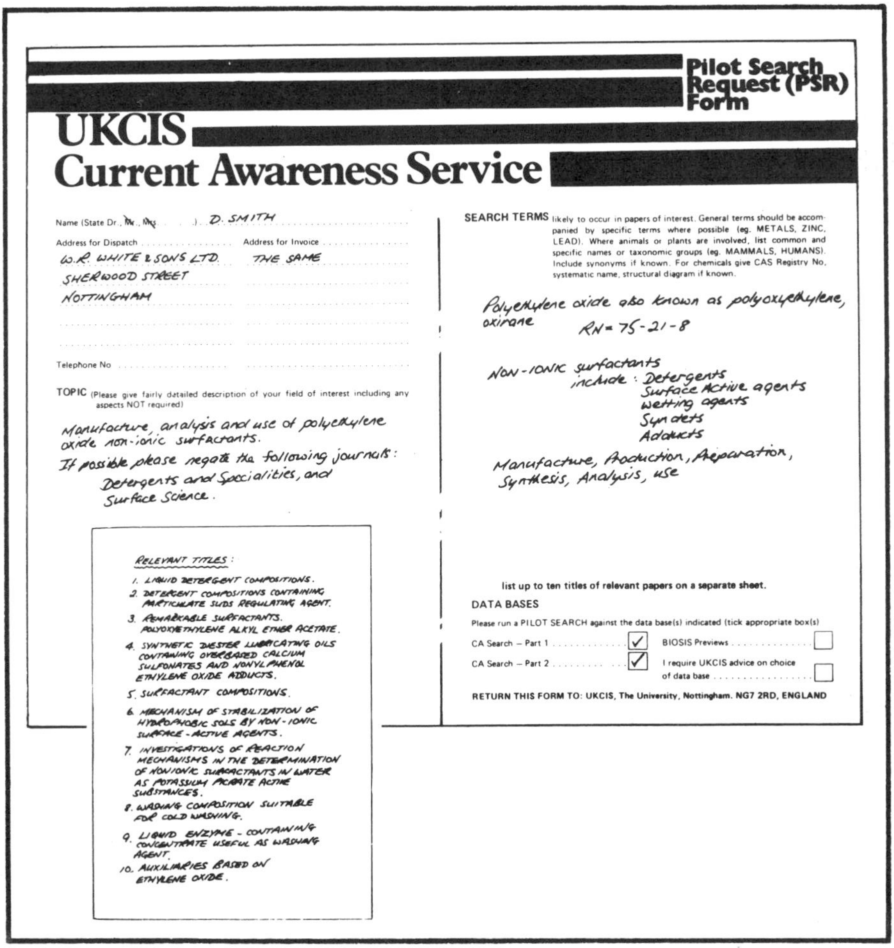

Pilot Search Request (PSR) Form

UKCIS
Current Awareness Service

Name (State Dr., Mr., Mrs ....) D. SMITH

Address for Dispatch: W.R. WHITE & SONS LTD. / SHERWOOD STREET / NOTTINGHAM

Address for Invoice: THE SAME

Telephone No

TOPIC (Please give fairly detailed description of your field of interest including any aspects NOT required)

Manufacture, analysis and use of polyethylene oxide non-ionic surfactants.
If possible please negate the following journals:
Detergents and Specialities, and
Surface Science.

RELEVANT TITLES:

1. LIQUID DETERGENT COMPOSITIONS.
2. DETERGENT COMPOSITIONS CONTAINING PARTICULATE SUDS REGULATING AGENT.
3. REMARKABLE SURFACTANTS. POLYOXYETHYLENE ALKYL ETHER ACETATE.
4. SYNTHETIC DIESTER LUBRICATING OILS CONTAINING OVERBASED CALCIUM SULFONATES AND NONYLPHENOL ETHYLENE OXIDE ADDUCTS.
5. SURFACTANT COMPOSITIONS.
6. MECHANISM OF STABILIZATION OF HYDROPHOBIC SOLS BY NON-IONIC SURFACE-ACTIVE AGENTS.
7. INVESTIGATIONS OF REACTION MECHANISMS IN THE DETERMINATION OF NONIONIC SURFACTANTS IN WATER AS POTASSIUM PICRATE ACTIVE SUBSTANCES.
8. WASHING COMPOSITION SUITABLE FOR COLD WASHING.
9. LIQUID ENZYME-CONTAINING CONCENTRATE USEFUL AS WASHING AGENT.
10. AUXILIARIES BASED ON ETHYLENE OXIDE.

SEARCH TERMS likely to occur in papers of interest. General terms should be accompanied by specific terms where possible (eg. METALS, ZINC, LEAD). Where animals or plants are involved, list common and specific names or taxonomic groups (eg. MAMMALS, HUMANS). Include synonyms if known. For chemicals give CAS Registry No, systematic name, structural diagram if known.

Polyethylene oxide also known as polyoxyethylene, oxirane RN = 75-21-8

NON-IONIC surfactants include: Detergents
Surface Active agents
wetting agents
Syndets
Adducts

Manufacture, Production, Preparation, Synthesis, Analysis, use

list up to ten titles of relevant papers on a separate sheet.

DATA BASES

Please run a PILOT SEARCH against the data base(s) indicated (tick appropriate box(s)

CA Search – Part 1 .... ✓
CA Search – Part 2 .... ✓
BIOSIS Previews .... ☐
I require UKCIS advice on choice of data base .... ☐

RETURN THIS FORM TO: UKCIS, The University, Nottingham. NG7 2RD, ENGLAND

Fig. 11.14. United Kingdom Chemical Information Service (UKCIS) current awareness service; pilot search request form.

and the 'electronic journal' operate. These latter two topics will be
described in the next chapter.

As an example of a computer based system for aiding the selective
dissemination of information UKCIS (Bla78) will be briefly outlined. This
is an example of a system that has been designed to support the conventional bibliographic journal system and acts as a current awareness
service to scientists. It is based upon the use of Chemical Abstracts
data bases. These are produced by scanning over 14,000 journals and
patents from 26 different countries. In addition, various conference
proceedings, government reports and a large number of books are abstracted
and indexed. Searches can be made based upon any of the following:-

Title and keyword phrases	CAS Registry Number
Author Names	Molecular Formula
Primary publication type	Chemical Abstracts Systematic Name
Original Language	Controlled language concept headings
Location of work	Free language concept modifiers
Journal	Chemical Abstract sections
Country of Patent Application	Patent Classification

When using the system the user formulates a specification of his/her
information interests and also presents any particular terms that are
known to be relevant to the chosen topic. In addition, a list of several
titles of papers or publications that are relevant to the field of
interest must be provided. These three pieces of information are then
converted into a 'search profile' by an information scientist. The search
profile is run against the data base and is used as a means of selecting
the required entries for its owner. In other words the profile acts like
a filter that permits only relevant material to 'slip' through. Figure
11.14 shows a typical search request.

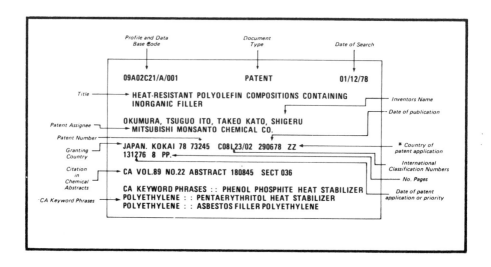

Fig. 11.15 Sample of output produced by the UKCIS
current awareness service.

The search profile that is constructed can be run on a weekly, fortnightly or monthly basis. Typical output produced by the system and returned to the user is illustrated in Fig. 11.15.

Usually, several references similar to that shown are returned. These contain all the information needed to obtain further details of the work concerned either from Chemical Abstracts or from the appropriate Chemical literature. The exact number of references retrieved by the search profile will depend upon its specificity. If it is too specific then it may not allow a sufficient number of references to be retrieved, while if it is too broad it may allow too many irrelevant references to be returned to the user. Consequently, the profile is usually 'tuned' in the light of experience with it.

Several other types of information dissemination systems exist. Many of these are very dependent upon the use of computer networks for their implementation. Some of them will be discussed later.

CONCLUSION

When a scientist starts a project it is important to know if any similar work has been done previously or is currently being conducted by any other laboratory or research group. To solve the problems associated with the large amount of chemical literature the scientist often uses computer based systems to help retrieve any information that is available. Once this 'literature search' has been completed the scientist then has to design the experiment(s) - often with the aid of a computer. This machine also helps to conduct the experiments and store the results that are produced as a result of experimentation. The computer can then be used as an aid to interpreting the results and, using word processing techniques, prepare papers/reports for publication. When disseminated to the scientific world through the journal system, details are abstracted and indexed and then entered into another computer system so that the results of the work can be made available to others through computer based searching and dissemination methods. King's cyclic model that was introduced at the beginning of this chapter therefore seems to be fulfilled. As computer networks become cheaper, more available and easier to use, so access to international data bases will become more widespread. Such networks should also enable the easier dissemination of scientific and technical information through the medium of the 'electronic journal' and 'computer conferencing' as will be outlined in the next chapter.

REFERENCES

All77   Allen, F.H. and Town, G.W., The Automatic Generation of Keywords from Chemical Compound Names: Preparation of a Permuted Name Index with KWIC Layout, Journal of Chemical Information and Computer Sciences, Volume 17, No. 1, 9-15, 1977.

Ant79   Antony, A., Guide to Basic Information Sources in Chemistry, John Wiley, ISBN: 0-470-26587-6, 1979.

Ash75   Ash, J.E. and Hyde, E., (Eds), Chemical Information Systems, John Wiley, ISBN: 0-470-03444-0, 1975.

Bag79   Baglow, G. and Bottle, R.T., Rate of Publication of British
         Chemists, Chemistry in Britain, Volume 15, No. 3, 138-141,
         March 1979.

Bar76   Barker, P.G., Information Retrieval - A Tape Slide
         PresentationPrismatron Productions Ltd., London, UK., 1976.

Bar79a   Barker, P.G., The BINSYS Batch System for Document Retrieval,
         Interactive Systems Research Group Working Paper, 1979.

Bar79b   Barker, P.G., The BINSYS On-line Retrieval System, BINSYS
         Documentation Volumes I and II, Interactive Systems Research
         Group, University of Durham, 1979.

Bat70   Batten, W., UKCIS - The United Kingdom Chemical Information
         Service, Chemistry in Britain, Volume 6, No. 10, 420-422,
         October 1970.

Ber79   Bernstein, H.J. and Andrews, L.C., The NIH/EPA Chemical
         Information System, DATABASE, Volume 2, Issue 1, 35-43 and
         46-49, March 1979.

Bir75   Bird, J.M., Substructure Search at the Connectivity Level Atom-
         by-Atom, Imperial Chemical Industries, Pharmaceutical Division,
         Data Services Section, January 1975.

Bla78   Blake, J.E., Mathias, V.J. and Patton, J., CA-Selects - A
         Specialised Current Awareness Service, Journal of Chemical
         Information and Computer Sciences, Volume 15, No. 1, 48-51,
         1975.

Bon63   Bonnett, H.T., Chemical Notations - A Brief Review, Journal of
         Documentation, Volume 3, 235-242, 1963.

Bot79   Bottle, R.T., Use of Chemical Literature - Information Sources
         for Research and Development, Butterworths, ISBN: 0-408-38452-2,
         1979.

Bur75   Burgess, M.T. and Eakin, D.R., The CROSSBOW Mark II Connectivity
         Table, Imperial Chemical Industries, Pharmaceuticals Division,
         Data Services Section, June 1975.

Caw70   Cawkell, A.E., Citations in Chemistry, Chemistry in Britain,
         Volume 6, 414-416, 1970.

Con80   Conger, L.D., Multiple System Searching: A Searcher's Guide to
         Making Use of the Real Differences Between Systems, ONLINE,
         Volume 4, No. 2, 10-21, April 1980.

Cra57   Crane, E.J., Patterson, A.M. and Marr, E.B., A Guide to the
         Literature of Chemistry, John Wiley, 1957.

Cua75   Cuadra, C.A., SDC Experiences with Large Data Bases, Journal of
         Chemical Information and Computer Sciences, Volume 15, No. 1,
         48-51, 1975.

Dav74   Davis, C.H. and Rush, J.E., Information Retrieval and Documenta-
         tion in Chemistry, Greenwood Press, ISBN: 0-8371-6364-1, 1974.

Dos80   Doszkocs, T.E., Rapp, B.A. and Schoolman, H.M., Automated
           Information Retrieval in Science and Technology, SCIENCE,
           Volume 208, 25-30, 4th April, 1980.

Eak74   Eakin, D.R., Hyde, E. and Palmer, G., The Use of Computers with
           Chemical Structural Information, Pestic. Sci., Volume 5,
           319-326, 1974.

Eak75   Eakin, D.R., The ICI CROSSBOW System, Chapter 14, 227-241 in Ash
           75.

Fre63   Freeman, R.R. and Dyson, G.M., Development and Production of
           'Chemical Titles', A Current Awareness Index Publication Prepared
           with the Aid of a Computer, Journal of Chemical Documentation,
           Volume 3, 16-20, 1963.

Gar64   Garfield, E., "Science Citation Indexing" - A New Dimension in
           Indexing, SCIENCE, New York, Volume 144, 649-654, 1964.

Has65   Hass, A.K., Internal Alerting with Keyword in Context Indexes,
           Journal of Chemical Documentation, Volume 5, No. 3, 160-163,
           August 1965.

Haw80   Hawkins, D.T. and Brown, C.P., What is an Online Search?,
           ONLINE, Volume 4, No 1, 12-18, January 1980.

Huf75   Huffenberger, M.A. and Wigington, R.L., Chemical Abstracts
           Service Approach to Management of Large Data Bases, Journal of
           Chemical Infromation and Computer Sciences, Volume 15, No. 1,
           43-47, 1975.

IBM62   IBM Corporation, Keyword-In-Context (KWIC) Indexing, Form:
           GE20-8091-0, 1962.

ISI78   Institute for Scientific Information, see ISI81, Index to
           Scientific and Technical Proceedings, 1978.

ISI81   Institute for Scientific Information, 325 Chestnut Street,
           Philadelphia, Pennsylvania 19106, USA.

Ken76   Kent, A.K., (Ed.), Techniques for the Retrieval of Chemical
           Information, Proceedings of the IUPAC Symposium 9th-10th
           November, 1976, ISBN: 0-08-021193-3, Pergamon Press, 1976.

Kin78   King, D.W., Statistical Indicators of Scientific and Technical
           Communication (1960-1980), 299-304 in Key Papers in the Design
           and Evaluation of Information Systems, (Ed. King, D.W.),
           American Society for Information Science, Knowledge Industry
           Publications, Inc., ISBN: 0-914236-31-8, 1978.

Kre78   Krentz., D.M., On-line Searching - Specialist Required, Journal
           of Chemical Information and Computer Sciences, Volume 18, No. 1,
           4-9, 1978.

Luh59   Luhn, H.P., Keyword in Context Index, Journal of Technical
           Literature (KWIC Index), IBM Advanced Systems Development
           Division, 1959.

Lyn74   Lynch, M.F., Computer-Based Information Services in Science and
         Technology - Principles and Techniques, Peter Peregrinus, ISBN:
         09-01223-55-7, 1974.

Mai79   Maizell, R.E., How to Find Chemical Information - A Guide for
         Practicing Chemists, Teachers and Students, John Wiley, ISBN:
         0-471-56531-8, 1979.

Mag78   Magrill, D., Information at the Touch of a Button, New
         Scientist, Volume 77, No. 1085, 76-79, 12th January, 1978.

Mar80   Marquaret, R.G., Marquaret, L.M., McDaniel, J.R., McGill, J.R.,
         Mintz, S.A., Heller, S.R. and Milne, G.W.A., The NIH/EPA CIS
         Federal Register Notices Search System, ONLINE, Volume 4, No.
         2, 45-49, April 1980.

Mel65   Mellon, M.G., Chemical Publications - Their Nature and Use,
         McGraw-Hill, 1965.

Mur67   Murdock, J.W. and Liston, D.W., A General Model of Information
         Transfer: Theme Paper 1968 Annual Convention, 287-298, in Key
         Papers in the Design and Evaluation of Information Systems, (Ed:
         King, D.W.), American Society for Information Science, Knowledge
         Industry Publications, Inc., ISBN: 0-914236-31-8, 1978.

Oko77   O'Korn, L.J., Algorithms in the Computer Handling of Chemical
         Information, Chapter 6, 122-148, in Algorithms for Chemical
         Computation, (Ed: Christopherson, R.E.), ACS Symposium Series
         No. 46, ISBN: 0-8412-0371-7, 1977.

Pal70   Palmer, G., Wiswesser Line-Formula Notation, Chemistry in
         Britain, Volume 6, No. 10, 422-426, October 1970.

Pet71.   Petrarca, A.E., Laitinen, S.V. and Lay, W.M., Use of the Double
         KWIC Coordinate Indexing Technique for Chemical Line Notations,
         Journal of Chemical Documentation, Volume 11, No. 3, 148-153,
         1971.

Ree76   van Ree, T., A Personal Reference Retrieval System, Journal of
         Chemical Information and Computer Sciences, Volume 16, No. 3,
         152-153, 1976.

Rog78   Rogalski, L., On-line Searching of the American Petroleum
         Institute's Data Bases, Journal of Chemical Information and
         Computer Sciences, Volume 18, No. 1, 9-12, 1978.

Row80   Rowlett, R.J., International Sharing of Chemical Information,
         Chemistry in Britain, Volume 16, No. 8, 425-427, August 1980.

Rud77   Rudman, R., A Coded Data Bank for Chemical Instrumentation,
         Journal of Chemical Information and Computer Science, Volume
         17, No. 4, 208-210, 1977.

Sko70   Skolnik, H., A Correlative Notation System for NMR Data, Journal
         of Chemical Documentation, Volume 10, No. 3, 216-220, 1970.

Smi68   Smith, E.G., The Wiswesser Line-Formula Chemical Notation,
         McGraw Hill, New York, 1968.

Sum75   Summit, R.K., Lockheed Experience in Processing Large Data Bases
          for its Commercial Information Retrieval Service, Journal of
          Chemical Information and Computer Sciences, Volume 15, No. 1,
          40-42, 1975.

War75   Warr, W.A., Substructure Search at the WLN Level, Imperial
          Chemical Industries Ltd., Pharmaceuticals Division, Data Services
          Section, January 1975.

Wei71   Weinstock, M., Citation Indexes, 16-40, in Volume 5 of
          Encyclopedia of Library and Information Science, Marcel Dekker,
          New York, 1971.

Wil80   Williams, P.W., Henry, W.M., Leigh, J.A. and Tedd, L.A., Online
          Searching - An Introduction, Butterworths, ISBN: 0-408-10696-4,
          1980.

Wip74   Wipke, W.T., Heller, S.R., Feldmann, R.J. and Hyde, E., Computer
          Representation and Manipulation of Chemical Information, John
          Wiley, ISBN: 0-471-95595-7, 1974.

Wis54   Wiswesser, W.J., A Line Formula Chemical Notation, Crowell, New
          York, 1954.

Woo74   Woodburn, H.M., Using the Chemical Literature - A Practical
          Guide, Marcel Dekker Inc., ISBN: 0-8247-6260-6, 1974.

12

Computer Networks and the Future

INTRODUCTION

Throughout this book considerable emphasis has been given to the various aspects of computing and their relevance to analytical chemistry. The main reasons for requiring access to computer systems derive from the continual need to be able to process data, store it and subsequently retrieve it. Additional and equally important reasons arise from the increasing requirement for the computer to be utilised as a control tool both in the laboratory and in the process control environment. In chapter 4, various types of computing system were described: micros, minis, mainframes and super-computers. The latter were based upon the highly parallel inter-connection of processing elements to produce multi-processing systems and array processors. The latest trend in this latter area is towards the use of ultra-computers (Sch80). These require a means of putting together computing assemblages consisting of thousands of interconnected elements. The practical realisation of such an arrangement depends upon the use of VLSI technology (see chapter 4). The motivation for connecting computing elements together lies in the fact that with such machines it is possible to achieve faster computational speeds (through parallelism) and greater reliability (through redundancy). Computers linked together in this way are usually located within fairly close proximity - often within the same room. Previously (chapter 4), the term multi-processing was introduced in order to describe this type of inter-connection.

There are many reasons for wanting to interconnect computer systems - particularly, in situations where the distances between the elements to be linked is geographically large - perhaps, thousands of miles. The term computer networking is used to describe this approach to digital systems interconnection. Figure 12.1 shows the topology of a typical network configuration.

In this diagram, individual computers in the network are referred to as nodes. The connecting lines between them represent data transmission links that enable data and information to flow between the various nodes. The term communications sub-network is often used to refer to

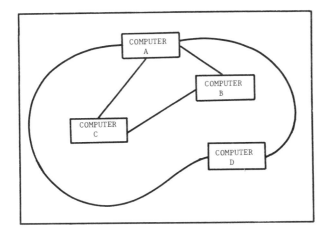

Fig. 12.1 Topology of a distributed computer network.

the underlying data transmission arrangements that support the integrated computer system. The geographical distribution of the nodes in the network will usually not influence its operation or the functions it is designed to perform. Thus, in Fig. 12.1, Computer A might be located in London, B in Paris, C in New York and D in Oslo. Sometimes there will be more than one direct link between given nodes (as in the case of A and D) in order to provide greater overall system reliability.

Closely dependent on the idea of computer networks and a concept of growing importance is that of distributed computing. Here the computational tasks to be performed in solving a problem are serviced by the resources of the computer network as a whole rather than those associated with any particular individual node. Thus, in the simple example network described above, it is feasible that laboratory data collected by a data acquisition system attached to computer D could be transmitted to computer A for processing; then, following processing, the results could be transmitted to computers B and C for storage. Retrieval requests for inspection of particular items of data might then arise from users of any of the four computers A, B, C or D. Distributed processing of this type can offer many advantages such as: high availability and greater reliability; improved work throughput and response time; distributed data processing, storage and retrieval; load levelling and resource sharing; greater security, integrity and privacy of data; and, system modularity and the implications that this offers for a highly structured approach to implementation.

Essentially, a distributed processing system may be thought of as an inter-connection of geographically distributed digital sub-systems each having certain processing capabilities and communicating with other sub-systems through the exchange of messages of various sorts - a more rigorous list of criteria has been given by Enslow (Ens78). Within such a distributed system processing elements may have their own local operating system (see chapter 4) and processing software which may be unique to that element. The various processing elements will communicate with each other

using common message transmission protocols. Two commonly used techniques
for transmission of information around a network are message switching and
packet switching, these will be described in more detail later. One
important feature of the network is that the route information takes from
an originating node to its destination node will not be guaranteed since
this will be influenced by the state of the network at any time. To the
user the system will present a common command language via the network
operating system. This will usually provide a set of high level commands
that enable the user to control the services and facilities that the
network offers - for example, CREATE, SEND, FETCH, FIND, to control the
manipulation of files of stored data; DATABASE XYZ, to establish
connection with a particular data base system, and so on.

Since their inception, computer networks have extended from simple in-
house affairs to systems that span both national and international
boundaries. They have extended across continents in order to inter-
connect major computing facilities around the world. Nowadays, with the
advent of inexpensive microcomputers and minicomputers (and equipment
based upon them), computer networks are extending between buildings and
along corridors to enable linking together of offices and laboratories on
a world-wide scale. Progress in this area is very rapid - particularly
since the advent of satellite links. In view of the importance of this
topic and its likely impact on analytical science this final chapter is
devoted to an introductory study of computer networks and the ways in
which they are likely to be used.

TYPES OF NETWORK AND THEIR PURPOSE

Computer networks may be classified by any of a wide range of possible
attributes: by their topology (star, tree, loop, etc); by the control
discipline used (central or distributed); by the type of information that
the network carries and the mechanism for transmitting it (message or
packet switching); by the communication links involved (cable, twisted
pair, radio, etc); and, the nature of the computers involved - these may
be of the same type or they may differ quite significantly (homogeneous
and heterogeneous networks, respectively). These attributes represent
just a few of those that are widely used to describe and classify the
different types of computer networks that currently exist. In this
section some of these attributes will be explained and examined in more
detail.

Network Topology

The term topology refers to the geometrical arrangement of links and nodes
of a network. Within these nodes it is possible to locate several
different types of hardware and software depending upon the function that
an individual node is to perform. When designing a network, many
different factors must be evaluated in order to choose the most suitable
topology. One major factor that is likely to strongly influence this
choice is the type of participation required by each of the nodes. Thus,
it is possible for a node to act

    (1)   solely as a consumer of resources,
    (2)   exclusively as a provider of resources, or,
    (3)   as both a consumer and provider of network resources.

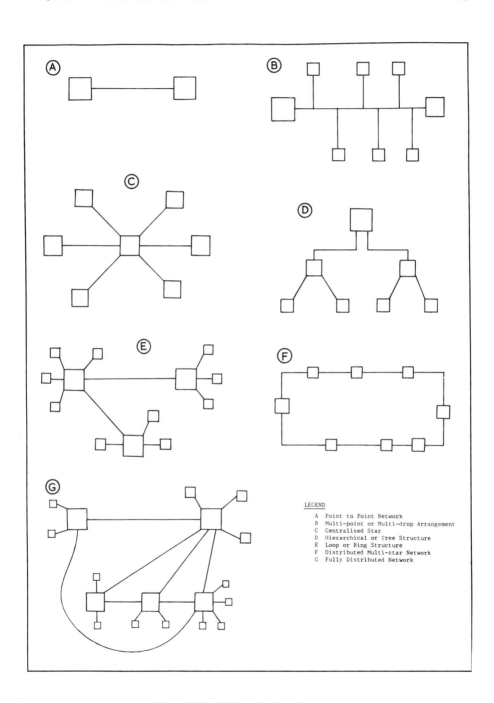

Fig. 12.2   Network structures.

Depending upon the likely resource utilisation and the way in which nodes need to communicate with each other, about half a dozen "standard" types of network topology are commonly used. These are summarised in the seven diagrams (A through G) contained in Fig. 12.2.

The simplest type of configuration is the point-to-point arrangement shown in diagram A. Here two nodes are interconnected by a single communication link. This may be a private line or a switched line as in the public dial-up network. A more complex arrangement of nodes is illustrated in the multi-point system depicted in diagram B. This requires that several nodes share the same communication line. One of the nodes in the network is designated as the controller and the others then become tributory stations. The control station controls network traffic by means of polling, that is, it invites tributary stations to send messages in turn. Multi-point networks are usually established over non-switched leased lines.

Diagram C shows another popular network arrangement called a centralised star system. In this type of topology all users communicate with a central point that has supervisory control over the entire system. Peripheral nodes can only communicate with each other via the central supervisory node. This thus provides a central message switching facility.

A typical hierarchical structure is shown in sketch D. Such an arrangement is often employed in industrial environments to supervise and control a variety of real-time, process control applications. A hierarchy of computers is used to control various processes, synchronise them and report their status. Both microcomputers and minicomputers are used as nodes in this type of arrangement. These occupy the lower levels of the tree structure with, perhaps, a mainframe or large minicomputer at the top.

Many organisations design their computer networks in the form of a loop or ring structure (diagram E). In an arrangement of this type there is a common communication loop to which all nodes are attached. The data to be transmitted is then looped around the nodes in turn. A loop or ring arrangement of this type is very economical when several remote stations and host processors are located near to each other - perhaps within the same building or distributed over a manufacturing plant. However, when the stations are geographically dispersed over long distances the line costs would probably be too expensive for a loop structure and a cheaper form of distributed network would probably be required. Two of these are described below.

The multi-star network similar to that shown in diagram F is an often used configuration in which there are several supervisory or exchange points each with their local cluster of attached nodes. The local hosts usually service the requirements of their attached nodes but also permit general communication between any nodes in the network. If properly designed, distributed networks can offer significant reliability advantages, since a failure at one node does not effect the rest of the network. Indeed, in applications where the reliability of continuous communication is important, a fully distributed network (diagram G) in which every point is connected to several neighbouring points may be preferred. The additional transmission paths provided by this type of structure improves the overall performance of the network. When using this type of topology, detailed traffic analysis must usually be performed in order to determine where the links are required.

The network structures described above and illustrated in Fig. 12.2
represent the most common types of discrete network architecture. It is
feasible, however, to use these as basic building blocks to construct even
more complex networks. Thus, two, three, or more, networks having
topologies similar to that shown in diagram G may be inter-connected to
form a highly distributed arrangement of nodes. Logically, the arrangement will appear as three separate networks linked at particular points.
Because the individual networks will require to retain various attributes
of autonomy, and, because they will differ considerably from each other in
their characteristics, special modes of interconnection are required.
Nodes that are used to interlink networks of different types in this way
are called gateway nodes. Their design has been described by a number of
authors (Hig75, Wal75). A description of one such gateway that connects
the University of Rochester to the ARPA network in the USA has been
presented by Ball et al (Bal76).

Circuit, Message and Packet Switching Networks

There are three basic methods for routing communications traffic from a
source to a destination within a computer network: circuit switching,
message switching and packet switching. In a circuit switching network -
similar to the public switched telephone system described in chapter 7 -
the role of the switching centre(s) is to establish a direct connection
between nodes in the network. Once established these may then carry on
one-way or two-way communication with minimal delay between transmission
of a message and its arrival at its destination. When communication is
complete, the switching centres disconnect the circuit and restore the
system in readiness for other connections. Circuit switching often
requires long connect times and ties up transmission capacity for long
periods. This arises because of a fundamental property of circuit
switching - once a path is determined through the network nodes, all
traffic between a source and destination pair then follows the same path.

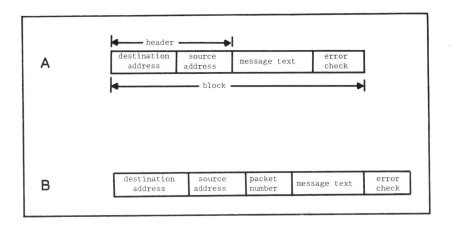

Fig. 12.3 Simple message (A) and packet structures (B) for
use in data routing networks.

An alternative mode of transmission which does not require a fixed route between source/destination could have many advantages. Using such an approach, two possibilities exist depending upon the volume of data to be transmitted: message switching and packet switching. In message switching each item of data is sent into the network as a discrete unit and is then routed to its destination. The format of the unit that is transmitted is illustrated in diagram A of Fig. 12.3.

A message makes its way through the network to the destination whose address is specified in the header. Each node in the network uses an appropriate routing algorithm in order to decide which node the message has to go to next in order to reach its destination. Since some stations may be busy, a message may often have to be stored at intermediate nodes before it is passed on. For this reason an arrangement such as this is often called a 'store and forward' system.

Packet switching is essentially similar to message switching and is used when large volumes of information are to be transmitted. At the source station a large message is sub-divided into a series of fixed length segments (called packets) of size 1,000 - 8,000 bits. Each packet has a unique number associated with it to enable the reconstruction of the complete message at the destination. The format of a packet is similar to that of a message and is shown in Fig. 12.3 (diagram B).

Packets are treated individually and are forwarded along the best avail- able route, that is, the route with the shortest transmission delay. Each packet is checked for errors at each node along the way by means of the error checking field contained in the packet. Because long messages are broken up and sent over different routes it is possible for them to arrive at their destination more quickly. Furthermore, since intermediate nodes in a packet switched network only have access to parts of messages they are unable to assemble the entire message. Thus, if data encryption is not being used, transmission is more secure.

Videotex Networks

Videotex networks (Bal80) were originally designed to provide low cost public data and information retrieval networks based upon broadcast TV signals or a switched telephone network. Intended primarily for use as public information utilities the systems were designed around the use of a single tree structured data base. Modified TV sets were used as user terminals through which could be implemented a variety of menu selection techniques in order to facilitate data/information retrieval operations. Figure 12.4 shows a simple videotex network.

There are two types of terminal: IPT - the information provider's terminal and UT - the user's terminal. The information provider is the person responsible for entering data into the data base and ensuring its accuracy. The arrangement of components is essentially a star network with the computer at the centre and the terminals and videotex data base (M) attached as peripheral nodes. This type of equipment is often used for the provision of in-house information systems - for a laboratory, operations room or sales office. In addition to their prime use as information retrieval tools, the two-way communication capability of many of these systems enables the implementation of a wide variety of elec- tronic mail and electronic journal facilities. On a large scale such systems are used to provide global or national information utilities. Typical examples of systems of this type include Prestel (UK),

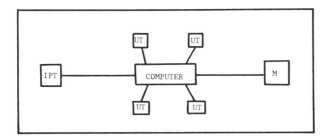

Fig. 12.4 Simple videotex network; IPT: information provider's
terminal; M: videotex data base; UT: user's terminal.

Bildschirmtext(Germany), Antiope (France), and Telidon (Canada). These
videotex networks are widely used for the provision of a variety of
commercial, scientific and technical information to the general public.

STANDARDS

One of the basic requirements necessary to enable the construction of a
viable computer network is the ability to connect various types of
computer equipment to a communication sub-system. In chapter 6 the
principles of interfacing were considered and some of the problems of
connecting one device (say, a computer) to another (such as a laboratory
instrument) were briefly outlined. An easy solution to many of the
problems involved in this operation could be avoided through the use of
standard interfaces. An analogous situation exists in the case of networking: when network components are inter-connected a variety of problems
can arise unless appropriate standards are available which specify how
components are to be linked together. Because of their much greater
complexity there will obviously be many more problems associated with the
formulation of standards for computer networks compared with those
necessary to enable simple interfacing of devices. Broadly, there will be
two classes of standard: hardware related - which specify how components
are to be physically connected with each other (typical considerations
might include electrical signal characteristics, signal speed, signal
types, etc); and, software related - which deal with the way in which
information is transmitted around the network (line protocols, error
checking, encryption methods and so on). Some of the different types of
standards used in networking will be briefly outlined in this section.
The treatment will not be comprehensive but will be sufficient to draw the
reader's attention to the type of problem involved when considering the
use of networks.

As in most other areas of technology standards originate from many
different sources - independent manufacturers, government bodies,
national/international standardisation committees and so on. Standards
produced by individual manufacturers are formulated for two reasons:
firstly, to ensure that the various components they manufacture will
ultimately fit together to form a variety of integrated operational
networks and, secondly, to enable customers to plan the details of the
particular configuration that they require. The large number of internal

standards involved in the description of network products such as IBM's
SNA (Systems Network Architecture - see IBM78, McF76), DEC's DECNET system
(DEC78) or Hewlett Packard's Distributed System Network (HP-DSN - see
Sch78) reflect the significant amount of detailed specification necessary
to achieve an operational network. Standards which are formulated by the
various standards organisations are not orientated towards any particular
vendor's product. Some of the international organisations involved in the
formulation of standards have been mentioned earlier (CCITT, ISO, EIA,
BSI, ANSI, etc). CCITT is probably one of the most well known of these.
Selected examples of some of the CCITT standards/recommendations employed
in computer networking are listed in Table 12.1.

TABLE 12.1 Some CCITT Standards used in Computer Networking

V3	International Alphabet Number 5
V15	Acoustic Coupling
V21	200 baud modem for use in switched telephone networks
V23	600/1200 baud modem for use in switched telephone networks
V24	Interface between data terminal equipment and data communication equipment
V26	2400 baud modem for 4 wire leased circuits
V26b	2400 baud modem for use in the switched telephone network
V27	4800 baud modem for leased circuits
V35	48 kilo-baud transmission on group band circuits
X1	User classes of service and data signalling rates for public data networks
X2	Recommended user facilities available in public data networks
X3	Defines facilities to be provided by a Packet Assembly and Disassembly (PAD) service for a packet switched network (Hou78)
X20	Interface between data terminal equipment and data circuit terminating equipment for start-stop services in user classes 1 and 2 on public data networks
X21	Interface between data terminal equipment and data circuit terminating equipment for synchronous operation on public data networks (circuit switching standard)
X25	Device independent interface between packet networks and user devices operating in the packet mode (Kir76, Ryb80)
X28	Defines the interface between asynchronous terminals and a packet assembly and disassembly unit (Mag79)
X29	Defines how a remote packet mode terminal (normally a host) communicates with an asynchronous terminal via a PAD
X75	Defines the interface between packet mode public data networks (Gro79)
X121	Proposed international numbering plan for public data networks (Hum 79)

Figure 12.5 illustrates how some of these standards might be employed in
the design of a typical packet switching network.

When data is exchanged between nodes in a network, transmission and
receipt of a block of data is achieved by handshaking. This concept was
introduced in chapter 6 to describe the various protocols necessary to
enable two devices to exchange data in an error free way - transmission is
considered to be complete only when the receiving device gives a positive
acknowledgement. The situation is entirely analogous in computer networks. A variety of handshaking protocols have been used but the most
popular of these has been Binary Synchronous Line Control (called BSC or
BISYNC - see DEC74) and its derivatives. BSC employs a rigorous set of
rules for establishing, maintaining and terminating a communications
sequence. It is a character orientated protocol and can be used on point-
to-point and multi-point lines. Another typical line protocol is
Synchronous Data Link Control (SDLC - see DEC74) which is similar to BSC
but is bit rather than character orientated. Both BSC and SDLC are line

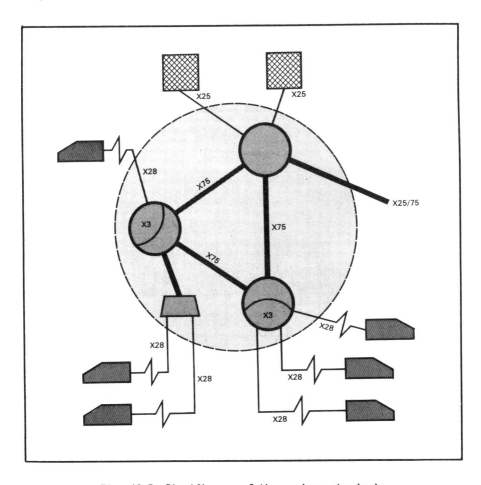

Fig. 12.5 Significance of the various standards
used in computer networks.

control standards that oiginate from IBM. An independent standard
developed by the International Standards Organisation (ISO) which is being
adopted by many manufacturers is High Level Data Link Control (HDLC - see
DEC74, Dav73, Hou78). This was originally developed to cover mainframe-
to-terminal applications and was specifically aimed at two-way simultaneous transmission. It is an important standard since it is closely
involved with the CCITT X25 recommendations for packet switching.

In most computer networks packet switching is becoming established as a
standard - probably because it offers the most cost effective approach to
networking (Ros76) on both a national and international level. Because of
the current interest in packet switching the X25 standard is of considerable importance since most large scale public networks (such as PSS in
the UK and TRANSPAC in France) and private networks (for example, TELENET,
TYMNET in USA and DATAPAC in Canada) seem to be adopting it. Furthermore,
the use of the X25 standard seems to be paving the way for the facile

construction of multi-vendor networks through the use of open systems
interconnection (Hou78) - see below. The X25 standard enables the inter-
connection of terminals of any type operating at any speed, that is, it
permits total device independence. The standard itself is sub-divided
into three levels. Level 1 stipulates the use of X21 interchange circuits
to provide the connection, maintenance and disconnection of the link
between a terminal and its local packet switching exchange (PSE). Level 2
is the link access procedure which manages the terminal - PSE link (via
HDLC) and the transfer of packets between the two, while level 3 concerns
the packet format, communications with other packet terminals and the
procedures for setting up calls. This last section lays down an
international standard packet size - a minimum of 128 characters for
information plus the address code. Although packet switching is currently
very popular and is likely to remain so for the near future it has technical limitations - particularly with respect to bandwidth for information
transfer. Furthermore, with the transition towards high speed digital
circuit switching technology for telecommunications applications there may
be severe migration and compatibility problems in future years.

The demand for computer networking facilities is growing substantially
within many organisations. However, until recently the inter-linking of
different types of computer has often presented many problems whenever the
machines to be connected have originated from different manufacturers.
The basic requirements of many organisations are often similar to those
depicted in Fig. 12.6. Here, two laboratory computers (LC) and a special
purpose data base machine (DBM) are inter-linked via the network to other
host and satellite processors within the organisation. In order to
overcome the problems involved in building this type of network and thus
mix vendor equipment with the minimum of difficulty the International
Standards Organisation has suggested the idea of an Open Systems Inter-
connection model (Hou78, Pia80, Bre80). Essentially, the model proposes
that any data processing system should be able to communicate readily with
any other by means of a set of inter-connection rules that are embodied in
an appropriate series of commonly agreed-upon standards. The ISO
recommendations are that a seven level model be used. This model is
illustrated in the diagram contained in Fig. 12.7.

The ISO model's lowest layer (physical) describes the lines between
network nodes; these could be RS-232-C or RS-449; V24 or V25; or, X21 for
interfacing to circuit switched networks. The second layer (link)
describes the passage of data packets over lines using HDLC. The third
layer (network control) describes the passage of entire messages between
nodes; network control routes traffic onto the correct physical network
circuit and is also able to support private networks as well as X25 nets.
These three lowest levels of the model are based essentially on the X25
definitions for packet switching systems. The other levels of the model
are awaiting definition. At level four, transport concerns the transfer
of messages from end-user to end-user - the sender and receiver of data.
The fifth layer (session control) establishes and manages the interaction
between two co-operating processes - these can be on the same host or
different machines. A connection protocol is used for setting up and
closing down a session and a dialogue protocol controls data flow between
processes; both are transparent to the application. The sixth layer of
the model (presentation) is required to carry out any necessary
reformatting or transformation of data thereby allowing a variety of
terminals and devices to be accessed transparently. Layer seven
(application) covers all other aspects - application software, system
software for all types of transaction processing, file management,

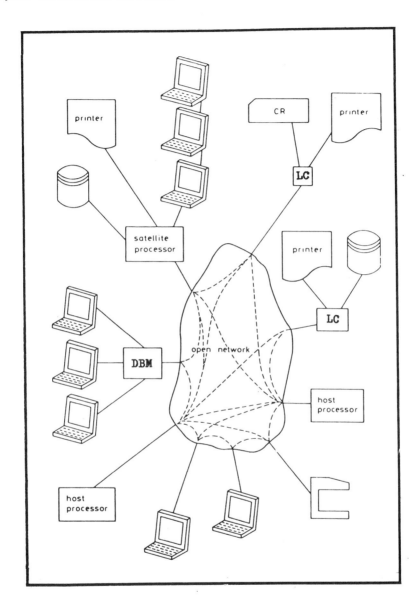

Fig. 12.6   Open systems network architecture; CR: card reader;
LC: laboratory computer; DBM: data base machine.

terminal concentration and so on. The way in which the levels of the
model are implemented will vary from manufacturer to manufacturer. All
the levels could be implemented in a single host system or they may be
distributed over two computers - as illustrated in the node in the top

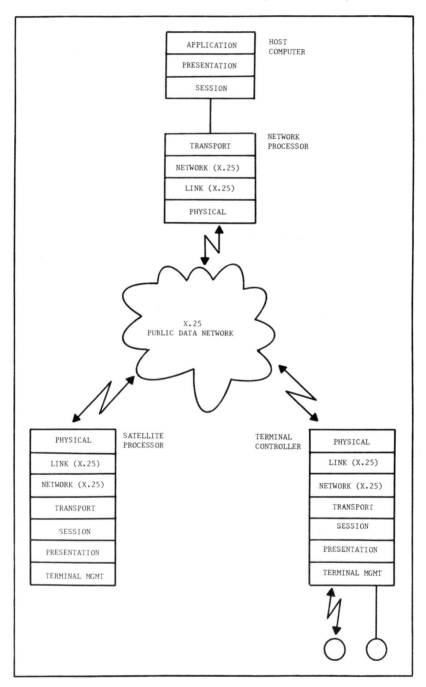

Fig. 12.7   Simple 'layered' model for open systems networking.

part of Fig. 12.7. Here both a mainframe (host system) and a network
interface processor are used. The particular approach shown illustrates a
clean separation between applications levels (host) and network functions
(on the network processor). A recent paper by Brenner (Bre80) describes
the use of Open Systems Interconnection standards for the design of an
open network system. It is likely that this approach to networking will
be of considerable value for designers of multi-vendor information
processing networks.

LOCAL AREA NETWORKS

Computer networks vary considerably in their coverage. As will be
described in subsequent sections national networks provide a variety of
services for individual countries.

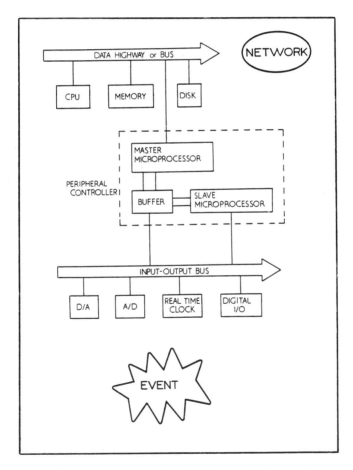

Fig. 12.8 Microprocessor based data acquistion unit used
to capture data for dissemination via a computer network.

International networks span different countries and continents.   In
contrast to these, local area networks are communication nets which have
only a relatively small geographical area of coverage compared with larger
regional, national and international systems. Typically, the area covered
might be:

     (a)   a number of different departments within an
            organisation which are spread over different
            floors within a building,
     (b)   a cluster of buildings distributed over a
            manufacturing or industrial site, or,
     (c)   a manufacturing plant with a distributed
            computer control system.

Such local networking environments are frequently found in many industrial, commercial and university areas with applications ranging from
simple time sharing services to complex data base and management information systems, transaction processing, process control and distributed
processing. A typical application is depicted in Figs. 12.8 and 12.9.
Figure 12.8 illustrates a simple microprocessor based data acquisition/
control unit being used to monitor and/or control some event. Acquired
data is transmitted via a local area computer network to a larger computer
for storage. Alternatively, control instructions may be loaded into a
remote microprocessor (or microprocessor based instrument) from a larger
machine via the network system. Figure 12.9 shows a typical local area
network topology to support activities such as this. An introduction to

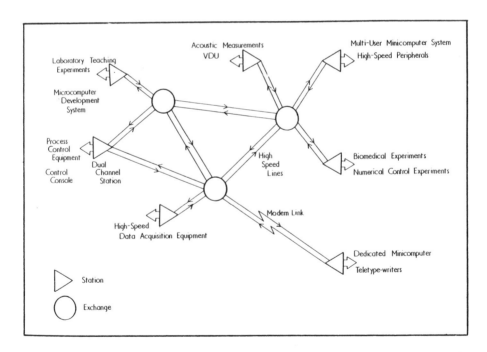

Fig. 12.9 Distributed support network for process
control and data acquisition.

the various ways of interconnecting computers and the likely applications of this type of networking to process control has been given by Harrison (Har78).

In recent years many local area computer networks have been developed using different topologies (star, hierarchy, bus, loop), control disciplines (central and distributed) and communication media (cable, twisted pair, radio, etc). Some of the more well known of these networks include:

ALOHANET - a packet radio network (Abr70),
ETHERNET - a bus broadcast network (Met76),
NWU   - a minicomputer star network (Len74),
MISS   - a three level hierarchical network (Ash75),
Z-NET   - a microcomputer based packet net (Est81),
CHIMPNET - a low cost microprocessor network (Kai79).

One of the earliest local area networks to be developed was the ALOHA system. This was produced by a group at the University of Hawaii and was designed to transmit data between network nodes using UHF (Ultra High Frequency) radio signals as the transmission medium. ETHERNET (ETH80) is a local area network architecture under joint development by DEC, Intel and Xerox Corporations. It is based upon the use of coaxial cable and allows transmission speeds up to 10 Mbits/sec. A variety of terminal devices can be connected and each may be located in any of the different offices of a building. Z-NET is the name of Zilog Corporation's local area network based upon microcomputers. It uses packet switching techniques and employs coaxial cable as the means of interconnection of the various network nodes. The prime intention of the network is to integrate DP, electronic mail and office automation traffic. The main idea behind Z-NET design is the distribution of a system's low cost elements such as CPU, memory and terminals while sharing such high cost elements as peripherals and data bases.

A variety of architectures thus exist for local area networks. Two of the most popular are the bus and ring (or loop) architectures. A bus configuration is essentially an extension of the computer/instrument bus ideas that have been previously described. Figure 12.10 shows a typical arrangement (Chl80).

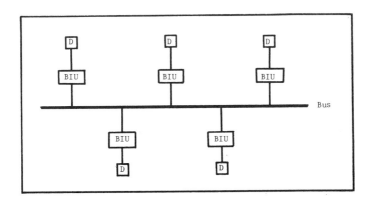

Fig. 12.10 Bus implementation of a local area network;
BIU: bus interface unit; D: attached device.

A bus configuration is attractive because it facilitates a fully connected
network where nodes can be easily added and deleted from the system
without affecting the connectivity and for many applications off-the-shelf
components can be used. Such an arrangement for a distributed system
requires bus interface units (BIU) with two faces. One face presents a
common interface to the bus and the other an interface that is unique to
the particular device (D) or equipment being added.

Another popular architecture is the closed loop or ring architecture. A
closed loop network consists of a very high speed (1-20 Mbits/sec) digital
communication channel that is arranged in a closed loop. Computers,
terminals and other peripheral devices are attached to the loop channel by
a special device called a loop interface unit (LIU). Figure
12.11 illustrates a simple ring architecture.

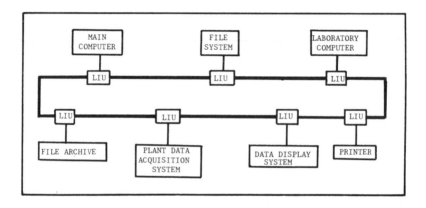

Fig. 12.11   Closed loop implementation of a local
area network; LIU: loop interface unit.

Messages in the form of programs, data or commands are placed onto the
loop in the form of addressed blocks or packets - sometimes referred to as
frames. When a message is to be sent from a local node to a remote one,
the local interface forms the frame and places in it the address of the
destination interface. The local interface then transmits the frame
around the ring. Each interface along the loop receives the frame, checks
its destination address and immediately relays it back onto the loop if
the proper destination has not been reached. When a receiving interface
recognises its own address as the destination of the incoming frame, it
receives the frame from the loop and delivers the message to its local
attached component. Some of the advantages of loop networks of this type
include (Liu78): easy message routing, simple node interfacing, high data
rates, low construction and expansion costs and the ease with which it
lends itself to distributed control.

An example of a loop architecture of this type is the Cambridge Digital
Communications Ring (Hop78, Hop80a, Hop80b, Wil80). This consists of a
monitor unit and a series of stations between which data is to be transmitted. Transmission of data is achieved by means of a number of short
fixed-length slots (or packets) circulating in a fashion similar to those
depicted in Fig. 12.12.

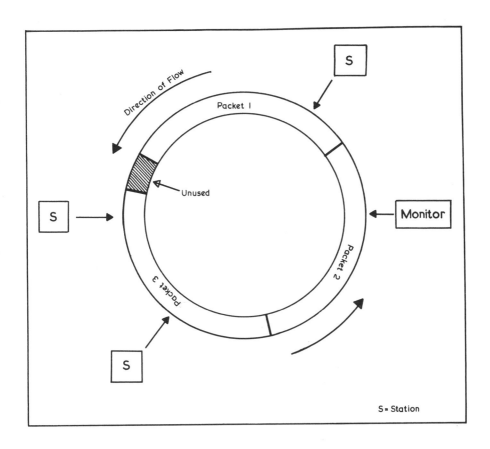

Fig. 12.12   Packet circulation in the Cambridge ring system.

Packets are passed serially around the loop and the number of packets in the ring depends upon the length of the wire. The purpose of the monitor is to create the packet structure and resolve any transmission errors that may arise. The monitor thus creates a series of continuously circulating packets into which data may be placed as they rotate. Each slot or packet contains a number of fields which show the source of the message, its destination and an indication of whether it is full or empty. The structure of these packets is shown in Fig. 12.13.

If a device connected to the ring wishes to send a message it examines the full/empty field of the packets as they go by. When it finds an empty one, it fills it with its message data (up to two bytes) and the required destination and then passes it on. The message will make its way around the ring to its designated destination where it will, in normal operation, be accepted and an acknowledgement flag set. The slot remains full until it has made a complete circuit of the ring back to its original source station. The control of the message is dealt with by the source so that

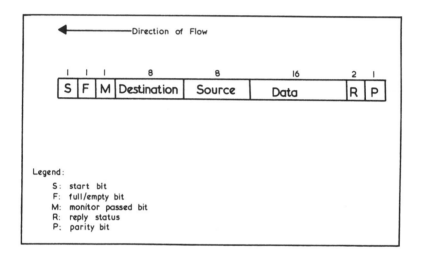

Fig. 12.13 Packet structure used in the Cambridge ring
system; numbers above individual fields specify number
of bits allocated to that field.

messages that go around the ring without being picked up (perhaps because
the destination station is in error) can be re-transmitted. A more
detailed description of the way in which these types of ring network
operate will be found in papers by Liu (Liu78) and Hopper (Hop80a,
Hop80b).

Local area networks are currently of considerable interest to many types
of organisation and product manufacturers. In order to produce some
degree of standardisation in this area and, hence, enable networks to be
more easily constructed, the IEEE has recently produced a standard
(Standard 802) for local area networks. Details of this will be found in
a recent paper by Clancy (Cla81).

NATIONAL NETWORKS

The ARPA network which currently operates in the USA was probably one of
the first large scale packet switching national computer networks to be
constructed. It was sponsored by the Advanced Research Projects Agency in
order to link together and coordinate the work of various research
establishments spread throughout the USA (Rob70). Construction of the
network commenced in 1967 and one of its early topologies is depicted
schematically in Fig. 12.14 (Car70).

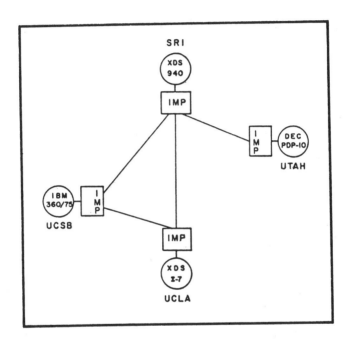

Fig. 12.14 The early topology (1970) of the Advanced
Research Project Agency's (ARPA) computer network.

During the time that has elapsed since its inauguration, the topology has
changed considerably - as can be seen from Fig. 12.15 (Kim75). Indeed,
today ARPANET should really be regarded as an international network since
there are now satellite links to Europe and other parts of the world.

Initially, the system consisted of two parts: a network of data processing
systems which were called HOSTs, and a communication sub-network of packet
switching node computers called IMPs (Interface Message Processors). In
1971 a new type of node was added called a TIP (Terminal Interface Processor). These were designed to provide network services at sites that did
not have their own large host system. The ARPANET topology continues to
expand and change as new developments are made. Overall, the network has
proved to be most effective in coordinating research at the centres it
connects - see, for example, the collaborative research involving the
DENDRAL programs described by Lederberger et al (Led75). Many of the

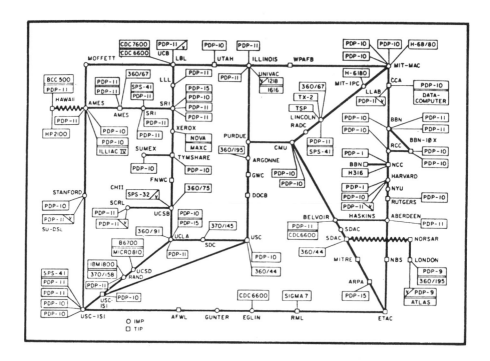

Fig. 12.15. A more recent (1975) site topology of the ARPA network.

centres are continuing to develop new techniques for using the network
effectively and several projects are aimed at providing new services and
facilities that would not be feasible without the sub-network. The inter-
connection of local area and national networks via gateways can produce
significant advantages. The University of Rochester's gateway node to the
ARPANET has been mentioned previously (Bal76). It is shown in Fig. 12.16.
Within the UK, the Packet Switching System (PSS) also uses appropriate
international gateways to enable trans-world data flow. This is depicted
in Fig. 12.17. In its initial configuration PSS contained nine packet
switching nodes situated throughout the UK with Network Management Centres
located in London. The switching nodes provide access facilities for
terminals and host computers. It has a capacity for 2,000 simultaneous
terminal users transmitting at speeds up to 1,200 bits/second and 150 host
computers operating at speeds up to 48,000 bits/second. PSS implements
the most up-to-date CCITT protocols and is capable of inter-connecting
with IPSS (International Packet Switching System) via a gateway node.
This node uses the CCITT X75 protocol for network-to-network communications to enable access to other national packet networks such as TELENET
in the USA (Rob75), DATAPAC in Canada (Dat80) and some of the other
European networks.

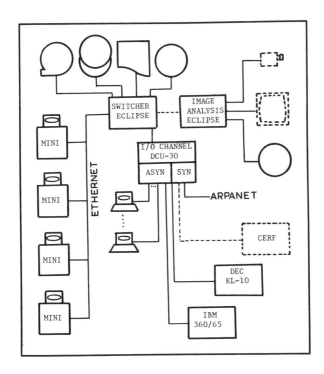

Fig. 12.16   University of Rochester's (USA)
gateway to the ARPA network.

Examples of other national computer communications networks are illustrated in the diagrams contained in Fig. 12.18. Within Europe several
public and private networks have been constructed or are in the process of
being built. The French PTT authority is constructing a public packet
switched network (called TRANSPAC) which is intended to offer a nationwide
service with a range of access speeds (50 - 48,000 bit/second) and with an
open systems architecture (Com80a). Similarly, CERNET is a private packet
switched network that operates at the CERN nuclear research establishment
on the French-Swiss border. CERNET handles over one million packets per
day (each of 300 bytes) and establishes transparent communication for any
CERN user to the research centre's mainframe (Com80b). A national packet
switching network similar to the UK's PSS system is being developed in
Australia and is scheduled to commence operation in 1982 (Com80c). AUSTPAC
is based on the CCITT X25 and X28 standards which are now being widely
accepted by the world's telecommunication authorities. An important
function of these networks is to enable the easy dissemination of scientific and technical information. Indeed, this is the sole purpose of the
JICST on-line network in Japan (Uch79). JICST (Japan Information Centre
for Science and Technology) provides an information processing facility
consisting of a data base production system, file management capability
and bibliographic retrieval systems. It has been available nationally
throughout Japan since 1976 via leased telephone lines and, more recently,
via dial-up lines. Most countries provide a variety of scientific

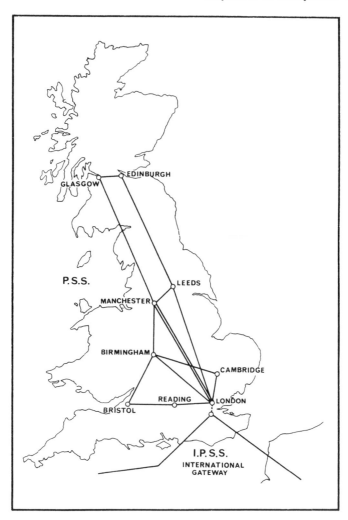

Fig. 12.17   The United Kingdom's packet switching
system (PSS) and its international gateway.

information dissemination systems of this type.   As was mentioned in
Chapter 10, the type of services that are available fall into two broad
categories: numerical and bibliographic.   Because there are fewer technical problems involved, progress with the development of techniques for
distributing bibliographic information seem to be at a more advanced stage
than those for numerical data.   Hawkins (Haw80) is one of many authors who
have recently described the need for further advancement in this area both
at a national and international level.   Carter (Car80) has given a description of many national and international data networks such as the

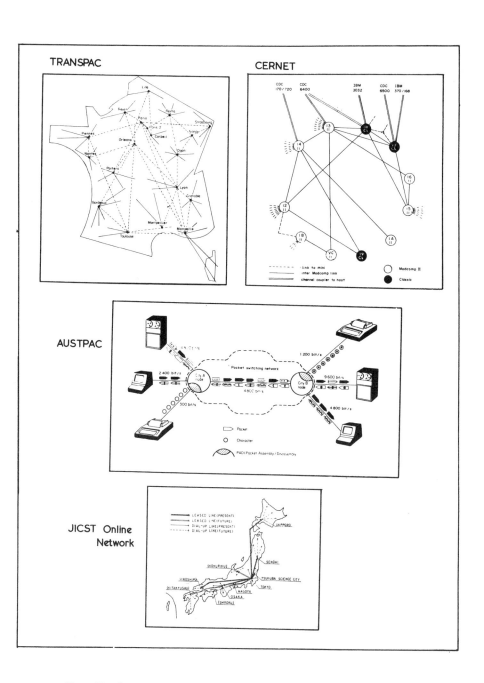

Fig. 12.18 Some examples of national computer networks.

Chemical Substances Information Network (of the Environmental Protection Agency in the USA) and EURONET. The first of these is aimed at providing on-line information on chemical substances to satisfy the legislative requirements for toxic substances, research institutions, industry and other needs. EURONET is an on-line information network that has been established for access from European countries. Through this network, a wide variety of data and information files, including many of US origin, can be queried. The network (as will be described later) will provide information and data for scientists, engineers, managers, documentalists, information scientists, environmentalists and legal and socio-economic data.

Even though national and international computer based networks now exist to enable the facile transmission of data/information between laboratories and between countries there are many problems which still have to be solved. A brief review of some of the more important issues has been given by Carter (Car80) while an in-depth discussion of problems relating to data exchange formats between files of information produced by different European countries has been described by Proctor et al (Pro78). This latter work is interesting since it has involved the design of data structures to enable the exchange of information on environmental chemicals between the members of an information sharing network - the member countries of the European Community. An environmental chemical may be defined as a substance which occurs in the environment as a result of human activity and which may be present in quantities capable of harming humans, other living species or the environment. Many of these chemicals enter the environment without any formal notification and, often, without sufficient toxicity testing. Because of their potentially harmful effects mechanisms are needed in order to record data and information on those chemicals that are likely to have an adverse effect on the environment. In 1973 the European Communities instigated a project aimed at constructing a data bank on environmental chemicals; this was called ECDIN - Environmental Chemicals Data and Information Network. It was designed as a tool which would rapidly enable all people engaged in environmental management and research to obtain reliable information on chemical products of environmental significance. Because different network partners had different file formats for their data appropriate methods of inter-conversion were necessary. These were achieved via the use of an exchange format. A file in exchange format is one which holds information from the processing files of the network partners in an agreed fashion. Any partner can access any required information from that file. Thus, in operation the system works as follows: any incoming file is converted to exchange format and is then passed on in that format to network partners who may then add the information to their own processing files. Any newly identified file (such as might arise from a new network partner) requires that software be written to convert the data supplied into the exchange format and from exchange format to its own format. Subsequently, the information exchange then becomes routine.

Obviously, as the requirement for data exchange between organisations increases as a result of more accessible data transmission networks that offer both national and world-wide coverage, so, the demand for file exchange schemes similar to that used in ECDIN is likely to increase. Schemes of this type are ideal since they enable individual users to maintain their own files in the forms most useful for them while at the same time encouraging the dissemination of data globally.

INTERNATIONAL NETWORKS

International computer communications networks are those which, by one means or another, cross national borders. Because of the satellite link between America and Norway (see Fig. 12.15) the US ARPA network may now be thought of as one of international significance even though when it originally commenced operation its domain was purely national. Many other examples of international networks exist. Their growth has been promoted by the rapidly growing volume of scientific and technical information that has been accumulating over the last decade. The need to be able to exchange and disseminate this material in a rapid and cost effective manner is of vital importance to economic development. The EURONET system (shown in Fig. 12.19) is an example of a packet switching network spanning many different countries within the European Economic Community (EEC). It was set up in order to satisfy the following three major objectives (Whi79):

(1) to develop data bases and information services in areas where gaps could be identified,
(2) to develop an international data telecommunications network, and,
(3) to improve the information environment to provide users with easy access to all major information services.

During the 1970's Europe was almost totally dependent on two large American companies (Lockheed Information Systems and System Development Corporation - see below) for access to scientific data bases and on-line information retrieval services. Thus, in implementing the network, top priority was given to establishing a data transmission system (EURONET) designed to provide access to data bases within the community and thereby decrease dependence of the EEC on US service.

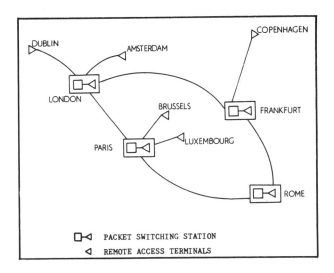

PACKET SWITCHING STATION
REMOTE ACCESS TERMINALS

Fig. 12.19 EURONET information dissemination network.

The basic EURONET network consists of four main nodes (packet switching exchanges) located in Frankfurt, London, Paris and Rome. Attached to these are five remote access facilities in Amsterdam, Brussels, Copenhagen, Dublin and Luxembourg. The control centre for the network is located in London. The initial configuration supports 140 simultaneous searchers though this is likely to increase substantially when the system expands. About sixteen data bases are supported - including Chemical Abstracts. It is anticipated that the biggest users of EURONET are likely to be chemical and pharmaceutical companies. The network currently permits only on-line retrieval of titles and bibliographic details. Unfortunately, for many applications this is not sufficient and so there is currently much interest in the electronic delivery of primary documents using the basic principles employed in electronic mail systems (Mar79).

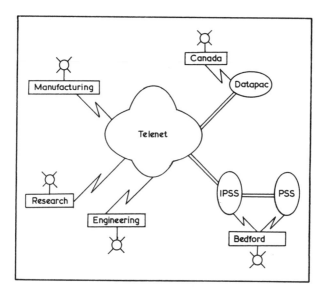

Fig. 12.20   An example of a corporate international network.

A more sophisticated example of an international network (Tag80) is illustrated in Fig. 12.20. This illustration depicts the essential features of a corporate network set up by Prime computers in order to co-ordinate their American and European activities. Obviously, the tech-niques involved and the benefits gained from using such a network are likely to apply to any international organisation. The network supports communication via local ring networks, point-to-point synchronous networks and X25 packet switching facilities. It uses an in-house product (PRIMENET) throughout, includes a US network linked to TELENET, a Canadian sub-net linked to DATAPAC and an IPSS/TELENET connection to the research and development centre in Bedford, UK.

PRIMENET users linked on a local ring use the same calls to link and transfer files between their own ring and a public data network. Local and remote communications facilities are provided, allowing communication between the computers themselves, with other manufacturer's computers and with terminals attached to packet networks in geographically different network configurations. Prime are currently working on the development of

an open network file transfer system. This will enable file transfer
between Prime and other manufacturer's computers via a packet switched
network. Such a facility is analogous to the requirements, outlined above
of ECDIN's file transfer arrangement and EURONET's proposed dissemination
system for primary documents.

Another example of a network system that started off as a national system
and which has grown to provide virtually world-wide coverage is the information services network offered by SDC (Systems Development Corporation).
This organisation operates a large data base system that permits users all
over the United States and in many foreign countries to search very large
bibliographic files interactively, that is, by means of a computer
terminal coupled through a telephone into a special nationwide communication network. The user is then able to interact with the various data
bases available using a time-shared retrieval program. The arrangement is
shown schematically in Fig. 12.21 (Cua75).

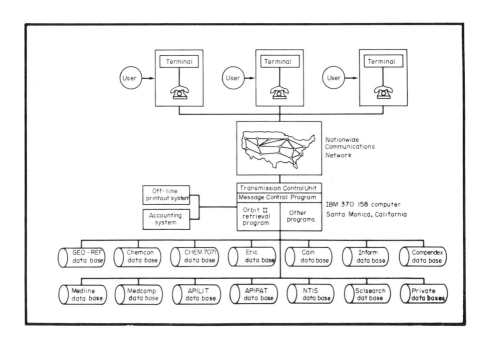

Fig. 12.21 The System Development Corporation (SDC) network
and associated data bases.

SDC developed its first interactive retrieval system in 1960. In 1965 the
first nationwide network was constructed as part of an experiment
sponsored by ARPA (Advanced Research Project Agency), a section of the US
government's Department of Defense. Today, SDC's facilities are available
anywhere in the world. This availability is made possible by a sophisticated web of international telecommunication networks. Access to services
of this type is being made increasingly easy as new forms of communication
channel, such as satellite transmission systems, become more commonly
available. Figure 12.22 gives an indication of the way in which satellites

are likely to integrate into the architecture of computer based communication networks (Iso80). In this diagram, TN denotes a traffic node on
the conventional terrestrial networks and ES denotes an earth station used
to beam information up to a satellite node and also receive information
back again. A more detailed discussion of satellite communication systems
will be found in the book by Martin (Mar78).

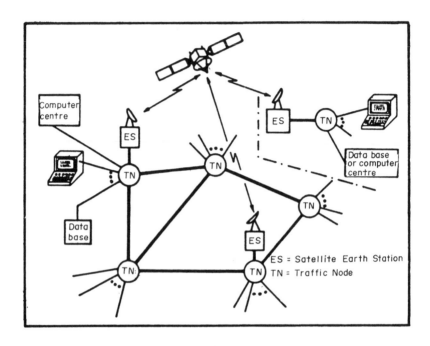

Fig. 12.22 Use of satellites for data and information transmission.

Many new kinds of service are likely to become available as these new
hybrid communication networks develop. Particularly important will be
their ability to support interactive person-to-person communication, and
hence, sophisticated forms of tele-conferencing (Val75, Hil78). There is
likely to be a wide variety of inexpensive electronic mail systems and a
number of other types of document transmission facility. The easy transmission of pictorial information (such as diagrams from primary documents,
spectra, 3-D crystal structures, etc) should also be possible in an
economic and rapid way. The advent of the electronic scientific journal
and its ability to make the results of scientific research instantaneously
available on a world-wide basis is likely to have a significant impact on
future scientific communities, industrial manufacturing and control of the
environment. Thus, automatic analytical equipment located anywhere on the
earth's surface performing an analytical function and transmitting the
results via satellite to a central control station is no longer a science
fiction concept. It is a feasible technique which reflects just one of
many ways new technology is likely to influence analysis.

CONCLUSION

The use of computer communication networks is growing considerably in many areas within manufacturing industry, management and research. This new technology enables a wide range of different approaches to computing to be realised. Most important of these, perhaps, are decentralisation of control, distribution of function, resource sharing and the use of computers as powerful tools for information dissemination. Decentralisation of control is useful in a variety of industrial situations, particularly in the process and manufacturing industries. Distribution of function can be usefully employed to remove bottle-necks within an organisation and greatly enhance smoothness of operation. Resource sharing is particularly important in areas relevant to computer software, experimental data and various forms of bibliographic information. The ability to disseminate information via computer networks has created a whole new information industry. None of these benefits of modern computer systems would be feasible without adequate computer based communication networks. The term distributed computing is now often used to collectively describe those modes of working with a computer that are in some way dependent upon a distributed communications system.

In this chapter many different aspects of computer networking have been discussed. The idea of network topology has been introduced and a description of the different types of basic modes of operation has been given - circuit, message and packet switching. Although they have some limitations, packet switching networks seem to be those which are being adopted by both private and public network producers throughout the world. In addition to the conventional approaches to networking, newer types of videotex systems have also been briefly described because of their growing future importance for both national and in-house use. As a consequence of the complexity of the inter-connection problems that are involved in networking, the need for internationally accepted standards is of vital importance. Some of the more well known CCITT standards have been mentioned and their significance outlined.

It was suggested that networks might be broadly classified into three basic types depending upon their coverage: local area, national and international. Regional networks (with a coverage between that of local area and national) are also important. However, because their mode of operation is essentially similar to that of a national network, a discussion of these has been omitted. The function of each of the three major network types was outlined and adequate illustrative examples given. Undoubtedly, the future of computer communication networks will be greatly influenced by satellite technology. This too, like the computer, automation and conventional communication technology, will have its impact on analytical chemistry. It is easy to predict, in general terms, some of the likely possibilities. However, the details of particular applications will be found with tomorrow's analyst, computer scientist and communication engineer. It is they, through the formation of multi-disciplinary teams, who will bring together the expertise necessary to harness these new technological advances and use them to improve the quality of life for future generations. The analytical scientist and technologist thus have complimentary parts to play since it is unlikely that either will succeed alone.

REFERENCES

Abr70   Abramson, N., The ALOHA System - Another Alternative for Computer
         Communications, Proc. AFIPS Fall Joint Computer Conference,
         281-285, 1970.

Ash75   Ashenhurst, R.A., Hierarchical Minicomputer Support as a Methodo-
         logical Aid to the Laboratory Investigator, Chapter 7, 108-117, in
         Computer Networking and Chemistry (Ed. P. Lykos), ACS Symposium
         Series No. 19, ISBN: 8412-0301-6, 1975.

Bal76   Ball, J.E., Feldman, J., Low, J.R., Rashid, R. and Rovner, P.,
         RIG: Rochester's Intelligent Gateway: System Overview, IEEE
         Transactions on Software Engineering, SE-2, 4, 321-328, December
         1976.

Bal80   Ball, A.J.S., Bochman, G.V. and Gecsei, J., Videotex Networks,
         IEEE Computer, Volume 13, No. 12, 8-13, December 1980.

Bre80   Brenner, J., Using Open Systems Interconnection Standards, ICL
         Technical Journal, Volume 2, No. 2, 106-116, ISSN: 0-142-155-7,
         May 1980.

Car70   Carr, C.S., Crocker, S.D. and Cerf, V.G., HOST-HOST Communications
         Protocol in the ARPA Network, 580-597, AFIPS Conference Proc.,
         (SJCC), Volume 36, 1970.

Car80   Carter, G.C., Numerical Data Retrieval in the US and Abroad, J.
         Chem. Inf. Comput. Sci., Volume 20, No. 3, 146-152, August 1980.

Chl80   Chlamtac, I. and Franta, W.R., Message-based Priority Access to
         Local Networks, Computer Communications, Volume 3, No. 2, 72-84,
         April.1980.

Cla81   Clancy, G., IEEE 802 Local Networks, Proc. of IEEE COMPCON 81,
         Conference on VLSI: in the Laboratory, the Office, the Factory,
         the Home, February 1981.

Com80a   Anon., International Review, Computer Communications, Volume 3,
         No. 1, 33-36, February 1980.

Com80b   Anon., Systems Review, Computer Communications, Volume 3, No. 5,
         page 247, October 1980.

Com80c   Anon., Austpac, Computer Communications, Volume 3, No. 6, page
         278, 1980.

Cua75   Cuadra, C.A., SDC Experiences with Large Data Bases, Journal of
         Chemical Information and Computer Sciences, Volume 15, No. 1,
         48-51, 1975.

Dat80   Anon., Canada - History of Communications, Computer Communica-
         tions, Volume 3, No. 1, 36-37, February 1980.

Dav73   Davies, D.W. and Barber, D.L.A., Communication Networks for
         Computers, John Wiley, ISBN: 0-471-19874-9, 1973.

DEC74   Digital Equipment Corporation, Introduction to Minicomputer
         Networks, 1974.

DEC78   Digital Equipment Corporation, Digital Data Communications
         Message Protocol (DDCMP), Digital Network Architecture (DECNET),
         Specification Version 4.0, Digital Equipment Corporation, Maynard,
         MA, USA, March 1978.

Ens78   Enslow, P.H., What is a "distributed" processing system?, IEEE
         Computer, Volume 11, No. 1, 13-21, January 1978.

Est81   Estrin, J., Architecture of Z-NET, a Microprocessor based Local
         Network, Proc. of the IEEE COMPCON '81 Conference on VLSI: in the
         Laboratory, the Office, the Factory, the Home, February 1981.

ETH80   Digital Equipment Corporation, Intel Corporation, Xerox Corpora-
         tion, The Ethernet: A Local Area Network, Data Link Layer and
         Physical Link Layer Specifications, Version 1.0, September 1980.

Gro79   Grossman, G.R., Hinchley, A. and Sunshine, C.A., Issues in Inter-
         national Public Data Networking, Computer Networks, Volume 3,
         259-266, 1979.

Har78   Harrison, T.J., Micros, Minis and Multiprocessing, Instrumenta-
         tion Technology, Volume 25, No. 2, 43-50, February 1978.

Haw80   Hawkins, D.T., Problems in Physical Property Data Retrieval, J.
         Chem. Inf. Comput. Sci., Volume 20, No. 3 143-145, August 1980.

Hig75   Higginson, P.L. and Hinchley, A.J., The Problems of Linking
         Several Networks with a Gateway Computer, 453-465 in Proc. of the
         European Computing Conference on Communications Networks, Online
         Conferences Ltd., Uxbridge, UK, ISBN: 0-201-03141-8, 1978.

Hil78   Hiltz, S.R. and Turoff, M., The Network Nation - Human Communica-
         tion via Computer, Adison-Wesley, ISBN: 0-201-03141-8, 1978.

Hop78   Hopper, A., Data Ring at the Computer Laboratory, University of
         Cambridge, p.11, Computer Science Technology: Local Area
         Networking, US National Bureau of Standards Special Publication
         500-31, 1978.

Hop80a   Hopper, A., The Cambridge Ring - a Local Network, 67-70, in
         Advanced Techniques for Microprocessor Systems (Ed. Hanna, F.K.)
         Peter Peregrinus Ltd., ISBN: 0-906048-31-1, 1980.

Hop80b   Hopper, A., Local Area Computer Networks, 103-111 in Business
         Telecommunications, Online Publications Ltd. ISBN: 0-903796-60-0,
         1980.

Hou78   Houldsworth, J., Standards for Open Network Operation, ICL
         Technical Journal, Volume 1, No. 1, 50-65, ISSN: 0-142-155-7,
         November 1978.

Hum79   Hummel, E., International Numbering Plan for Public Data Networks
         Provisionally Approved, Computer Networks, Volume 3, p.136,
         1979.

IBM78   IBM Corporation, SNA Format and Protocol Reference Manual: Archi-
          tecture and Logic, Form: SC30-3112-01, IBM Corporation, Research
          Triangle Park, N.C., June 1978.

Iso80   Isotta, N.E.C., The Future Impact on Telecommunications on
          Information Science, Journal of Information Science, Volume 1,
          No. 5, 249-258, January 1980.

Kai79   Kain, R.Y., Franta, W.R. and Jelatis, G.D., CHIMPNET: A Network
          Testbed, Computer Networks, Volume 3, 447-457, 1979.

Kim75   Kimbleton, S.R. and Schneider, G.M., Computer Communications
          Networks: Approaches, Objectives and Performance Considerations,
          ACM Computing Surveys, Volume 7, No. 3, 129-173, September 1975.

Kir76   Kirstein, P.T., Planned New Public Data Networks, Computer
          Networks, Volume 1, 79-94, 1976.

Led75   Lederberger, J., Carhart, R.E., Johnson, S.M., Smith, D.H.,
          Buchanan, B.G. and Dromey, G., Networking and a Collaborative
          Research Community: A Case Study Using the DENDRAL Programs,
          Chapter 13, 192-218, in Computer Networking and Chemistry (Ed.
          Lykos, P.), American Chemical Society Symposium Series, No. 19,
          ISBN: 8412-0301-6, 1975.

Len74   Lennon, W.J., A Minicomputer Network for Support of Real Time
          Research, Proc. ACM Annual Conference, 595-604, 1974.

Liu78   Liu, M.T., Distributed Loop Computer Networks, 163-221, in Volume
          17 of Advances in Computers, (Ed. Yovits, M.C.), Academic Press,
          ISBN: 0-12-012117-4, 1978.

Mag79   Magnee, F., Endrizzi, A. and Day, J., A Survey of Terminal
          Protocols, Computer Networks, Volume 3, 299-314, 1979.

Mar78   Martin, J., Communication Satellite Systems, Prentice-Hall,
          ISBN: 0-13-153163-8, 1978.

Mar79   Marsh, P., Who Gains from the Great Data Boom?, New Scientist,
          Volume 81, No. 1139, 239-242, January 1979.

McF76   McFadyen, J.H., Systems Network Architecture - An Overview, IBM
          Systems Journal, Volume 15, No. 1, 4-23, 1976.

Met76   Metcalfe, R.M. and Boggs, D.R., ETHERNET: Distributed Packet
          Switching for Local Computer Networks, Comm. ACM., Volume 19,
          No. 7, 395-404, 1976.

Pia80   Piatkowski, T.F., The ISO - ANSI Open Systems Reference Model - A
          Proposal for a Systems Approach, Computer Networks, Volume 4,
          No. 3, 111-124, June 1980.

Pro78   Proctor, D.J., Robson, A., Veal, M.A., Petrie, J.H. and Town,
          W.G., Development of an Exchange Format for the European Environ-
          mental Chemical Data and Information Network (ECDIN), Information
          Processing and Management, Volume 14, 429-443, 1978.

Rob70   Roberts, L.G. and Wessler, B.D., Computer Network Development to Achieve Resource Sharing, 543-549, in AFIPS Conference Proceedings, (SJCC), Volume 36, 1970.

Rob75   Roberts, L.G., Telenet: Princples and Practice, Proceedings of the European Computing Conference on Communications Networks, 315-329, Online Conferences Ltd., Uxbridge, UK., ISBN: 0-903796-05-8, 1975.

Ros76   Rosner, R.D. and Springer, B., Circuit and Packet Switching - A Cost and Performance Tradeoff Study, Computer Networks, Volume 1, No. 1, 7-26, 1976.

Ryb80   Rybczynski, A.M. and Palframon, J.D., A Common X25 Interface to Public Data Networks, Computer Networks, Volume 4, No. 3, 97-110, June 1980.

Sch78   Schwager, A.O., The Hewlett Packard Distributed System Network, Hewlett-Packard Journal, Volume 29, No. 7, 2-6, March 1978.

Sch80   Schwartz, J.T., Ultracomputers, ACM Transactions on Programming Languages and Systems, Volume 2, No. 4, 484-421, October 1980.

Tag80   Taggart, A., Interactive Networking, Computer Communications, Volume 3, No. 3, 136-142, June 1980.

Uch79   Uchida, H., Information System, Data Bases and On-line Services of the Japan Information Centre of Science and Technology (JICST), J. Chem. Inf. Comput. Sci., Volume 19, No. 4, 199-201, 1979.

Val75   Vallee, J., Lipinski, H., Johansen, R. and Wilson, T.H., Computer Conferencing, SCIENCE, Volume 188, page 203, April 1975.

Wal75   Walden, D.C. and Rettberg, R.D., Gateway Design for Computer Network Interconnections, 113-128, in Proc. of the European Computing Conference on Communication Networks, Online Conferences Ltd., Uxbridge, UK, ISBN: 0-903796-05-8, 1975.

Whi79   White, M., Why Europe is caught up in the Data Net, New Scientist, Volume 81, No. 1140, 310-312, February 1979.

Wil80   Wilkes, M.V., The Impact of Wide-Band Local Area Communication Systems on Distributed Computing, IEEE COMPUTER, Volume 13, No. 9, 22-25, September 1980.

Author Index

Subject Index